Recent Advances in Photovoltaic Systems

Recent Advances in Photovoltaic Systems

Editor: Fraser Cox

RCALLISTO
REFERENCE

www.callistoreference.com

Callisto Reference,
118-35 Queens Blvd., Suite 400,
Forest Hills, NY 11375, USA

Visit us on the World Wide Web at:
www.callistoreference.com

ISBN: 978-1-64116-339-2 (Hardback)

Trademark Notice: Registered trademark of products or corporate names are used only for explanation and identification without intent to infringe.

Cataloging-in-Publication Data

Recent advances in photovoltaic systems / edited by Fraser Cox.
    p. cm.
Includes bibliographical references and index.
ISBN 978-1-64116-339-2
1. Photovoltaic power generation. 2. Solar energy. I. Cox, Fraser.
TK1087 .R43 2020
621.312 44--dc23

Table of Contents

Permissions

List of Contributors

Index

Preface

Every book is a source of knowledge and this one is no exception. The idea that led to the conceptualization of this book was the fact that the world is advancing rapidly; which makes it crucial to document the progress in every field. I am aware that a lot of data is already available, yet, there is a lot more to learn. Hence, I accepted the responsibility of editing this book and contributing my knowledge to the community.

The power systems that are designed to supply usable solar power using photovoltaics are referred to as photovoltaic systems. There are various types of modern photovoltaic systems such as grid-connected photovoltaic systems, stand-alone photovoltaic systems, building integrated systems, rack-mounted systems, residential and utility systems, roof-top and ground-mounted systems and many more. The photovoltaic system is made up of an arrangement of various components. It uses solar panels to absorb the sunlight and convert it into electricity. It also uses a solar inverter to convert DC into AC, as well as other electrical accessories such as mounting and cabling to set up a working system. Photovoltaic systems use a solar tracking system to improve the overall performance of the system. This book elucidates the concepts and innovative models around prospective developments with respect to photovoltaic systems. It presents researches and studies performed by experts across the globe. Those in search of information to further their knowledge will be greatly assisted by this book.

While editing this book, I had multiple visions for it. Then I finally narrowed down to make every chapter a sole standing text explaining a particular topic, so that they can be used independently. However, the umbrella subject sinews them into a common theme. This makes the book a unique platform of knowledge.

I would like to give the major credit of this book to the experts from every corner of the world, who took the time to share their expertise with us. Also, I owe the completion of this book to the never-ending support of my family, who supported me throughout the project.

Editor

Rapid, Chemical-Free Generation of Optically Scattering Structures in Poly(ethylene terephthalate) Using a CO$_2$ Laser for Lightweight and Flexible Photovoltaic Applications

Simon D. Hodgson and **Alice R. Gillett**

Department of Mechanical Engineering, University of Chester, Chester, UK

Correspondence should be addressed to Simon D. Hodgson; s.hodgson@chester.ac.uk

Academic Editor: Yanfa Yan

Highly light scattering structures have been generated in a poly(ethylene terephthalate) (PET) film using a CO$_2$ laser. The haze, and in some cases the transparency, of the PET films have been improved by varying the processing parameters of the laser (namely, scanning velocity, laser output power, and spacing between processed tracks). When compared with the unprocessed PET, the haze has improved from an average value of 3.26% to a peak of 55.42%, which equates to an absolute improvement of 52.16% or a 17-fold increase. In addition to the optical properties, the surfaces have been characterised using optical microscopy and mapped with an optical profilometer. Key surface parameters that equate to the amount and structure of surface roughness and features have been analysed. The CO$_2$ laser generates microstructures at high speed, without affecting the bulk properties of the material, and is inherently a chemical-free process making it particularly applicable for use in industry, fitting well with the high-throughput, roll to roll processes associated with the production of flexible organic photovoltaic devices.

1. Introduction

The photovoltaic (PV) industry is both incredibly valuable and growing rapidly. In 2017, more capacity was added from solar PV than any other technology and can be considered amongst the leading contenders for replacing fossil fuels as the major provider of energy worldwide [1]. Traditional first-generation solar cells have reached efficiencies of 26.7%, very close to the theoretical maximum of ~33% [2, 3]. Thin-film PV, also known as second-generation PV, has closed the gap significantly in recent years. The two dominant technologies are cadmium telluride (CdTe) and copper indium gallium diselenide (CIGS) which have verified records of 21.0% (with the standard active area of ≥ 1 cm^2, at active areas below this, the record is 22.1%) and 22.9%, respectively [2]. Thin-film PV is designed using materials that have very high absorption coefficients. This enables the devices to have absorber layer thicknesses of the order of ~10 μm and below, a considerable material saving over the 100–500 μm thick active layers in first-

generation PV [4]. Third-generation PV (3GPV) covers a broad area of different technologies. 3GPV primarily focuses on overcoming the Shockley-Queisser limit through the application of one or more methods such as hot-carrier absorption, photon management techniques, multiple p-n junctions, and concentration [5–8]. Despite this, many novel new material types are also classed as 3GPV. This includes dye-sensitised (DSC), organic (OPV), and perovskite-based devices [9–11]. These new technologies offer the potential for even lower production costs and have applications to a range of niche markets otherwise inaccessible to classic module design. Of these, perovskites have the highest recorded efficiency of 20.9% to date (standard active area, 22.7% on smaller) [2, 12]. One advantage of these new technologies is the ability to be deposited on polymer substrates. This is particularly the case for OPV, something that not only provides a reduction in cost, but allows devices to be flexible [13].

The engineering of structures at the micro- and nano-scale to increase the photon collection of PV cells is well established. These have found particular application to thinner absorber materials where maximising the absorption over a small optical path length is key to increasing performance. The structures range from the texturing of surfaces to reduce reflection (e.g., through the generation of black silicon and similar structures or through the creation of both ordered and randomly generated structures designed to create a high degree of light scattering (also known as haze) or induce light trapping) [14–16]. Another route used to increase photon capture is the application of plasmonic layers [17]. The methods used to create these types of surfaces revolve around either the growth of structures on top of a PV layer (e.g., through PVD- or CBD-based methods) or through the removal of material to create the structures (i.e., etching) [18, 19]. In nearly all cases, additional and normally highly toxic chemicals are needed as reaction precursors, or as part of an etching process. Laser processing, by contrast, is normally free of the use of toxic material beyond the use of common solvents for cleaning [20]. Lasers have been used in many fields. These range from a machining tool in cutting and welding applications [21] to the engineering of surfaces in biomedical or orthopaedic applications [22, 23], as well as microfluidics and as a part of the deposition of thin films [24, 25]. Carbon dioxide (CO_2) based lasers are amongst the most well-used systems today. They are fairly efficient, low in cost, and highly flexible for use in a range of applications. CO_2 lasers are the highest power continuous wave systems around and are also q-switchable to yield very high power densities per pulse [26–28].

Poly(ethylene terephthalate) (PET) is a common polymer that is used in food packaging, in medical devices, and in PV as either a substrate or as part of the encapsulation of a module [29–32]. It is low in cost, lightweight and has good mechanical and optical properties [29]. This work presents the first steps towards the development of a chemical-free (beyond the initial clean) CO_2 laser-based process to engineer a PET layer for the production of light scattering structures designed to increase the performance of flexible PV devices at high speeds.

2. Materials and Methods

2.1. Surface Engineering. The CO_2 laser system used was a 60 W Synrad Firestar system fitted with a galvanometric scanning head producing a FWHM spot size of 171 μm. Each sample was produced as a series of parallel lines in a 10 mm^2 area. Three processing parameters have been varied throughout the set of experiments. These are the laser power, scanning speed, and the line spacing. The parameters were set according to Table 1, with every possible combination tested, resulting in a total of 64 samples created. Prior to experimentation, the laser power was measured using a Mahoney laser power probe. The measured values have been included in Table 1 next to the percentage power used. The transparent 0.25 mm thick a-PET film was purchased from Goodfellows Inc., and prior to laser treatment, samples were briefly cleaned with isopropyl alcohol wipes to remove surface contamination and minimise additional solvent use. All processing was performed in ambient air.

TABLE 1: The range of setting sued for laser engineering structures onto the PET surface. Every combination was tested, resulting in 64 total sample types.

Laser power (%)	Scanning speed (mm/s)	Line spacing (μm)
5 (3.75 W)	100	300
10 (6.33 W)	200	400
15 (11.25 W)	300	500
20 (14.67 W)	400	600

2.2. Surface Characterisation. Two types of surface characterisation have been performed: surface profilometry and optical microscopy. Surfaces have been mapped using a STIL Micromeasure 2 confocal chromatic imager (CCI). An area of 1×1 mm was scanned on each sample with a step size of 10 μm. Data processing was done using MountainMaps software, and all raw data was comparably filtered and processed. Multiple 3D surface parameters have been calculated and analysed, namely, S_a (arithmetic mean roughness), S_z (10-point height), S_{ku} (kurtosis of height distribution), and S_{sk} (skewness of height distribution). Optical microscopy was performed on a Leica DM2700 microscope.

2.3. Optical Measurements. An Ocean Optics QE-Pro series UV-VIS spectrometer has been used to measure the optical properties of the laser-processed PET samples. All samples, including an unprocessed as-received (AR) sample, were measured using an integrating sphere. Two reference measurements have been taken: T_1, the equivalent to 100% transmission, and T_3, identical to T_1 with the exit port of the integrating sphere open to allow the direct beam of light to pass through the sphere so only stray light from the natural bloom of the beam is detected. For each sample, two further measurements have been taken. These are T_2, the equivalent to T_1 but with the sample in place, and T_4, the equivalent to T_3 with the sample in place. Equation 1 can then be used to calculate the total transmission (T_T) of each sample, equation 2 can be used to calculate the diffuse transmission (T_d), and equation 3 can be used to calculate the total haze (T_H).

$$T_T = \frac{T_2}{T_1}, \tag{1}$$

$$T_d = \frac{T_4 - T_3(T_2/T_1)}{T_1}, \tag{2}$$

$$T_H = \frac{T_d}{T_T}. \tag{3}$$

3. Results and Discussion

3.1. Surface Analysis. Analysis of the surfaces produced shows a range of different structures produced on the PET samples. Interestingly, two of the samples, both processed at 20% power and 100 mm/s scanning speed whilst having spacings of 300 and 400 μm, respectively, caused the samples to melt through, resulting in their destruction due to power densities greater than the PET being able to withstand.

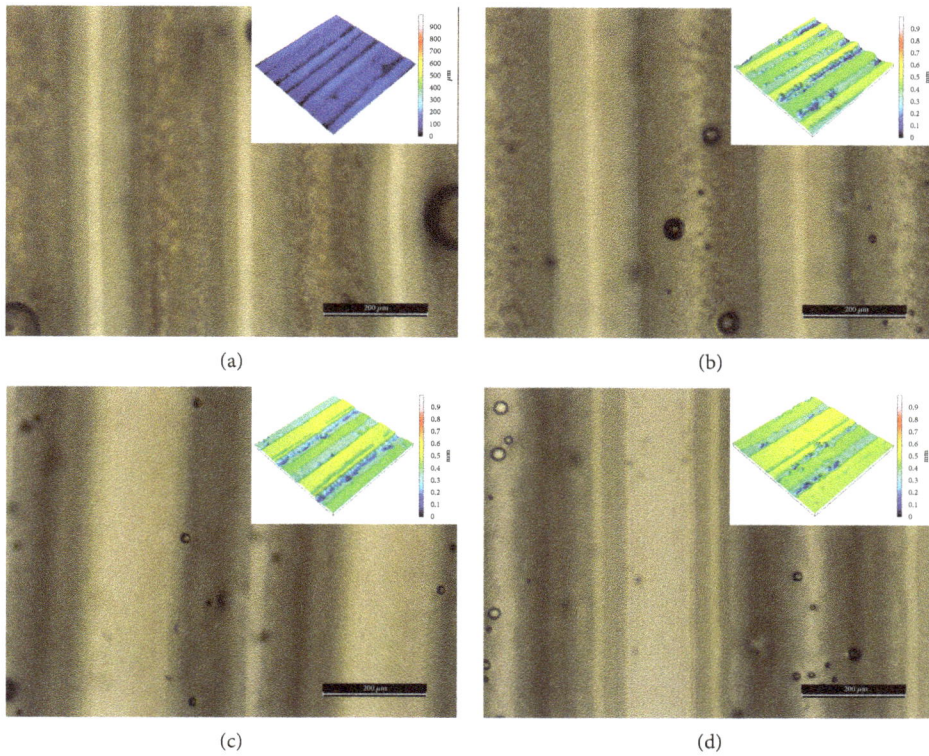

FIGURE 1: Optical micrographs of 4 laser-processed PET samples with increasing spacing between the parallel laser processing lines. (a) 300 μm; (b) 400 μm; (c) 500 μm; (d) 600 μm. All samples have been processed at 15% power and a scanning speed of 200 mm/s. The insets are CCI surface maps that have been set to the same z scale, and horizontally the scale bar is equal to 200 μm.

Optical microscopy of the samples demonstrates in detail the structure of the surfaces. Figure 1 displays a series of samples with increasing spacing from 300 to 600 μm. In each image, the power and scanning speed are constant. As would be expected, the wider the spacing, the greater the area of AR-like material. Some bubbling is visible in the samples, caused by air becoming trapped in molten PET as it cools rapidly following laser processing. From the CCI surface maps, it can be observed that the sample with the smallest spacing has a smaller z range than the wider-spaced samples. This has been caused by heat transfer within the sample from successive line scans that are in close enough proximity to affect the previous scan. Here, the effect is to reduce the effective depth of the process by remelted PET partly filling any trenches previously engraved in the surface. The FWHM of the laser yields a spot size of 171 μm, which is the equivalent of a $1/e^2$ spot size of 290 μm. As polymers do not resist heat particularly effectively, it can be seen that the proposed remelting effect is likely occurring due to spot size and spacing (for the smallest spacing) being similar in scale. Once the spacing reaches a sufficiently wide level, this effect reduces and disappears. By analysing the surface roughness data further, support for this is found. By averaging the S_a, S_z, S_{sk}, and S_{ku} for all the samples with the same spacing, we find that the average roughness (S_a) increases as spacing increases. The 300 μm spacing displays an average roughness of ~10.8 μm which increases to 12.8 μm by 600 μm spacing. 400 and 500 μm spacing samples have comparable values. S_a alone is a comparatively

useless value, however, as samples with a large z range and a small z range can have a similar average roughness despite vastly different surfaces. Comparing S_a with S_z provides a better demonstration of the scale of the surface roughness. S_z gives a measure of the scale of surface roughness, i.e., from the lowest valleys to the highest peaks. The same trend is observed, showing that not only does the average roughness of the samples increase so but also does the scale of the roughness. There is approximately a 100 μm increase in surface roughness from 300 to 600 μm spacing. Other key parameters, S_{sk} and S_{ku}, also provide useful information about the roughness features on the surface. S_{sk} is the skewness of the surface and relates to the relative dominance of peaks or valleys in the surface. S_{ku} is the kurtosis and is an indication of how high or not the peaks and valleys are. Both S_{sk} and S_{ku} are unitless. All the sample spacings demonstrate a negative skewness, indicating a predominance of valleys. This would be expected as laser processing is typically an ablative removal process that generates valleys. There is a slightly larger average negative skewness value for 300 μm spacing (-1.60), vs. the increasing spacing (-1.51, -1.43, and -1.33, respectively). Whilst the values are, of course, close, one explanation could be due to increasing spacing resulting in a relative decrease in the volume of processed area—reducing the relative amount of valleys. S_{ku} values demonstrate an increase in the lowest level of kurtosis—meaning the lower level of deep valleys—for the smallest spacing. This increases from 8.2 to 9.8 and 9.4 for 400 and

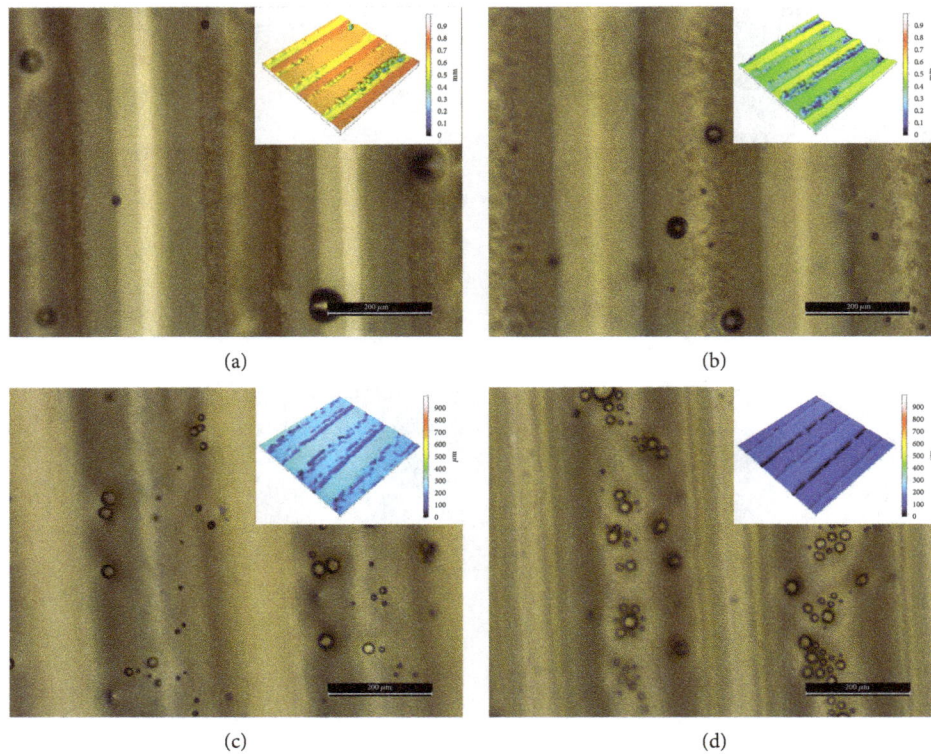

FIGURE 2: Optical micrographs of 4 laser-processed PET samples with increasing scanning speed between the parallel laser processing lines. (a) 100 mm/s; (b) 200 mm/s; (c) 300 mm/s; (d) 400 mm/s. All samples have been processed at 15% power and at 400 μm spacing. The insets are CCI surface maps that have been set to the same z scale, and horizontally the scale bar is equal to 200 μm.

500 μm spacings, very similar values, and then 10.2 for the largest spacing. This would seem to support the idea of a remelt at lower spacings causing some of the depth of process to be removed. This also explains the reduced number of bubbles within the 300 μm spaced sample, due to slower cooling rates (caused by the remelting) allowing more trapped air to escape the sample.

Figure 2 shows optical micrographs illustrating the impact of scanning speed on the surface of PET. Scanning speed ranges from 100 to 400 mm/s. All other parameters have been held constant. Interestingly, at faster scanning speeds there appears to be an increase in air bubbles trapped within the polymer. The likely explanation for this is that due to the faster scanning speeds resulting in less laser power per second striking an area of the surface, that whilst providing sufficient energy to process the PET, the temperature the PET reaches will likely be lower. This would result in an increase in the speed any molten polymer cools at, reducing the likelihood of any trapped air escaping. From the CCI images, it can be observed that the z scale of the surfaces decreases with increasing speed. Again, this relates to the laser power per second striking a surface. The longer the laser beam interacts with a spot on the surface, the greater the level of melting and ablation that occurs. Indeed, it can also be seen on the optical micrographs that the laser tracks appear to be significantly deeper at 100 mm/s than at 400 mm/s where there is comparatively little surface processing. Analysis of the roughness data supports this. Whilst the average

roughness shows no particular trend, the S_z of 100 mm/s is the largest (252 μm), with 300 and 400 mm/s showing the lowest values (165 and 192 μm). S_{sk} shows negative values for all samples, with the greater negative skewness, indicating a higher proportion of valleys, at the slowest scanning speeds. Again, the S_{ku} supports the theory of slower speeds yielding deeper valleys, with the average S_{ku} of 100 mm/s samples being over 12, whilst for all other speeds it is less than 10.

Figure 3 contains optical micrographs displaying the impact of laser processing power on PET surfaces. Laser powers range from 5 to 20%; all other parameters have been held constant. At a laser power of 5%, very little impact has been observed on the surface, whereas at higher powers tracks are visible. Bubbling is observed significantly at 10% power, a small amount at 15% power, and not at all at 20%. This relates to the aforementioned higher power per second causing hotter melted PET resulting in a longer cooling time allowing air pockets time to escape. There is no significant bubbling at 5% power due to the limited laser material interaction taking place. As demonstrated in the CCI maps, the width and depth of engraving increase significantly with laser power. The average of the mean and scale of the surface roughness is at its lowest for 5% power ($S_a = 9.5 \mu$m, $S_z = 132 \mu$m), compared to the other samples which all exhibit S_a values $> 10.7 \mu$m, and S_z values > 210 μm. The average skewness and kurtosis of the 5% power processed samples are also significantly lower, indicating fewer and smaller valleys being created in the PET surface

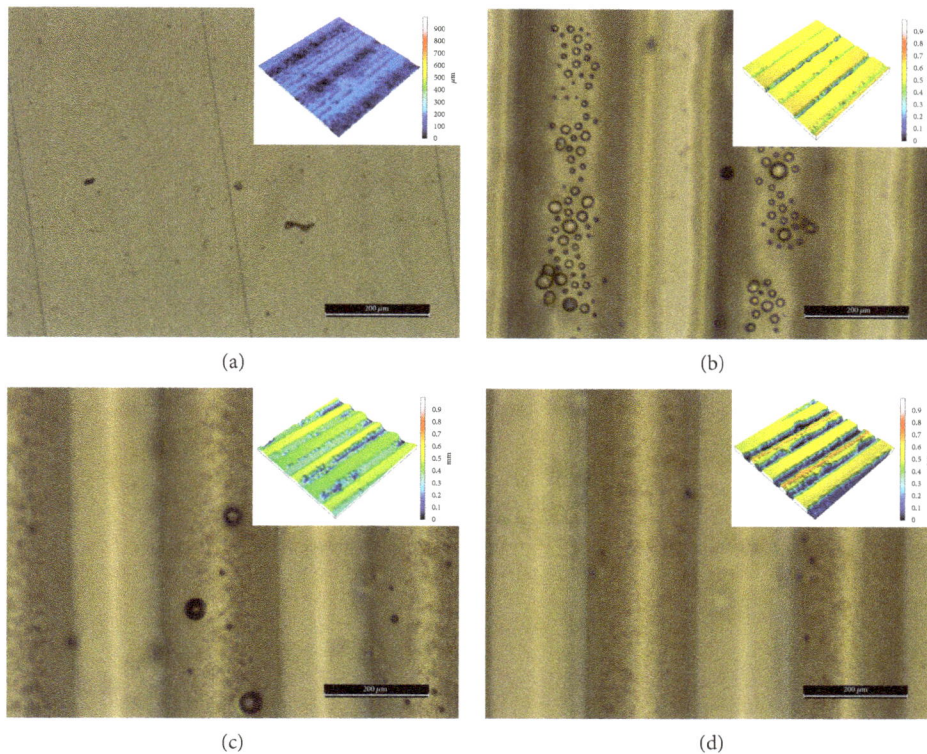

FIGURE 3: Optical micrographs of 4 laser-processed PET samples with increasing laser power. (a) 5%; (b) 10%; (c) 15%; (d) 20%. All samples have been processed at a scanning speed of 200 mm/s and at 400 μm spacing. The insets are CCI surface maps that have been set to the same z scale, and horizontally the scale bar is equal to 200 μm.

of the 5% sample which are −0.77 and 6.0, respectively. The other powers all exhibit an S_{sk} of <−1.45 and S_{ku} of >10.0. The caveat with all of these averaged roughness values is that each parameter has its own individual effect, which can confound other results. Despite this, there are visible trends and observations from this data that make sound logical sense when considering how an infrared laser interacts with PET.

3.2. Optical Properties. After characterising the surfaces of the samples, the optical properties have been measured. Considering the location of the PET layer in the PV cell, as either a substrate or part of the encapsulation system, and therefore incident to the incoming light, the two most important properties are the transmissivity, i.e., how much of the incoming light will pass through to the device, and the haze, i.e., how much the light is scattered. PET has a refractive index of ~1.57, which yields an approximate reflection loss of around 4–5% per surface interface. For analysis of the optical data, a wavelength range between 400 and 750 nm has been selected, although longer was measured. This range coincides with the majority of photon energy available to all PV active materials for absorption not only for the most common semiconductor materials, i.e., silicon, CdTe, and CIGS, but also and more importantly for the target PV materials for this light scattering technology, OPV and perovskites which often do not absorb towards the near-infrared [2]. Laser processing the samples appears to have a small impact on the transmissivity of the PET when compared to the AR sample. Figure 4(a) displays the average transmission of samples as a function of

laser parameter, whereas Figure 4(b) displays example transmission spectra of the AR PET sample and compares it to the sample that produced the highest haze value. In Figure 4(a), the values for each run with a particular setting have been averaged, in order to view the effect that a particular parameter shows. Many of the settings show a slight increase in transmission, ~1–2%. This could be because of two primary reasons. The first is that due to material removal due to the laser-material interaction. Less material will cause a slight reduction in absorption. However, due to the very low absorption that occurs within PET at these wavelengths, this would appear to have minimal impact. Another possibility is that the surface structures are acting in some way to partially trap light. The textured surface is nonflat, which allows for an increase in total internal reflection due to the change in angle of incidence relative to the surface. Another possibility is a pseudolensing effect caused by the regular ray of structures that have been created that may partially concentrate the light at a very small scale. Perhaps the most interesting effect is that of changing laser power on the transmissivity of the PET film. The lowest laser power demonstrates an average transmission value equivalent to that of the unprocessed PET, whereas the higher powers all display a slight increase to transmission. This would imply that at the lowest power, any structures created on the surface are comparatively small and having a minimal effect on the incident light.

Figure 5 shows the average haze of laser-processed PET films, again averaged between the wavelength range of 400–750 nm. The laser processing has generated impressive levels

(a)

(b)

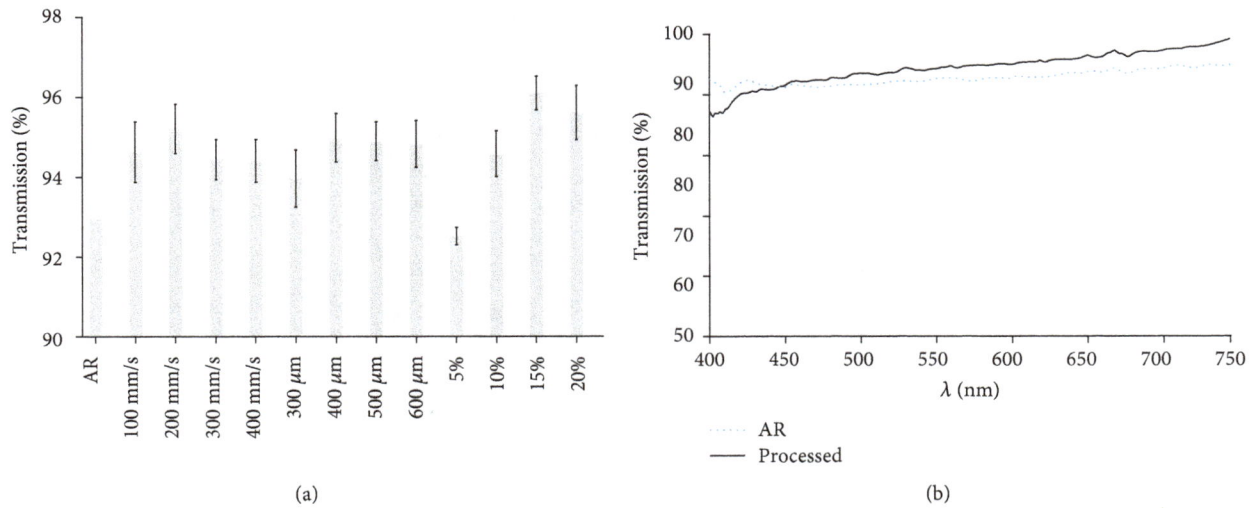

FIGURE 4: (a) The averaged transmission values from 400 to 750 nm for each of the laser parameters tested and the unprocessed AR sample. The error bars show the standard error of the mean. (b) A comparison between the transmission spectra of the as-received PET sample and those of the sample that produced the highest haze.

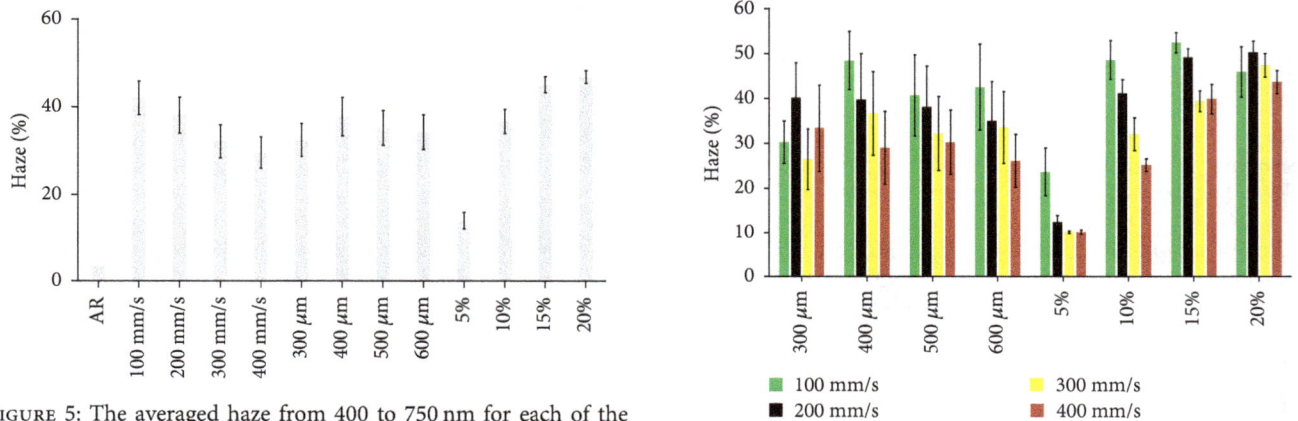

FIGURE 5: The averaged haze from 400 to 750 nm for each of the laser parameters tested and the unprocessed AR sample. The error bars show the standard error of the mean.

FIGURE 6: The averaged haze from 400 to 750 nm as a function of the scanning speed. The colours relate to the results separated by processing speed: green: 100 mm/s, black: 200 mm/s, yellow: 300 mm/s, and red: 400 mm/s. The error bars show the standard error of the mean.

of haze in the PET film. The AR PET has a very low level of native haze, ~3%, whereas the laser-engineered surfaces, in some cases, exhibit average haze levels in excess of 50%. Evaluating the light scattering properties of the laser-textured films also suggests that the power of the beam is the most important property for generating high haze surfaces. As the laser power increases, there appears to be a clear trend towards increased average haze. A deeper analysis is required, however, as laser parameters often have a similar effect. In particular (and as discussed previously), the scanning speed also heavily affects the power density per unit time of a laser, with a slower scanning speed resulting in a greater effective power density hitting a single spot. This is also visible to an extent in the haze levels produced.

Figure 6 shows the average haze of the PET films as a function of the laser scanning speed. It has previously been observed in this paper that the slower the scanning speed, the greater the feature, and thus the greater the scattering; however, this can confound the effect of the laser power on

the haze. At the slowest scanning speed, 100 mm/s, the average haze of the lowest power, 5%, is at its highest. The haze levels of the other three powers are very comparable at this speed, suggesting possibly the upper limit in the amount of haze creatable in these films by this method. As the scanning speed increases, the haze at a power of 5% decreases dramatically from >20% to <10%. Laser powers of 10% and 15% show their highest total haze values at the slowest scanning speed, with decreasing average haze as speed increases. However, at the highest power the levels of Haze remain quite constant. The line spacing of the samples appears to have limited effect on the haze of the sample once the spacing is sufficient to limit the effect of subsequent process lines on prior ones. The highest haze value obtained in this work is

Rapid, Chemical-Free Generation of Optically Scattering Structures in Poly(ethylene terephthalate) Using a CO2...

7

(a)

(b)

(c)

(d)

(e)

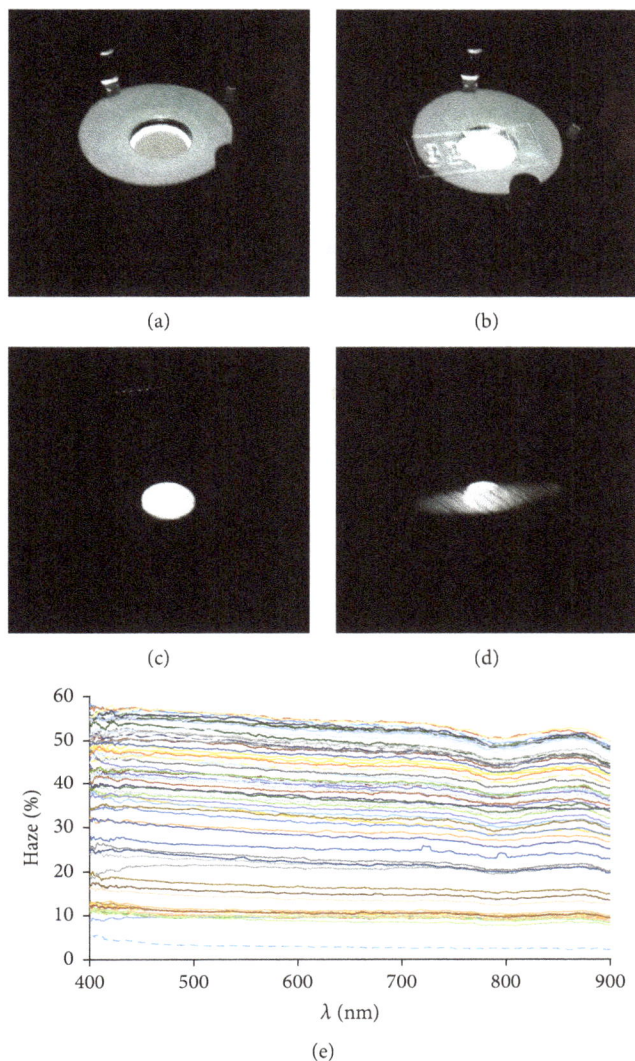

FIGURE 7: (a–d) Photographs showing the directional scattering caused by the laser-processed PET (b, d) relative to having no sample (a, c). The sample used for this image has a speed of 400 mm/s, a spacing of 400 μm, and a power of 20%. The average haze (400–750 nm) is 43.5%. The graph displays the measured haze of all 64 samples and demonstrates the range of haze values created. The dashed line in (e) is the AR sample.

from a sample with 100 mm/s scanning speed, 400 μm spacing and a power of 15% (the equivalent sample with 20% power was one of the two that destroyed the sample).

Another interesting observation is that of the highly directional scattering of light by the high haze structures. Figure 7 displays photographs showing unscattered light compared with the presence of a structured PET film. Instead of a wider circle of light leaving the integrating sphere, which would indicate uniform scattering in both the x and y directions, the light is scattered perpendicular to the directions of the laser marks. This is intriguing and gives the potential for combinations of differentially oriented structures to maximise light scattering effects. The haze measurements

for all the samples can also be observed in Figure 7, demonstrating the range of Haze values created by varying only three parameters. It should be noted that there is no consistent trend between the roughness values and the overall Haze. This would suggest that the light scattering is dependent on the production of features and patterning rather than microroughness alone.

3.3. Potential for Enhancing Photovoltaics and the Advantages of Laser Processing. Laser-structured PET clearly has the ability to improve the efficiency of solar cells by utilising light scattering mechanisms. The increased haze is comparable to many high haze films that have been demonstrated to increase some cell efficiencies by over 10% [33–36]. The advantages in the process described in this paper have already been discussed somewhat in the introduction; however, they bear repeating when combined with the discoveries within this work. The laser process is, most importantly for any industrial process, very fast. Films demonstrating up to 50% haze have been generated at scanning speeds of 400 mm/s using laser powers that are very low considering that CO_2 lasers in the kW levels of power output are commonly available. This leads to the chance of significantly faster processing speeds to counteract the possible power increases and certainly at levels sufficient to be incorporated as part of any batch or roll-to-roll process. This is particularly important as roll-to-roll production is one of the key selling points of OPV [13]. Perhaps equally important to the processing speed is the lack of complexity. There are many excellent publications demonstrating surface texturing (often at the sub 10-micron scale) of surfaces for a similar end goal [19, 37–40]. Many processes used for this can be time-consuming when compared to the laser process and are certainly more complicated than the "point and shoot" approach of industrial laser usage. The high levels of scattering observed here also demonstrate that complex nanostructures are unnecessary for the production of highly light scattering surfaces. This process can also be completely chemical-free, which offers the opportunity for both production savings and increased safety with the removal of potentially toxic chemicals from the production line. By texturing PET, the ability to improve highly novel forms of PV that use flexible polymer substrates has been demonstrated. Although a contender for this role, PET is not always the substrate used for flexible applications. It does, however, behave comparably to several other transparent polymer materials when irradiated with a CO_2 laser source making it a useful reference material. Indeed, there is evidence in the literature of CO_2 laser processing and/or machining of other polymers that are considered of importance as potential substrates for PV technologies including polyimide and polydimethylsiloxane [41–45].

Another important factor to consider is the effect the process may have on the transparent conducting layers (typically oxides like indium doped tin oxide, TCOs, or polymer mixtures like PEDOT:PSS, TCPs) that coat the substrate. These highly conductive layers act as the front contact of the PV device and as such are vital to attaining high levels of cell performance. The morphology of the organic films in an OPV cell has a significant effect on cell properties. For example, thicker films, whilst naturally absorbing a greater portion of the incident light, may lead to increased series resistance within the cell, and voids or other defects can increase the likelihood of short circuits leading to a reduction in the shunt resistance (both effects resulting in a reduced cell fill factor) [46–48]. A potential issue could be adverse heat transfer from the PET substrate to the TCO/TCP layer, caused by the laser process, distorting the layer and causing an increase in defects where the organic layer meets the front contact. Whilst this does have the potential to cause problems, these should be minor for two key reasons. Firstly, the processing takes place on the external facing side of the PET substrate and only the more extreme end of the processing conditions tested within this work also affects the internal side (see relative feature size on the inset surfaces on Figures 1–3). Secondly, the conductive coating can be applied after the PET has been laser processed. This would allow for an even TCO/TCP layer to be applied, reducing the likelihood of introducing microshorts. There is also the possibility of gaining a potentially advantageous texturing to the TCO/TCP layer through a templating effect, as it has been demonstrated in the past that texturing of or adding scattering centres into these layers can result in enhanced photon capture and/or superior junction formation, albeit primarily on glass substrates [46–49]. In order to adequately test this, further studies into processing precoated substrates and on coating pretextured substrates are required.

4. Conclusions

This paper has demonstrated a high-speed, simple, and chemical-free method for the production of highly light scattering surfaces for use in flexible PV applications. The parameters of laser power and scanning speed have the largest effect on the level of haze a textured surface produces. It has been observed that at higher powers and slower scanning speeds, there is significantly reduced trapping of air pockets, which is likely due to slower cooling allowing more time for air to escape. Films exhibit an increase in valleys, demonstrated by negative S_{sk} values, which can be considered as very large roughness features due to the high S_{ku}. S_a and S_z values demonstrate that when spacing gets to levels close to those of the spot size, heat transfer/remelting occurs which results in smaller feature sizes. The roughness values do not directly relate to the scattering of light, indicating that the light scattering is more dependent on feature size than explicit microroughness. The light scattering occurs in a direction perpendicular to the direction of the processed laser lines, allowing for the potential for directional applications or novel multidirectional structures. Finally, a brief discussion of how this process may result in improved PV cells has been included discussing both the advantages and some potential issues that may be caused and how they may be overcome. The sample that produced the highest measured haze between 400 and 750 nm had a scanning speed of 100 mm/s, a spacing of 400 μm, and a laser power of 15%. The next step in realising this technological approach to enhanced PV devices requires extensive modelling of the light scattering and testing on functional PV devices.

Conflicts of Interest

The authors declare that there is no conflict of interest regarding the publication of this paper.

Acknowledgments

The authors would like to thank the members of the Faculty of Science and Engineering at the University of Chester for useful conversations whilst undertaking this work, particularly the members of the departments of Natural Sciences and Mechanical Engineering. The work performed here was funded internally through the University of Chester.

References

[1] REN21, *Renewables 2018: Global Status Report*, REN21 Secretariat, Paris, 2018.

[2] M. A. Green, Y. Hishikawa, E. D. Dunlop, D. H. Levi, J. Hohl-Ebinger, and A. W. Y. Ho-Baillie, "Solar cell efficiency tables (version 52)," *Progress in Photovoltaics*, vol. 26, no. 7, pp. 427–436, 2018.

[3] W. Shockley and H. J. Queisser, "Detailed balance limit of efficiency of p-n junction solar cells," *Journal of Applied Physics*, vol. 32, no. 3, pp. 510–519, 1961.

[4] V. Fthenakis, "Sustainability of photovoltaics: the case for thin-film solar cells," *Renewable & Sustainable Energy Reviews*, vol. 13, no. 9, pp. 2746–2750, 2009.

[5] J. F. Yang, Y. Feng, R. Patterson, S. Huang, S. Shrestha, and G. Conibeer, "Theoretical investigation of plasmon enhanced optically-coupled hot carrier extraction," in *2015 IEEE 42nd Photovoltaic Specialist Conference (PVSC)*, New Orleans, LA, USA, 2015.

[6] S. D. Hodgson, G. Kartopu, S. L. Rugen-Hankey, A. J. Clayton, V. Barrioz, and S. J. C. Irvine, "Accessing the quantum palette: quantum-dot spectral conversion towards the BIPV application of thin-film micro-modules," *Journal of Optics*, vol. 17, no. 10, p. 105905, 2015, no. 105905.

[7] M. Yamaguchi, T. Takamoto, K. Araki, and N. Ekins-Daukes, "Multi-junction III-V solar cells: current status and future potential," *Solar Energy*, vol. 79, no. 1, pp. 78–85, 2005.

[8] V. M. Andreev, V. A. Grilikhes, V. P. Khvostikov et al., "Concentrator PV modules and solar cells for TPV systems," *Solar Energy Materials and Solar Cells*, vol. 84, no. 1–4, pp. 3–17, 2004.

[9] M. Gratzel, "Recent advances in sensitized mesoscopic solar cells," *Accounts of Chemical Research*, vol. 42, no. 11, pp. 1788–1798, 2009.

[10] M. M. Lee, J. Teuscher, T. Miyasaka, T. N. Murakami, and H. J. Snaith, "Efficient hybrid solar cells based on meso-superstructured organometal halide perovskites," *Science*, vol. 338, no. 6107, pp. 643–647, 2012.

[11] M. A. Green, A. Ho-Baillie, and H. J. Snaith, "The emergence of perovskite solar cells," *Nature Photonics*, vol. 8, no. 7, pp. 506–514, 2014.

[12] A. Isakova and P. D. Topham, "Polymer strategies in perovskite solar cells," *Journal of Polymer Science Part B: Polymer Physics*, vol. 55, no. 7, pp. 549–568, 2017.

[13] J. E. Carle and F. C. Krebs, "Technological status of organic photovoltaics (OPV)," *Solar Energy Materials and Solar Cells*, vol. 119, pp. 309-310, 2013.

[14] I. M. Peters, H. Hauser, N. Tucher, and B. Blasi, "Optical modeling of honeycomb textures for multicrystalline silicon solar cells," *IEEE Journal of Photovoltaics*, vol. 6, no. 6, pp. 1480–1487, 2016.

[15] H. Melkonyang, S. Saylan, A. Heidelberg et al., "Submicron texturing for broadband light management in thin-film PV," in *Proceedings Volume 8823, Thin Film Solar Technology V*, vol. 8823, p. 2013, San Diego, CA, USA, 2013.

[16] S. Kontermann, T. Gimpel, A. L. Baumann, K. M. Guenther, and W. Schade, "Laser processed black silicon for photovoltaic applications," *Energy Procedia*, vol. 27, pp. 390–395, 2012.

[17] M. Batmunkh, T. J. Macdonald, W. J. Peveler et al., "Plasmonic gold nanostars incorporated into high-efficiency perovskite solar cells," *ChemSusChem*, vol. 10, no. 19, pp. 3750–3753, 2017.

[18] G. Zhang, S. Finefrock, D. Liang et al., "Semiconductor nanostructure-based photovoltaic solar cells," *Nanoscale*, vol. 3, no. 6, pp. 2430–2443, 2011.

[19] D. J. Rogers, V. E. Sandana, S. Gautier et al., "Core–shell GaN–ZnO moth-eye nanostructure arrays grown on a-SiO$_2$ /Si (1 1 1) as a basis for improved InGaN-based photovoltaics and LEDs," *Photonics and Nanostructures - Fundamentals and Applications*, vol. 15, pp. 53–58, 2015.

[20] T. E. Lizotte and T. R. Okeeffe, "Chemical free cleaning using excimer lasers," *Conference on Lasers as Tools for Manufacturing of Durable Goods and Microelectronics*, 1996, pp. 279–287, San Jose, Ca, USA, 1996.

[21] P. Waurzyniak, "Laser cutting and welding technologies advance," *Manufacturing Engineering*, vol. 147, no. 2, p. 69, 2011.

[22] A. Gillett, D. Waugh, J. Lawrence, M. Swainson, and R. Dixon, "Laser surface modification for the prevention of biofouling by infection causing Escherichia coli," *Journal of Laser Applications*, vol. 28, no. 2, article 022503, 2016.

[23] R. Narayan and P. Goering, "Laser micro- and nanofabrication of biomaterials," *MRS Bulletin*, vol. 36, no. 12, pp. 973–982, 2011.

[24] Y. Cheng, K. Sugioka, and K. Midorikawa, "3D integration of microoptics and microfluidics in glass using femtosecond laser direct writing," in *Proceedings SPIE 5662, Fifth International Symposium on Laser Precision Microfabrication*, Nara, Japan, October 2004.

[25] R. Cristescu, G. Socol, I. N. Mihailescu et al., "New results in pulsed laser deposition of poly-methyl-methacrylate thin films," *Applied Surface Science*, vol. 208-209, pp. 645–650, 2003.

[26] U. Bielesch, J. Uhlenbusch, and W. Viol, "Q-switched low pressure CO2 laser," in *Proceedings Volume 2502, Gas Flow and Chemical Lasers: Tenth International Symposium*, pp. 31–37, Friedrichshafen, Germany, 1995.

[27] T. Sakai and N. Hamada, "A high-power Q-switchedco2laser using intense pulsed RF discharge excitation," *Japanese*

Journal of Applied Physics, vol. 35, Part 1, No. 6A, pp. 3428–3435, 1996.

[28] A. K. Dubey and V. Yadava, "Laser beam machining - a review," *International Journal of Machine Tools and Manufacture*, vol. 48, no. 6, pp. 609–628, 2008.

[29] D. E. Duvall, "Environmental degradation of pet and its potential effect on long-term mechanical-properties of oriented pet products," *Polymer-Plastics Technology and Engineering*, vol. 34, no. 2, pp. 227–242, 1995.

[30] S. Ebadian and P. Servati, "Reducing the roughness of transparent electrodes in organic photovoltaic devices on plastic substrate by PEDOT:PSS treatment," *2009 IEEE Nanotechnology Materials and Devices Conference*, 2009, pp. 221–224, Traverse City, MI, USA, June 2009.

[31] C. Roldán-Carmona, O. Malinkiewicz, A. Soriano et al., "Flexible high efficiency perovskite solar cells," *Energy & Environmental Science*, vol. 7, no. 3, pp. 994–997, 2014.

[32] G. Oreski and G. M. Wallner, "Aging mechanisms of polymeric films for PV encapsulation," *Solar Energy*, vol. 79, no. 6, pp. 612–617, 2005.

[33] J.-m. Liu, X.-l. Chen, J. Fang, Y. Zhao, and X.-d. Zhang, "High-haze and wide-spectrum hydrogenated MGZO TCO films on micro-textured glass substrates for thin-film solar cells," *Solar Energy Materials and Solar Cells*, vol. 138, pp. 41–50, 2015.

[34] A. Hongsingthong, T. Krajangsang, A. Limmanee, K. Sriprapha, J. Sritharathikhun, and M. Konagai, "Development of textured ZnO-coated low-cost glass substrate with very high haze ratio for silicon-based thin film solar cells," *Thin solid films*, vol. 537, pp. 291–295, 2013.

[35] K. Li, Y. Zhang, H. Zhen et al., "Versatile biomimetic haze films for efficiency enhancement of photovoltaic devices," *Journal of Materials Chemistry A*, vol. 5, no. 3, pp. 969–974, 2017.

[36] P. Vincent, D. S. Song, H. B. Kwon et al., "Towards maximizing the haze effect of electrodes for high efficiency hybrid tandem solar cell," *Applied Surface Science*, vol. 432, pp. 262–265, 2018.

[37] M. G. Salvaggio, R. Passalacqua, S. Abate et al., "Functional nano-textured titania-coatings with self-cleaning and antireflective properties for photovoltaic surfaces," *Solar energy*, vol. 125, pp. 227–242, 2016.

[38] C. Trompoukis, A. Herman, O. El Daif et al., "Enhanced absorption in thin crystalline silicon films for solar cells by nanoimprint lithography," in *Proceedings Volume 8438, Photonics for Solar Energy Systems IV*, Brussels, Belgium, 2012.

[39] Y. F. Liu, G. J. Blayney, and O. J. Guy, "Rapid microwave growth of ZnO nanowires for low cost photovoltaics cells using reclaimed silicon substrates," in *2012 IEEE International Conference on Electron Devices and Solid State Circuit (EDSSC)*, Bangkok, Thailand, December 2012Chulalongkorn Univ.

[40] K. Watanabe, T. Inoue, H. Sodabanlu, M. Sugiyama, and Y. Nakano, "Thin-film solar cells with InGaAs/GaAsP multiple quantum wells and a rear surface etched with light trapping micro-hole array," *Japanese Journal of Applied Physics*, vol. 54, no. 8S1, article 08KA13, 2015.

[41] K. Znajdek, M. Sibinski, A. Strakowska, and Z. Lisik, "Polymer substrates for flexible photovoltaic cells application in personal electronic system," *Opto-Electronics Review*, vol. 24, no. 1, 2016.

[42] X. Jin, D. Zhu, J. Liu, and X. Zeng, "Mechanism and process of fabricating fluorinated polyimide optical waveguide by CO_2 laser direct-writing," *Optical and Quantum Electronics*, vol. 43, no. 11–15, pp. 163–174, 2012.

[43] K. Coupland, P. R. Herman, and B. Gu, "Laser cleaning of ablation debris from CO_2-laser-etched vias in polyimide," *Applied Surface Science*, vol. 127-129, pp. 731–737, 1998.

[44] B. A. Fogarty, K. E. Heppert, T. J. Cory, K. R. Hulbutta, R. S. Martin, and S. M. Lunte, "Rapid fabrication of poly(dimethyl-siloxane)-based microchip capillary electrophoresis devices using CO_2 laser ablation," *The Analyst*, vol. 130, no. 6, pp. 924–930, 2005.

[45] M. T. Khorasani and H. Mirzadeh, "Laser surface modification of silicone rubber to reduce platelet adhesion in vitro," *Journal of Biomaterials Science, Polymer Edition*, vol. 15, no. 1, pp. 59–72, 2004.

[46] J. Xu, Z. Hu, K. Zhang, L. Huang, J. Zhang, and Y. Zhu, "Enhancement in photocurrent through efficient geometrical light trapping in organic photovoltaics," *Energy Technology*, vol. 4, no. 2, pp. 314–318, 2016.

[47] R. M. Howden, E. J. Flores, V. Bulovic, and K. K. Gleason, "The application of oxidative chemical vapor deposited (oCVD) PEDOT to textured and non-planar photovoltaic device geometries for enhanced light trapping," *Organic Electronics*, vol. 14, no. 9, pp. 2257–2268, 2013.

[48] C. Cho, H. Kim, S. Jeong et al., "Random and V-groove texturing for efficient light trapping in organic photovoltaic cells," *Solar energy materials and solar cells*, vol. 115, pp. 36–41, 2013.

[49] Y. Park, L. Muller-Meskamp, K. Vandewal, and K. Leo, "PEDOT:PSS with embedded TiO_2 nanoparticles as light trapping electrode for organic photovoltaics," *Applied Physics Letters*, vol. 108, no. 25, article 253302, 2016.

Comparative Performance Analysis of Grid-Connected PV Power Systems with Different PV Technologies in the Hot Summer and Cold Winter Zone

Chong Li ⑩ [1,2]

[1]*School of Electronic Engineering, Nanjing Xiaozhuang University, Nanjing 211171, China*
[2]*College of Economics and Management, Nanjing University of Aeronautics and Astronautics, Nanjing 211106, China*

Correspondence should be addressed to Chong Li; chongli630@163.com

Academic Editor: Giulia Grancini

The objective of this paper is to establish the performance of $8\,kW_p$ grid-connected photovoltaic (PV) power systems based on different PV module technologies in Nanjing, China. Nanjing has a hot summer and a cold winter which are considered based on monthly average solar irradiation and ambient temperature specifically for the deployment of grid-connected PV systems. The study focuses on performance assessment of grid-connected PV systems using typical PV modules made of monocrystalline silicon (m-Si), polycrystalline silicon (p-Si), edge-defined film-fed growth silicon (EFG-Si), cadmium telluride (CdTe) thin film, copper indium selenide (CIS) thin film, heterojunction with intrinsic thin layer (HIT), and hydrogenated amorphous silicon single-junction (a-Si:H single-PV) installed on location. The yearly average energy output, PV module and system efficiency, array yield, final yield, reference yield, performance ratio, monthly average array capture losses, and system losses of seven PV module technologies are all analyzed. The results show that grid-connected PV power system performance depends on geographical location, PV module types, and climate conditions such as solar radiation and ambient temperature. In addition, based on energy output and efficiency, the HIT PV power technology can be considered as the best option and CdTe and p-Si as the least suitable options for this area. The monthly average performance ratio of the CdTe technology was higher than those of other technologies over the monitoring period in Nanjing.

1. Introduction

Thermal power generation has a number of shortcomings, including consumption of coal, oil, and natural gas resources, serious environmental pollution, and low utilization rate. Solar energy is universal, nonpolluting, vast, noiseless, safe, and inexhaustible. So far, solar energy has experienced vigorous growth around the world and is one of the most prominent renewable energy sources. Therefore, more attention has been paid recently to solar PV power generation to reduce greenhouse gas emissions. The global cumulative installed capacity of solar PV power systems has increased rapidly over the past decade [1–3].

Nanjing ($32°02'38''$N, $118°46'43''$E, approximately 67.9 m above sea level) is the capital of Jiangsu province of China and is located in the southwestern part of Jiangsu

province, close to Anhui province, as shown in Figure 1 [4, 5]. The city covers a total land area of 6597 square kilometers with an estimated population of 8.23 million. Nanjing has a hot summer and a cold winter with a north subtropical monsoon climate with distinct seasons, plentiful sunshine, and rain. The four seasons are distinct in this region, with hot summers and cold winters. The average annual temperature is about 17.8°C, and precipitation is about 1106 millimeters per year. The annual sunshine is about 2200–3000 hours in Nanjing, and the city has good light conditions, making this region suitable to build a certain capacity of PV power stations. The chemical industry is the main component of heavy industry in Nanjing, creating high energy demand and environmental pressures. In the long term, Nanjing should develop renewable energy sources such as solar energy [6–8].

FIGURE 1: Geographical location of Nanjing, China [5].

In recent years, many researchers have studied the performance of grid-connected solar PV power systems in different locations. Kymakis et al. [9] analyzed the performance of a grid-connected PV park on the island of Crete. The performance parameters that were calculated included reference yield, array yield, final array yield, capture losses, system losses, and performance ratio. Ayompe et al. [10] analyzed the performance of a 1.72 kW$_p$ grid-connected PV system installed in Dublin, Ireland, and its performance parameters were evaluated on a monthly, seasonal, and annual basis. Wittkopf et al. [11] assessed the performance of a 142.5 kW$_p$ grid-connected BIPV system on the roof of a zero-energy building in Singapore. They found that the performance ratio was lowest for clear days, with high irradiance fluctuations due to higher capture rates and system losses. Al-Sabounchi et al. [12] evaluated the performance of a distributed PV generation system rated at 36 kW and installed at ground level in the Abu Dhabi industrial area. Results showed that the amount of accumulated dust deposition affected the system seriously. Padmavathi and Daniel [13] examined the performance of a 3 MW grid-connected solar PV plant located in Karnataka State, India. Energy yields, system losses, and component efficiencies were discussed in detail. Farhoodnea et al. [14] presented the performance of an in-campus 3 kW$_p$ grid-connected polycrystalline silicon PV system using experimental data. Six-month performance data for the system installed at the Universiti Kebangsaan Malaysia campus were used. Sundaram and Babu [15] analyzed the performance of a 5 MW$_p$ grid-connected PV system using the RETscreen software in Sivagangai, India. The annual average daily array yield, reference yield, final yield, module efficiency, inverter efficiency, system efficiency, capture loss, and system loss were discussed. Okello et al. [16] analyzed the performance of a 3.2 kW$_p$ grid-connected PV system at the Nelson Mandela Metropolitan University, South Africa. System performance was simulated using PVsyst software with measured and Meteonorm-derived climate data sets. Mpholo et al. [17] evaluated the performance of a newly installed 281 kW$_p$ first grid-connected solar PV farm in Lesotho. The results showed that the area was suitable for grid-connected PV systems. Sidi et al. [18] analyzed the performance of the first large-scale solar PV plant in

Mauritania. Some performance indices such as reference yield, array yield, system yield, performance ratio, capacity factor, and array capture losses were calculated. The results showed that the PV plant performance depended on both insolation and environmental conditions. Dabou et al. [19] presented the effect of weather conditions on the performance of a 1.75 kW$_p$ grid-connected PV system installed in southern Algeria. The final yield, reference yield, performance ratio, and system efficiency were discussed. Allouhi et al. [20] assessed the performance of two 2 kW$_p$ grid-connected PV systems in Meknes. Monthly and annual performance indicators such as total energy generated, final yield, capacity factor, and overall system efficiency were evaluated and compared for mono-Si and poly-Si PV technologies using the PVsyst software. Maammeur et al. [21] investigated the performance of a grid-connected PV system for family farms in northwestern Algeria. They found that 49% of total on-farm electricity consumption came from renewable energy. Lima et al. [22] carried out a performance analysis of a 2.2 kW$_p$ PV system installed at the State University of Ceará, Fortaleza, Brazil. The results showed that the PV systems performed well.

Up to now, few researchers have focused on the comparative performance of grid-connected solar PV power systems with different PV modules under hot summer and cold winter conditions. There are two main research methodologies available to study the performance of grid-connected PV systems: numerical simulation and experimental investigation methods. The advantages of the numerical simulation method are that it is convenient, quick, low cost, etc. [14, 23–25]. Therefore, this study compares the performance of seven PV technologies under this climate condition using the PVsyst V5.74 tool [26]. Simulations were carried out on the performance of different PV modules, taking into account the solar radiation and ambient temperature of Nanjing.

2. Description of the Grid-Connected Solar PV Power System

Figure 2 shows the proposed grid-connected solar PV system. Its main components are PV arrays, DC/DC converters,

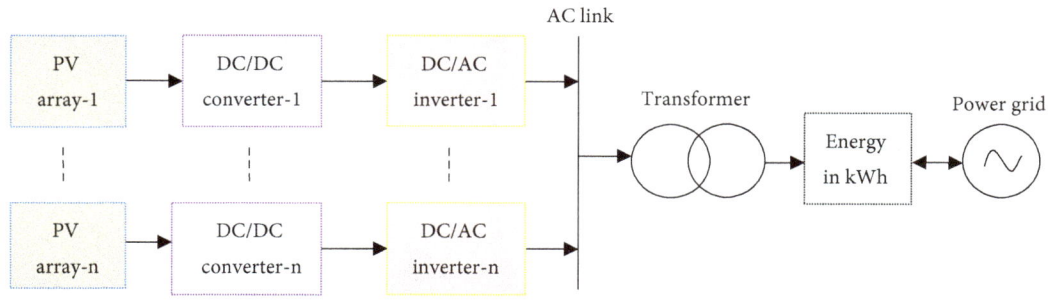

FIGURE 2: Grid-connected PV power system.

DC/AC inverters, a transformer, DC and AC cables, and other accessory devices. In this study, the grid-connected PV system was used without batteries. The PV system consisted of 806 modules covering a total area of $77 \, m^2$ with an installed capacity of $8 \, kW_p$. HB-1210 $10 \, W_p$ modules were used. The modules were tilted at a fixed angle of $23°$ facing south. The Sunway TG 10-600V inverter was used to transform the voltage from DC to AC and was connected to the utility grid. The inverter had a rated maximum efficiency of 95.3% and maximum AC power of 8.7 kW [27–29].

2.1. PV Modules. The m-Si, p-Si, EFG-Si, CdTe thin-film, CIS thin-film, HIT, and a-Si:H single-PV modules were chosen from the PVsyst product database. Table 1 shows the main specifications of the PV module types at standard testing conditions (STC): $1 \, kW/m^2$ solar irradiance, 25°C cell temperature, and 1.5 air mass (AM) [18].

3. Methodology

3.1. Meteorological Data for Nanjing. Local meteorological data, such as solar irradiation and ambient temperature, are important factors for estimating solar PV power system performance. Monthly meteorological data for Nanjing were available with the help of the METEONORM V7.0 software, which was used to generate hourly synthetic meteorological data [30].

Figure 3 shows monthly average solar irradiation parameters for horizontal global irradiation, horizontal diffuse irradiation, global incident irradiation in the collector plane (tilt angle = 23°, azimuth angle = 0°, albedo = 20%), and ambient temperature for Nanjing. Monthly average horizontal global irradiation varied from $61 \, kWh/m^2$ in January and December to $145 \, kWh/m^2$ in July. Monthly average horizontal diffuse irradiation varied from $37 \, kWh/m^2$ in January to $95 \, kWh/m^2$ in May and July. Monthly average direct normal radiation varied from $35 \, kWh/m^2$ in May to $72 \, kWh/m^2$ in July. Monthly average ambient temperature varied from 3°C in January to 28.7°C in July, and the average ambient temperature over the whole year was 16.4°C.

3.2. Performance Parameters

3.2.1. Energy Generated by the Solar PV System. The total daily, monthly, and yearly alternating current (AC) energy

generated by the solar PV system over a given period can be obtained as [22, 31]

$$E_{(AC,d)} = \sum_{h=1}^{24} E_{(AC,h)},$$

$$E_{(AC,m)} = \sum_{d=1}^{n} E_{(AC,d)}, \quad (1)$$

$$E_{(AC,y)} = \sum_{m=1}^{12} E_{(AC,m)},$$

where $E_{(AC,h)}$ is the hourly AC energy output at time h (kWh), $E_{(AC,d)}$ is the daily AC energy output (kWh), $E_{(AC,m)}$ is the monthly AC energy output (kWh), $E_{(AC,y)}$ is the yearly AC energy output (kWh), and n is the number of days in one month.

3.2.2. Array Yield (Y_A). The array yield is the ratio of daily, monthly, or yearly direct current (DC) energy output from a PV array to the rated PV array power and is given by [10]

$$Y_A = \frac{E_{DC}}{P_{PV,rated}}, \quad (2)$$

where E_{DC} is the total DC energy output from the PV arrays (kWh) and $P_{PV,rated}$ is the rated output power of the PV system (kW_p).

3.2.3. Final Yield (Y_F). The final yield can be defined as the total AC energy during a given period divided by the rated PV array power and is given by [18]

$$Y_F = \frac{E_{AC}}{P_{PV,rated}}, \quad (3)$$

where E_{AC} is the total AC energy output from the inverter generated by the PV power system for a specific period (kWh).

3.2.4. Reference Yield (Y_R). The reference yield is the ratio of total in-plane solar radiation to the reference irradiance at standard test conditions (STC). It represents the total in-

TABLE 1: Main characteristics of the PV module types in the proposed system.

PV module type	m-Si	p-Si	EFG-Si	CdTe	CIS	HIT	a-Si:H single-PV
Model	GES-5 M5	HB_1210	DSM 24	CX-50	WSK 0001	VBHN230SE51	Flexcell 27 W
Manufacturer	GESOLAR	HBL Power Systems Ltd	Van de Loo	Calyxo	Wurth Solar	Panasonic	VHF Technologies
Nominal power at STC (W_P)	5	10	24	50	5.5	230	27
Module dimensions ($L \times W \times T$)	$0.306 \times 0.216 \times 0.018$ m	$0.28 \times 0.34 \times 0.022$ m	$0.546 \times 0.452 \times 0.052$ m	$1.2 \times 0.6 \times 0.069$ m	$0.305 \times 0.205 \times 0.031$ m	$1.58 \times 0.798 \times 0.035$ m	$1.22 \times 0.548 \times 0.0012$ m
Area (m^2)	0.066	0.095	0.247	0.72	0.063	1.261	0.669
Weight (kg)	0.8	2	3.5	12	1	15	1.5
Modules in series	29	31	30	7	29	12	27
Modules in parallel	55	26	11	23	50	3	11
Short circuit current (A)	0.32	0.67 A	1.57	1.06	0.37	5.78	2.4
Open circuit voltage (V)	21.6	21	21	86.5	22	52.1	23
Maximum power current (A)	0.29	0.588	1.43	0.83	0.34	5.42	1.8
Maximum power voltage (V)	17.4	17	16.8	60	16.5	42.5	15
Temperature coefficient of maximum power (%/°C)	−0.031	−0.666	−0.06	−0.021	−0.051	−0.03	−0.07

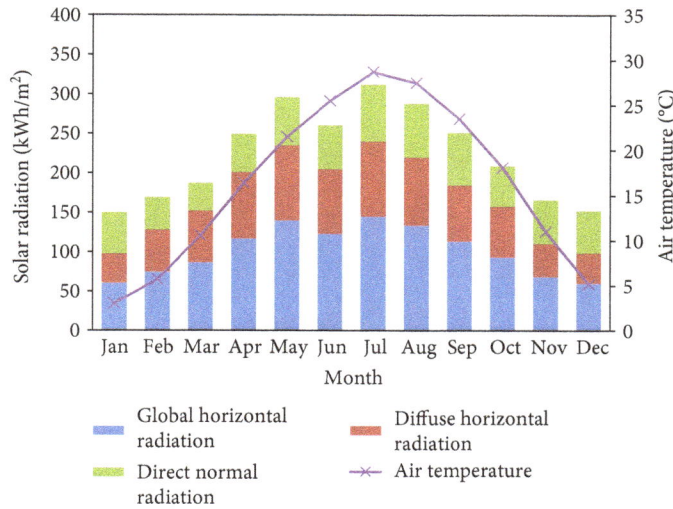

FIGURE 3: Mean monthly solar radiation and ambient temperature for Nanjing.

plane solar radiation or an equivalent number of hours at the reference irradiance and is given by [10, 18]

$$Y_R = \frac{S_R}{H_R}, \tag{4}$$

where S_R is the total in-plane solar radiation (kWh/m²) and H_R is the array reference irradiance at STC (1 kW/m²).

3.2.5. Performance Ratio (PR). The performance ratio (PR) is the ratio of the final energy yield of the PV system to the reference yield. It provides information about the overall losses incurred in converting DC to AC power. Therefore, it represents the percentage of energy actually available after deducting energy losses [29, 32, 33]:

$$PR = \frac{Y_F}{Y_R}. \tag{5}$$

3.2.6. PV Module Efficiency (η_{PV}). The PV module efficiency is calculated as [10]

$$\eta_{PV} = \frac{E_{DC}}{S_R A_{PV}} \times 100\%, \tag{6}$$

where A_{PV} is the PV module total area (m²).

3.2.7. PV System Efficiency (η_S). The overall PV system conversion efficiency is defined as the energy output from a PV array divided by the total in-plane solar insolation and is given as [10]

$$\eta_S = \frac{E_{AC}}{S_R A_{PV}} \times 100\%. \tag{7}$$

3.2.8. Array Capture Loss. Array capture loss (L_A) is due to PV array losses and is given as [10, 18]

$$L_A = Y_R - Y_A. \tag{8}$$

3.2.9. System Loss. System loss (L_s) is due to inverter inefficiencies and is calculated as [10, 18]

$$L_S = Y_A - Y_F. \tag{9}$$

4. Results and Discussion

In this section, the results of yearly average energy output, PV module and system efficiency, array yield, final yield, reference yield, performance ratio, monthly average array capture losses, and system losses for seven different PV module technologies are discussed.

Figure 4 shows the effective energy output as well as PV array and system efficiency. The yearly average energy outputs of the systems were 8831.1 kWh, 9556 kWh, 8584.4 kWh, 9570 kWh, 9411.2 kWh, 9606 kWh, and 8424.1 kWh for m-Si, p-Si, EFG-Si, CdTe, CIS, HIT, and a-Si:H systems, respectively. The energy output from the p-Si PV system was similar to those from the CdTe and HIT PV systems. The maximum PV module and system efficiencies were 15.7% and 14.73%, respectively, for the HIT PV module, whereas the minimum efficiencies were 3.15% and 2.95%, respectively, for the a-Si:H single-PV module system. Therefore, based on energy output and efficiency, HIT PV systems can be considered as the best option and CdTe and p-Si the least suitable options for this location [34].

Figure 5 shows that the monthly array yields for different PV module technologies were different for different months. The array yields of these power systems varied from a minimum value of 2.17 h/d in January for an a-Si:H single-PV module system to a maximum value of 4.18 h/d in August for a p-Si PV module system.

Figure 6 shows the monthly average final yield of the PV module technologies tested. The AC energy fed to the grid is indicated by the final yield [29]. The a-Si:H single-PV technology had the lowest monthly average final yield in January, with a value of 2.03 h/d, and a high value of 3.38 h/d in

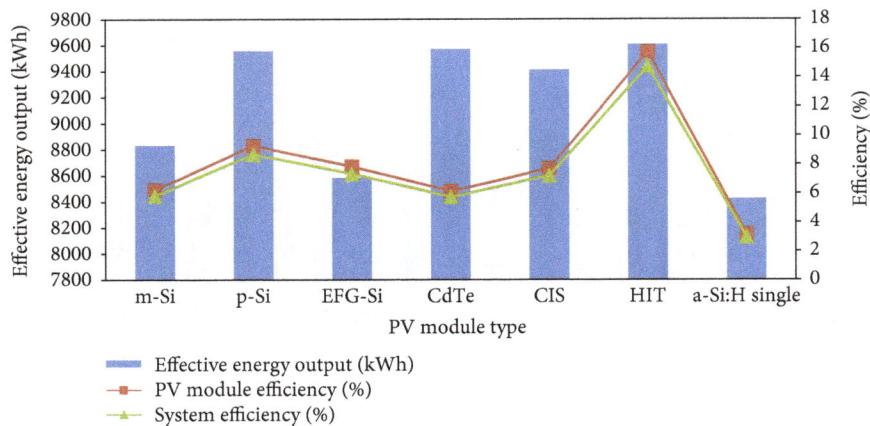

FIGURE 4: Yearly average energy output, PV array, and system efficiency of seven PV module technologies.

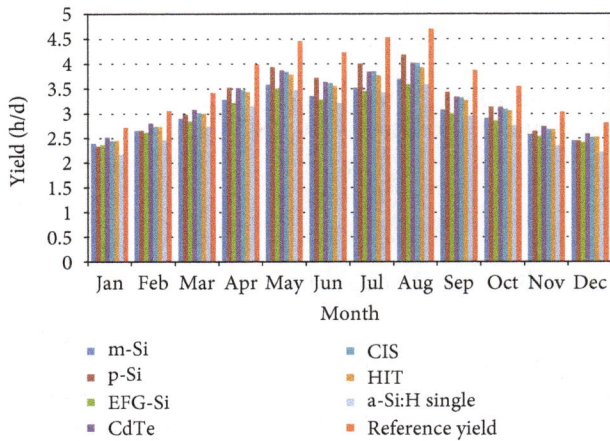

FIGURE 5: Monthly average array yield of seven PV module technologies compared with the reference yield.

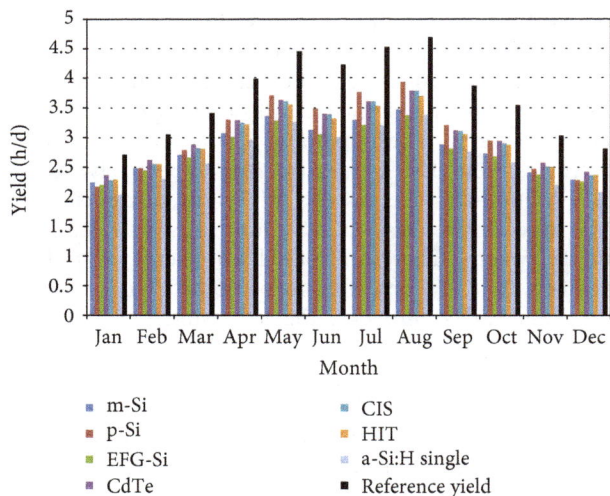

FIGURE 6: Monthly average final yield of seven PV module technologies compared with the reference yield.

August. The p-Si PV technology has the highest monthly average final yield in August, with a value of 3.93 h/d, and a low value of 2.17 h/d in January.

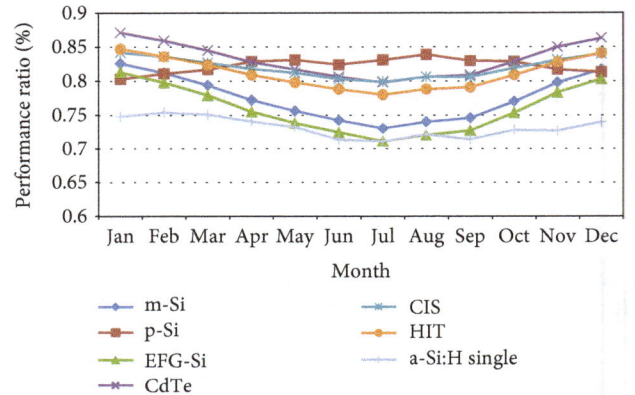

FIGURE 7: Monthly average performance ratio of seven PV module technologies.

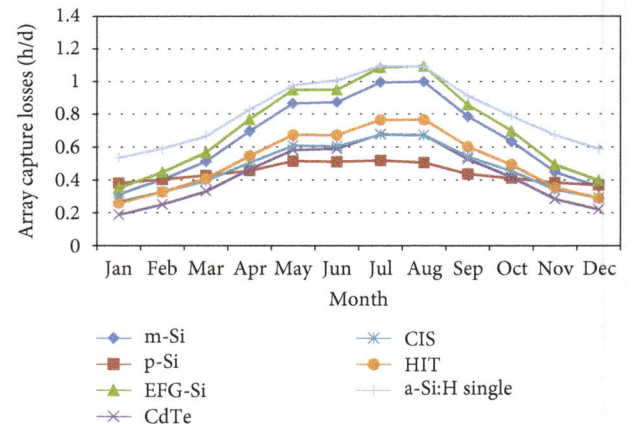

FIGURE 8: Monthly average array losses of seven PV module technologies.

Figure 7 shows the monthly average performance ratio of each PV module technology. The monthly average performance ratio of the CdTe technology was higher than that of the other technologies over the monitored period. However, the monthly average performance ratio of the a-Si:H single-PV technology was lower than that of the other technologies

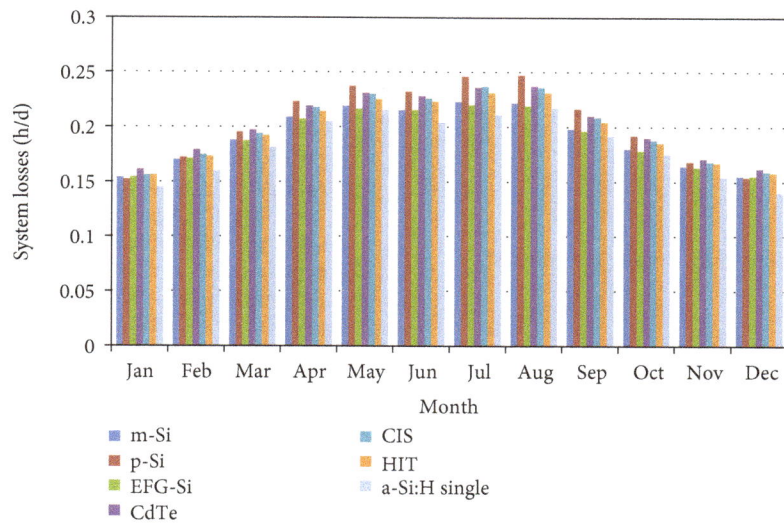

FIGURE 9: Monthly average system losses of seven PV module technologies.

over the twelve months. Note that the monthly average performance ratios of the m-Si, EFG-Si, CdTe, CIS, and HIT modules decreased from January to July but increased from July to December. It is clear that the PRs of the m-Si, EFG-Si, CdTe, CIS, and HIT were lowest in July. The variation in the monthly average performance ratio of the a-Si:H single-PV module is similar to the variation in the m-Si, EFG-Si, CdTe, CIS, and HIT modules. However, the monthly average performance ratio of the p-Si increased from January to May, remained basically identical from May to July, and then decreased from August to December.

Figure 8 shows the monthly average array capture losses of each PV module technology. The monthly average array capture losses of the a-Si:H single-PV module were higher than those of the other technologies over the monitored period. Monthly array capture losses were relatively higher in summer (June to August) than in other seasons. This may have been due to high solar irradiation and air temperature.

Figure 9 shows the monthly average system losses of each PV module technology. The monthly average system losses of the p-Si module were higher than those of the other technologies over the monitored period. Monthly array system losses were also higher in summer and lower in winter (December to February). The maximum monthly array system loss for the p-Si module due to the temperature effect was 0.247 h/d, which occurred in August, whereas the minimum value for the a-Si:H single-PV module was 0.14 h/d, which occurred in December.

5. Conclusions

The performance of 8 kW$_p$ grid-connected solar PV power systems based on seven PV module technologies (m-Si, p-Si, EFG-Si, CdTe, CIS, HIT, and a-Si:H single-PV) installed in Nanjing, China, has been investigated.

The yearly average energy output of the seven PV systems was 8831.1 kWh, 9556 kWh, 8584.4 kWh, 9570 kWh,

9411.2 kWh, 9606 kWh, and 8424.1 kWh for m-Si, p-Si, EFG-Si, CdTe, CIS, HIT, and a-Si:H single-PV, respectively. The maximum PV module and system efficiencies were 15.7% and 14.73%, respectively, for the HIT PV module, whereas the minimum efficiencies were 3.15% and 2.95%, respectively, for the a-Si:H single-PV module system. The array yield of these PV systems varied from a low value of 2.17 h/d in January for the a-Si:H single-PV system to a high value of 4.18 h/d in August for the p-Si PV system. The a-Si:H single-PV technology has the lowest monthly average final yield in January, with a value of 2.03 h/d, and a high value of 3.38 h/d in August. The p-Si technology had the highest monthly average final yield in August, with a value of 3.93 h/d, and a low value of 2.17 h/d in January.

The monthly average performance ratio of the CdTe technology was higher than those of the other technologies over the monitored period. However, the monthly average performance ratio of the a-Si:H single-PV technology was lower than those of the other technologies over the twelve months. The monthly average array capture losses of the a-Si:H single-PV module were higher than those of the other technologies over the monitored period. Monthly array capture losses were higher in summer (June to August) than in other seasons. The maximum monthly array system loss for the p-Si module due to temperature was 0.247 h/d and occurred in August, whereas the minimum value for the a-Si:H single-PV module was 0.14 h/d and occurred in December.

Data Availability

Monthly meteorological data for Nanjing were available with the help of the METEONORM V7.0 software, which was used to generate hourly synthetic meteorological data. This study compares the performance of grid-connected solar PV power systems with seven PV technologies under hot summer and cold winter conditions using the PVsyst V5.74 tool. Simulations were carried out on the performance of different PV modules, taking into account the solar radiation and ambient temperature of Nanjing. I have finished writing this article aided by the METEONORM V7.0 and PVsyst V5.74 simulation software. Data is done through software. There is no experimental data.

Conflicts of Interest

The author declares that there is no conflict of interest.

Acknowledgments

The author would like to thank the editors and anonymous reviewers for their valuable comments and suggestions to improve the quality of the paper. The author is also grateful to the scientific research project of Nanjing Xiaozhuang University (No. 2017NXY41) and the talent introduction project of Nanjing Xiaozhuang University (No. 4177020).

References

[1] K. Kawajiri, T. G. Gutowski, and S. B. Gershwin, "Net CO_2 emissions from global photovoltaic development," *RSC Advances*, vol. 4, no. 102, pp. 58652–58659, 2014.

[2] M. Asif, "Urban scale application of solar PV to improve sustainability in the building and the energy sectors of KSA," *Sustainability*, vol. 8, no. 11, pp. 1–11, 2016.

[3] E. R. Shouman, E. T. E. Shenawy, and N. M. Khattab, "Market financial analysis and cost performance for photovoltaic technology through international and national perspective with case study for Egypt," *Renewable and Sustainable Energy Reviews*, vol. 57, pp. 540–549, 2016.

[4] H. Wu, X. J. Wang, S. Shahid, and M. Ye, "Changing characteristics of the water consumption structure in Nanjing city, southern China," *Water*, vol. 8, no. 8, p. 8, 2016.

[5] "Nanjing maps," https://www.chinahighlights.com/nanjing/map.htm.

[6] "Characteristics of Nanjing geographical climate," http://www.cma.gov.cn/2011xzt/2014zt/20140730/2014073002/20140730 0201/201408/t20140802_254423.html.

[7] "Nanjing climate & weather," https://www.topchinatravel.com/nanjing/nanjing-climate-and-weather.htm.

[8] Z. H. Gu, Q. Sun, and R. Wennersten, "Impact of urban residences on energy consumption and carbon emissions: an investigation in Nanjing, China," *Sustainable Cities and Society*, vol. 7, pp. 52–61, 2013.

[9] E. Kymakis, S. Kalykakis, and T. M. Papazoglou, "Performance analysis of a grid connected photovoltaic park on the island of Crete," *Energy Conversion and Management*, vol. 50, no. 3, pp. 433–438, 2009.

[10] L. M. Ayompe, A. Duffy, S. J. McCormack, and M. Conlon, "Measured performance of a 1.72 kW rooftop grid connected photovoltaic system in Ireland," *Energy Conversion and Management*, vol. 52, no. 2, pp. 816–825, 2011.

[11] S. Wittkopf, S. Valliappan, L. Y. Liu, K. S. Ang, and S. C. J. Cheng, "Analytical performance monitoring of a 142.5 kW$_p$ grid-connected rooftop BIPV system in Singapore," *Renewable Energy*, vol. 47, no. 6, pp. 9–20, 2012.

[12] A. M. Al-Sabounchi, S. A. Yalyali, and H. A. Al-Thani, "Design and performance evaluation of a photovoltaic grid-connected system in hot weather conditions," *Renewable Energy*, vol. 53, no. 9, pp. 71–78, 2013.

[13] K. Padmavathi and S. A. Daniel, "Performance analysis of a 3 MW$_p$ grid connected solar photovoltaic power plant in India," *Energy for Sustainable Development*, vol. 17, no. 6, pp. 615–625, 2013.

[14] M. Farhoodnea, A. Mohamed, T. Khatib, and W. Elmenreich, "Performance evaluation and characterization of a 3-kWp grid-connected photovoltaic system based on tropical field experimental results: new results and comparative study," *Renewable and Sustainable Energy Reviews*, vol. 42, pp. 1047–1054, 2015.

[15] S. Sundaram and J. S. C. Babu, "Performance evaluation and validation of 5 MW$_p$ grid connected solar photovoltaic plant in South India," *Energy Conversion and Management*, vol. 100, pp. 429–439, 2015.

[16] D. Okello, E. E. V. Dyk, and F. J. Vorster, "Analysis of measured and simulated performance data of a 3.2 kW$_p$ grid-connected PV system in Port Elizabeth, South Africa," *Energy Conversion and Management*, vol. 100, pp. 10–15, 2015.

[17] M. Mpholo, T. Nchaba, and M. Monese, "Yield and performance analysis of the first grid-connected solar farm at Moshoeshoe I International Airport, Lesotho," *Renewable Energy*, vol. 81, pp. 845–852, 2015.

[18] C. E. B. E. Sidi, M. L. Ndiaye, M. E. Bah, A. Mbodji, A. Ndiaye, and P. A. Ndiaye, "Performance analysis of the first large-scale (15 MW$_p$) grid-connected photovoltaic plant in Mauritania," *Energy Conversion and Management*, vol. 119, no. 1, pp. 411–421, 2016.

[19] R. Dabou, F. Bouchafaa, A. H. Arab et al., "Monitoring and performance analysis of grid connected photovoltaic under different climatic conditions in south Algeria," *Energy Conversion and Management*, vol. 130, pp. 200–206, 2016.

[20] A. Allouhi, R. Saadani, T. Kousksou, R. Saidur, A. Jamil, and M. Rahmoune, "Grid-connected PV systems installed on institutional buildings: technology comparison, energy analysis and economic performance," *Energy and Buildings*, vol. 130, no. 15, pp. 188–201, 2016.

[21] H. Maammeur, A. Hamidat, L. Loukarfi, M. Missoum, K. Abdeladim, and T. Nacer, "Performance investigation of grid-connected PV systems for family farms: case study of north-west of Algeria," *Renewable and Sustainable Energy Reviews*, vol. 78, pp. 1208–1220, 2017.

[22] L. C. D. Lima, L. D. A. Ferreira, and F. H. B. D. L. Morai, "Performance analysis of a grid connected photovoltaic system in northeastern Brazil," *Energy for Sustainable Development*, vol. 37, pp. 79–85, 2017.

[23] H. M. Bahaidarah, B. Tanweer, P. Gandhidasan, N. Ibrahim, and S. Rehman, "Experimental and numerical study on non-concentrating and symmetric unglazed compound parabolic photovoltaic concentration systems," *Applied Energy*, vol. 136, pp. 527–536, 2014.

[24] H. M. Bahaidarah, P. Gandhidasan, A. A. B. Baloch, B. Tanweer, and M. Mahmood, "A comparative study on the effect of glazing and cooling for compound parabolic concentrator PV systems-experimental and analytical investigations," *Energy Conversion and Management*, vol. 129, pp. 227–239, 2016.

[25] A. Raghoebarsing and A. Kalpoe, "Performance and economic analysis of a 27 kW grid-connected photovoltaic system in Suriname," *IET Renewable Power Generation*, vol. 11, no. 12, pp. 1545–1554, 2017.

[26] "PVSYST," http://www.pvsyst.com/.

[27] A. M. Humada, M. Hojabri, H. M. Hamada, F. B. Samsuri, and M. N. Ahmed, "Performance evaluation of two PV technologies (c-Si and CIS) for building integrated photovoltaic based on tropical climate condition: a case study in Malaysia," *Energy and Buildings*, vol. 119, pp. 233–241, 2016.

[28] H. A. Kazem, T. Khatib, K. Sopian, and W. Elmenreich, "Performance and feasibility assessment of a 1.4 kW roof top grid-connected photovoltaic power system under desertic weather conditions," *Energy and Buildings*, vol. 82, pp. 123–129, 2014.

[29] B. Tripathi, P. Yadav, S. Rathod, and M. Kumar, "Performance analysis and comparison of two silicon material based photovoltaic technologies under actual climatic conditions in Western India," *Energy Conversion and Management*, vol. 80, pp. 97–102, 2014.

[30] "Meteonorm software," http://www.meteonorm.com/.

[31] V. Sharma, A. Kumar, O. S. Sastry, and S. S. Chandel, "Performance assessment of different solar photovoltaic technologies under similar outdoor conditions," *Energy*, vol. 58, no. 9, pp. 511–518, 2013.

[32] A. Balaska, A. Tahri, F. Tahri, and A. B. Stambouli, "Performance assessment of five different photovoltaic module technologies under outdoor conditions in Algeria," *Renewable Energy*, vol. 107, pp. 53–60, 2017.

[33] I. Jamil, J. Q. Zhao, L. Zhang, R. Jamil, and S. F. Rafique, "Evaluation of energy production and energy yield assessment based on feasibility, design, and execution of 3×50 MW grid-connected solar PV pilot project in Nooriabad," *International Journal of Photoenergy*, vol. 2017, Article ID 6429581, 18 pages, 2017.

[34] D. A. Quansah, M. S. Adaramola, G. K. Appiah, and I. A. Edwin, "Performance analysis of different grid-connected solar photovoltaic (PV) system technologies with combined capacity of 20 kW located in humid tropical climate," *International Journal of Hydrogen Energy*, vol. 42, no. 7, pp. 4626–4635, 2017.

3

Modeling of Photovoltaic System with Modified Incremental Conductance Algorithm for Fast Changes of Irradiance

Saad Motahhir ⓘ, Abdelaziz El Ghzizal, Souad Sebti, and Aziz Derouich

Laboratory of Production Engineering, Energy and Sustainable Development, Higher School of Technology, SMBA University, Fez, Morocco

Correspondence should be addressed to Saad Motahhir; saad.motahhir@usmba.ac.ma

Academic Editor: Philippe Poggi

The first objective of this work is to determine some of the performance parameters characterizing the behavior of a particular photovoltaic (PV) panels that are not normally provided in the manufacturers' specifications. These provide the basis for developing a simple model for the electrical behavior of the PV panel. Next, using this model, the effects of varying solar irradiation, temperature, series and shunt resistances, and partial shading on the output of the PV panel are presented. In addition, the PV panel model is used to configure a large photovoltaic array. Next, a boost converter for the PV panel is designed. This converter is put between the panel and the load in order to control it by means of a maximum power point tracking (MPPT) controller. The MPPT used is based on incremental conductance (INC), and it is demonstrated here that this technique does not respond accurately when solar irradiation is increased. To investigate this, a modified incremental conductance technique is presented in this paper. It is shown that this system does respond accurately and reduces the steady-state oscillations when solar irradiation is increased. Finally, simulations of the conventional and modified algorithm are compared, and the results show that the modified algorithm provides an accurate response to a sudden increase in solar irradiation.

1. Introduction

The energy generated by the PV systems depends on various parameters, either environmental as temperature and irradiance or internal parameters of the PV panel, namely, the series and shunt resistors [1, 2]. Thus, the load imposes its own characteristic on the output power [3]. Therefore, in order to predict and analyze the effect of these parameters on the PV power, the model of the PV panel should be previously studied and achieved, and this model should be in accordance with the real comportment of the PV panel. Therefore, different models were proposed in the literature. In [4], a single diode model is used; in [5], a two-diode model is proposed to illustrate the influence of the recombination of carriers; and in [6], a model of three diodes is used to present the effects which are neglected by the two-diode model. However, the single-diode model is the most adopted due to its good simplicity and accuracy [7]. Moreover, manufacturers of PV panels offer only some characteristics. But other characteristics required to model PV panel are missed in the

datasheet, as the photocurrent, the diode saturation current, the series and shunt resistors, and the ideality factor [8]. Hence, in [7, 9, 10], researchers have proposed different methods to extract the lacking characteristics based on the datasheet values, but these methods require an implementation and this can increase the time spent in the development of a PV application. Therefore, this paper aims firstly to extract lacking parameters in the manufacturers' datasheet by using a simple tool provided by MathWorks [11] and then model the PV panel. The single-diode model is used in this work because it gives a high compromise between accuracy and simplicity [12], and several researchers have used it in their works [13, 14]. In addition, this work shows the effect of parameters that may change the performance of the PV panel. Then, this model will be a platform to design a PV array.

On the other side, the maximization of the PV power always remains a major challenge. Researchers have proposed different MPPT algorithms to maximize PV power, namely, FSCC, FOCV, fuzzy logic, neural network, P&O,

and INC [15, 16]. FSCC and FOCV are the simplest MPPT algorithms, which are based on the linearity of short-circuit current or open-circuit voltage to the maximum power point current or voltage. However, these techniques isolate the PV panel to measure the short-circuit current or open-circuit voltage. Therefore, the loss of energy is increased due to the periodic isolation of the panel [17]. Alternatively, fuzzy logic and neural network obtain a consistent MPPT technique due to their ability to treat the nonlinearity of the PV panel. But the main drawback of fuzzy logic is that the effectiveness of this algorithm depends a lot on choosing the right error computation and an appropriate rule base [18]. In addition, neural network presents many disadvantages like the fact that the data needed for the training process have to be specifically acquired for every PV array and location, and also, the PV characteristics change with time, so the neural network has to be periodically trained [18]. On the other hand, P&O and INC are mostly used. These techniques use the (P-V) characteristic of the PV panel. For P&O, steady-state oscillations occur after the MPP is found due to the perturbation made by this technique to maintain the MPP, which in turn increases the loss of power [19]. For INC, it is founded in the fact that slope of P-V characteristic is zero at the maximum power, and theoretically, there is no perturbation after the MPP is found. Therefore, oscillations are minimized. However, during implementation, the zero value is hardly found on the slope of the P-V characteristic due to the truncation error in digital processing. Thus, the INC technique can make an inaccurate response when the irradiation is suddenly increased [20]. Therefore, this work aims also to propose and implement a modified INC algorithm, which can overcome the wrong response made by the conventional INC algorithm when the irradiance is suddenly increased. Therefore, this work proposes a new technique to detect the increase in solar irradiation. The variation of voltage (ΔV) and current (ΔI) are used to identify the increase in irradiation instead of the slope ($\Delta P/\Delta V$) of the P-V characteristic. The modified algorithm detects the increase of irradiance and makes a correct decision. Moreover, a mini error is accepted to admit that the slope is near to zero and minimize the steady-state oscillations.

This paper is structured as follows. Following the introduction, Section 2 presents the modeling of PV panel and array and presents the impact of different environmental and internal parameters. Section 3 presents the design of the boost converter, the conventional algorithm, and the proposed algorithm.

2. Modeling of PV Panel and Array

2.1. Model of PV Panel. As shown in Figure 1, the single-diode model of PV panel can be presented by a photocurrent source and a diode connected with series and shunt resistances. The mathematical model of the PV panel can be presented by the following equations [19]:

$$I = I_{ph} - I_S \left(\exp \frac{q(V + R_s I)}{aKTN_s} - 1 \right) - \frac{(V + IR_s)}{R_{sh}}, \quad (1)$$

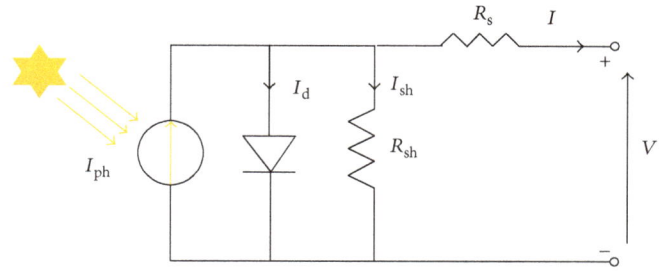

FIGURE 1: PV panel equivalent circuit.

TABLE 1: Characteristics of MSX-60 PV panel at STC.

PV panel parameters	Values
Maximum power, P_{max}	60 W
Maximum power voltage, V_{mp}	17.1 V
Maximum power current, I_{mp}	3.5 A
Short-circuit current, I_{sc}	3.8 A
Open-circuit voltage, V_{CO}	21.1 V
Voltage/temp. coefficient, K_V	−0.38%/°C
Current/temp. coefficient, K_I	0.065%/°C
The number of cells, N_s	36

$$I_{ph} = (I_{SC} + K_i(T - 298.15)) \frac{G}{1000}, \quad (2)$$

$$I_S = \frac{I_{SC} + K_i(T - 298.15)}{\exp(q((V_{OC} + K_V(T - 298.15))/aKTN_s)) - 1}. \quad (3)$$

Hence, the physical behavior of the PV panel depends on the shunt and series resistances, solar irradiation, and temperature. Therefore, in this work, the impact of these parameters on the output of the PV panel is investigated.

The panel used in this work is MSX-60 panel, and as presented in Table 1, the datasheet of PV panel provides only some characteristics of PV panel. Thus, other characteristics required to model PV panel are missed in the datasheet, as the photocurrent, the diode saturation current, the series and shunt resistors, and the ideality factor. Hence, researchers have proposed different methods to extract the lacking characteristics based on the datasheet [7, 9, 10]. But these methods require an implementation and this can increase the time spent in the development of a PV system. Therefore, this work aims firstly to extract these parameters by using a simple tool provided by MathWorks which is "PV array;" the latter is available in Simulink 2015 or later [11]. Hence, as shown in Figure 2, we only have to set the datasheet values and automatically it will generate the lacking parameters.

Equations (1), (2), and (3) are modeled using PSIM (software for power electronics simulation), and Figure 3 presents the PSIM model.

Figure 4 shows the I-V and P-V curves of experimental and PSIM model under STC. The experimental data $P(V)$ and $I(V)$ are taken from the manufactured datasheet [21].

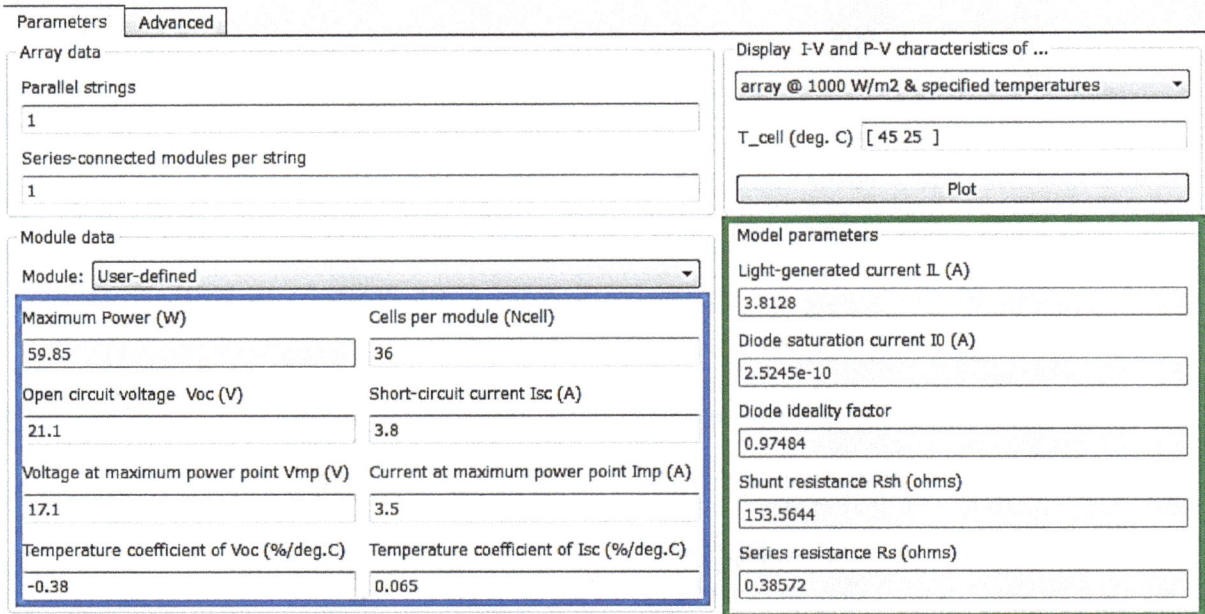

FIGURE 2: PV array tool.

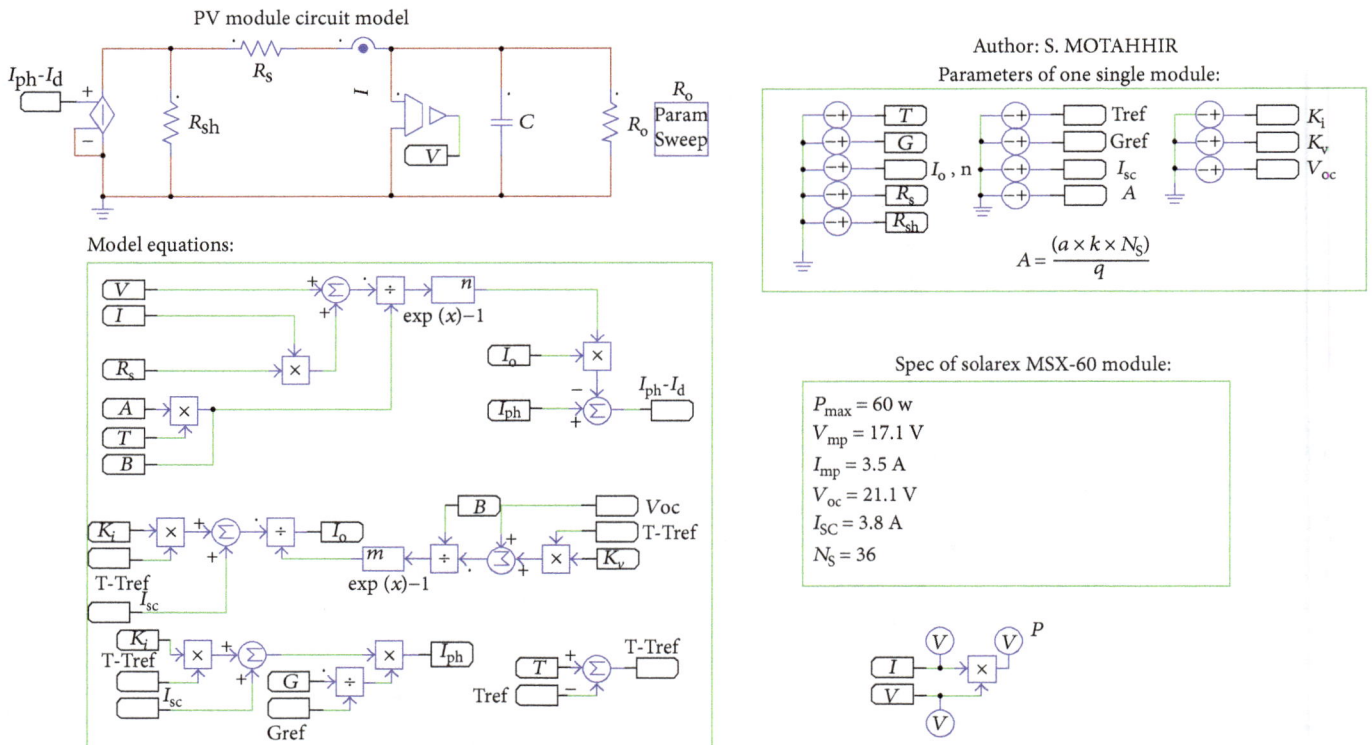

FIGURE 3: PV panel PSIM model.

And as presented, the model data are in accordance with the experimental data both in the current and power curves.

2.2. Effect of Solar Irradiation Variation. Figure 3 contains the model of the three equations: one of these equations computes the photocurrent based on temperature and irradiance (2). The model of this equation is presented in Figure 5, and Figure 6 presents I-V and P-V curves for different values of solar irradiation.

As presented in Figure 6, the PV panel current depends heavily on solar irradiation. However, the voltage increases just by 1 V once the irradiance is increased from $400\,W/m^2$

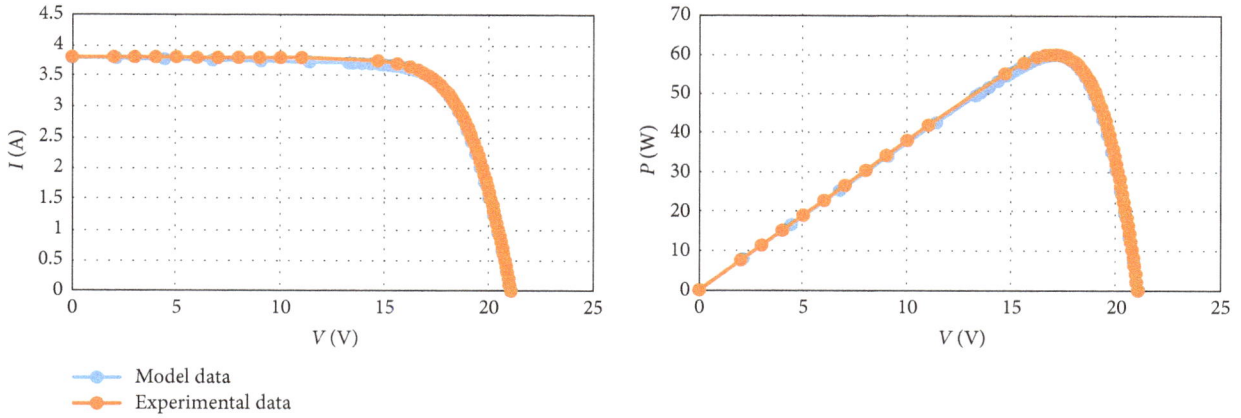

FIGURE 4: *I-V* and *P-V* characteristics of model and experimental data.

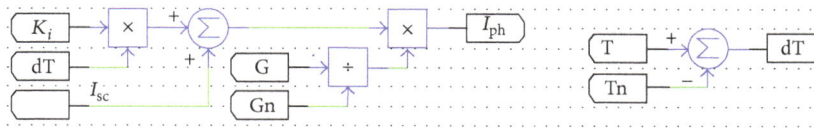

FIGURE 5: Model of (2).

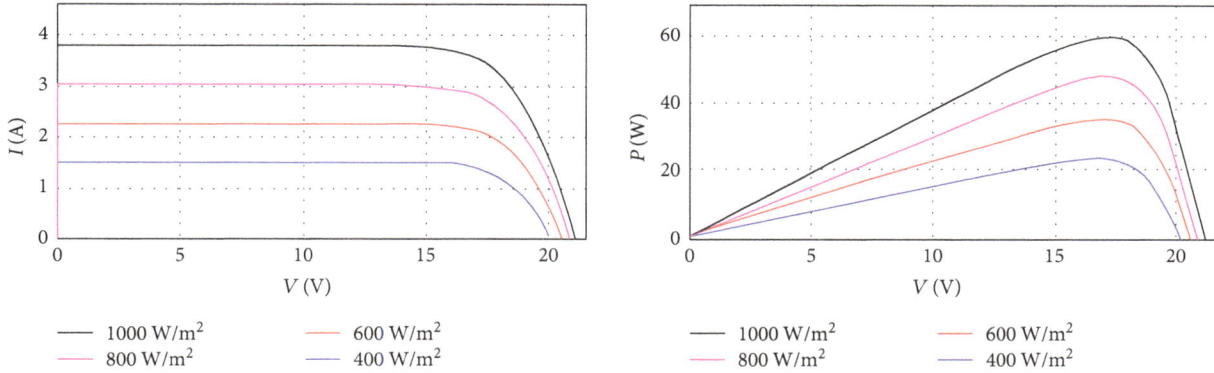

FIGURE 6: *I-V* and *P-V* curves for different values of irradiance.

to $1000 \, \text{W/m}^2$. Therefore, the irradiance change affects heavily the PV panel current.

2.3. Effect of the Temperature Variation.
Figure 3 contains also the modeling of (3), which computes the diode saturation current based on the temperature. The model of this equation is shown in Figure 7, and Figure 8 shows the *I-V* and *P-V* curves for different values of temperature.

Generally, as shown in Figure 8, for a fixed solar irradiation and when the temperature increases, the open-circuit voltage decreases and the short-circuit current increases with a little value. Therefore, the temperature change affects strongly the PV panel voltage.

2.4. Effect of Series Resistor Variation.
The series resistor value is very small, and it may be neglected in some cases. Nevertheless, to make the appropriate model for any PV panel, it is recommended to make a variation of this resistor

and show its effect on the PV panel output. As shown in Figure 9, the change of the series resistor results on the deviation of the MPP.

The simulation was made for three values of series resistance ($1 \, \text{m}\Omega$, $4 \, \text{m}\Omega$, and $8 \, \text{m}\Omega$). Moreover, as shown in Figure 9, the upper values of series resistance decrease the output power. In addition, the fill factor presented by (4) decreases as series resistance increases [22].

$$\text{FF} = \frac{P_{\max}}{V_{\text{OC}} I_{\text{SC}}}. \tag{4}$$

2.5. Effect of Shunt Resistor Variation.
As presented in Figure 10, R_{sh} should be quite large for a good fill factor. In fact, when R_{sh} is small, the current collapses more strongly, then the loss of power is high, and the fill factor is low. Therefore, R_{sh} of any PV panel should be large enough for a good efficiency.

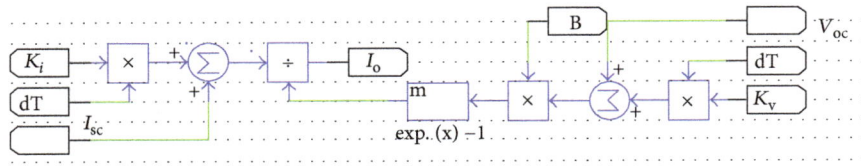

FIGURE 7: Model of (3).

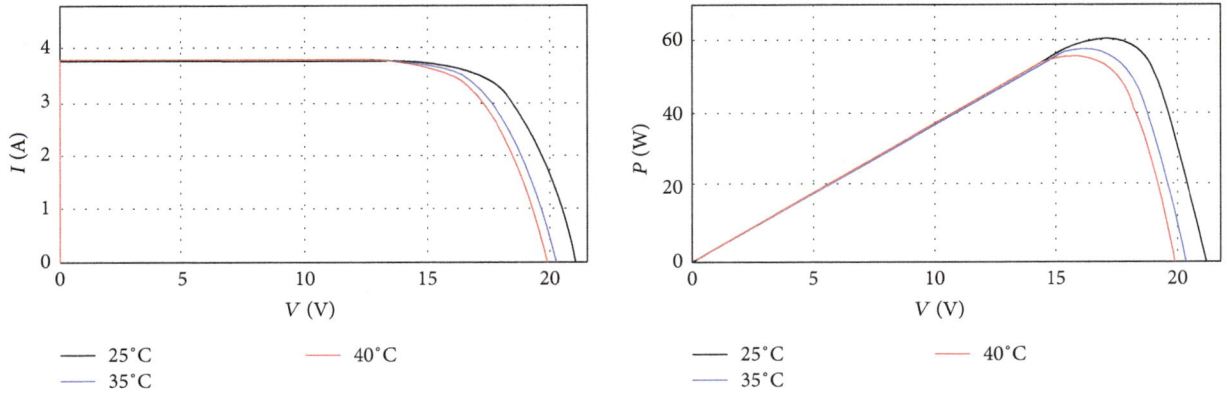

FIGURE 8: I-V and P-V curves for different values of temperature.

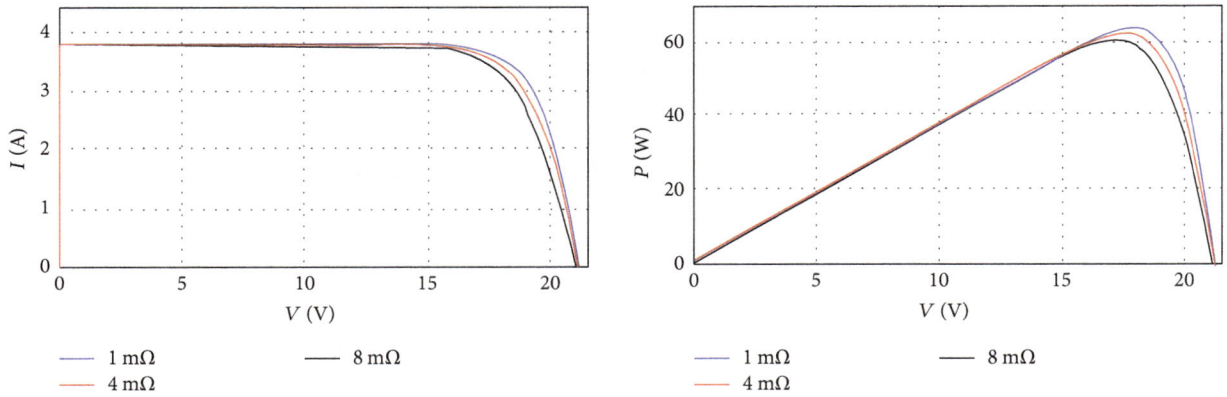

FIGURE 9: I-V and P-V curves for different values of R_s.

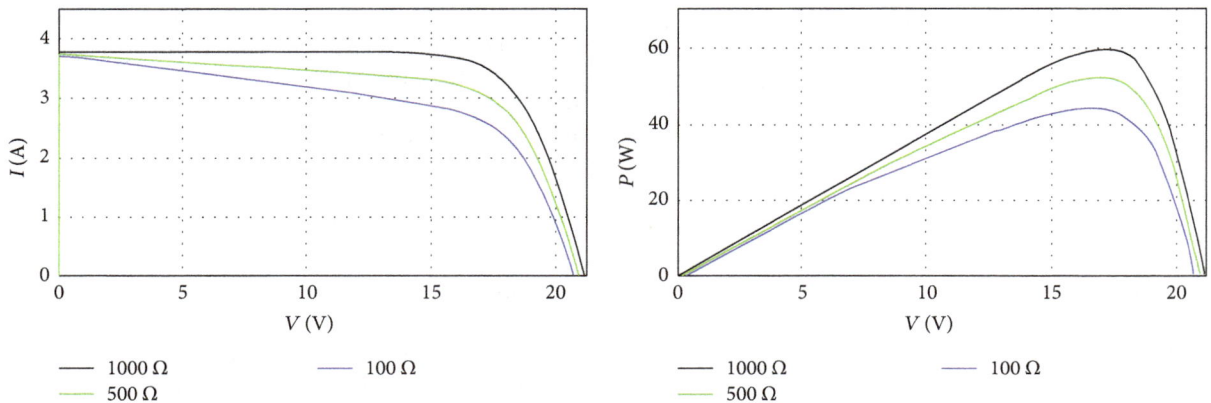

FIGURE 10: I-V and P-V curves for different values of R_{sh}.

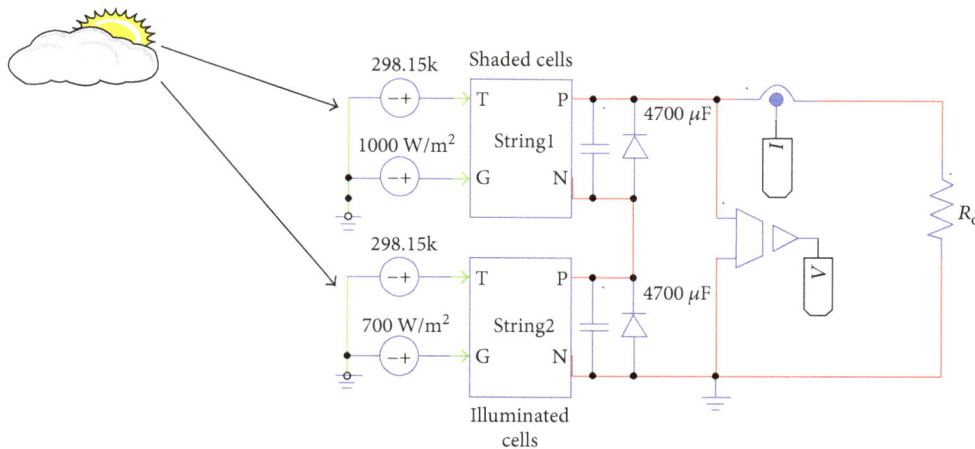

FIGURE 11: PV panel under nonuniform irradiation.

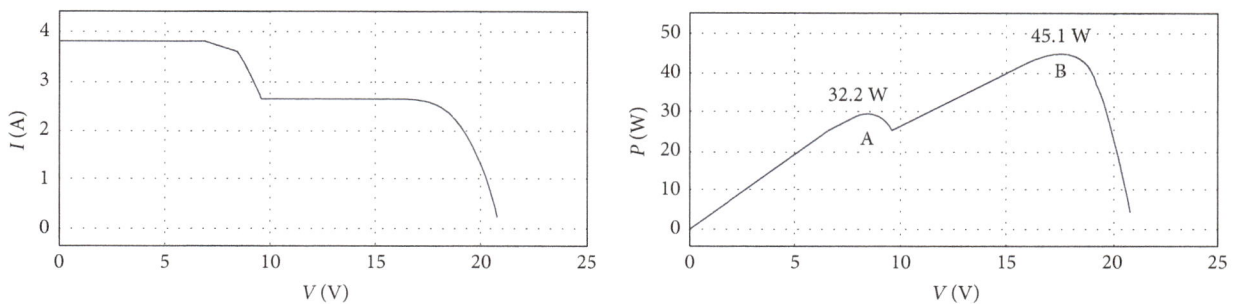

FIGURE 12: I-V and P-V characteristics of the PV panel under nonuniform irradiation.

2.6. Effect of Shading. Partial shading also presents a major impact on PV output power. When the insolation received by a part of the PV panel (shaded cells) is less than the insolation received by another part (illuminated cells), the current generated by the illuminated cells is greater than the current produced by the shaded cells; this mismatch makes the diode of shaded cells reverse biased; consequently, the power will be lost in the shaded cells and that may cause a hot spot problem which is the reason of permanent damage to the PV panel [19]. Hence, in order to overcome this problem, the bypass diodes can be connected in parallel with PV cells [19].

To simulate the effect of shading, a bypass diode is associated with each string of the panel, and it should be mentioned that the panel used includes two strings and each string is a set of 18 cells. Thus, as shown in Figure 11, the first string is exposed by $1000 \, \text{W/m}^2$ and the second string by $700 \, \text{W/m}^2$.

Under uniform irradiation, the bypass diodes have no impact because they are reversely biased. But under shading, the current flows through the bypass diode instead of the shaded string because the bypass diode is directly biased, and as a result, no power will be lost in the shaded cells and only the illuminated cells generate power. Figure 12 shows the effect of bypass diodes on the PV panel characteristics and as presented, multiple peaks may occur on the P-V curve; as in this case, there are two peaks, point A which is the global peak and point B which is the local peak.

Therefore, conventional MPPT algorithms are unable to track the global peak which is the real MPP [19].

2.7. Photovoltaic Array. To get benefit from the model developed, a PV array of 18 PV panels has been built in order to supply a solar pumping station, not studied in this paper. Therefore, as shown in Figure 13, three strings of six PV panels have been linked in parallel and each group is composed of six panels in series.

The model of the PV array has been achieved on PSIM, and simulation result obtained is presented in Figure 14.

As presented in Figure 14, the panels connected in parallel increase the current and the panels connected in series increase the voltage. However, as discussed in the previous section, this connection between panels can lead to hot spot problem when the insolation received by a part of the PV array (shaded panels) is less than the insolation received by another part (illuminated panels).

3. Modified INC Algorithm with Boost Converter

PV panel provides I-V and P-V curves presented in Figure 15; these curves highlight one point where the power is maximum. As presented above, this point depends on solar irradiation and temperature. Moreover, as presented in Figure 15, in general, load's characteristic is different from the MPP. Therefore, the boost converter controlled by a duty

PV array
Author: S. MOTAHHIR

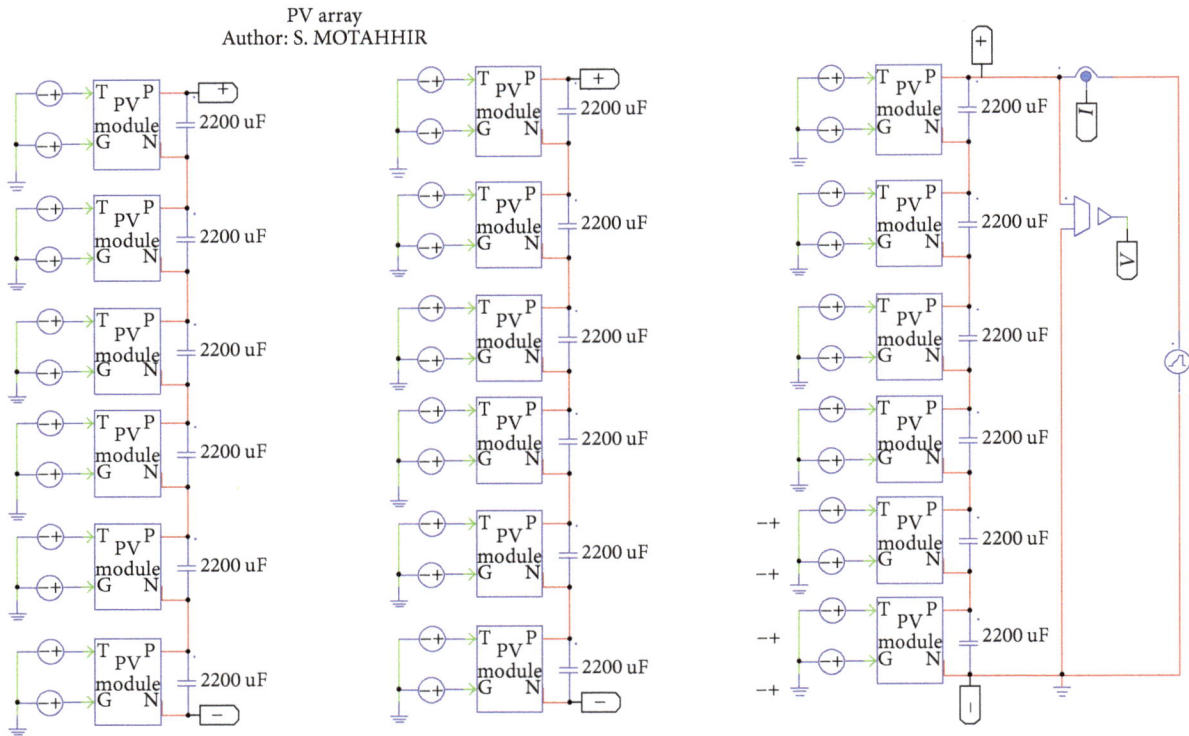

FIGURE 13: PV array model.

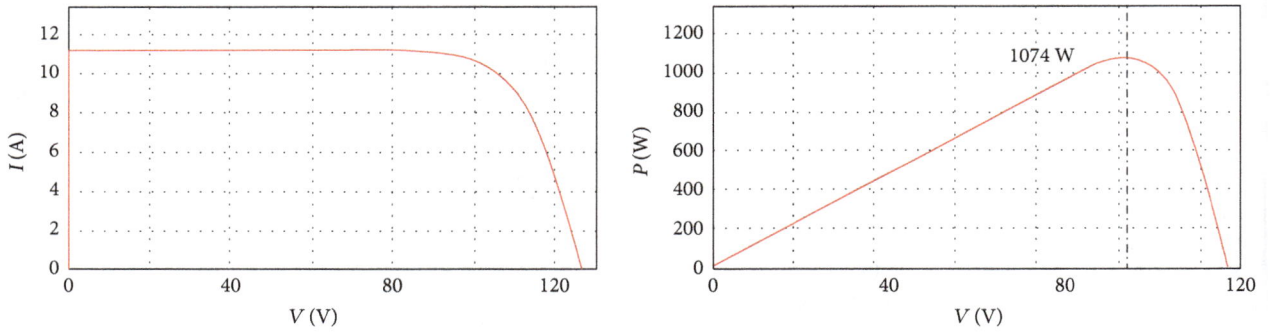

FIGURE 14: *I-V* and *P-V* curves of the PV array.

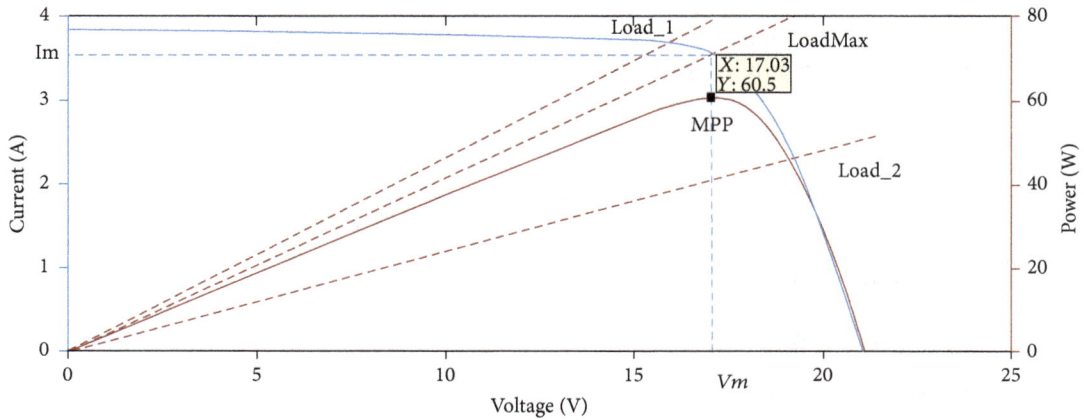

FIGURE 15: The impact of load on *I-V* and *P-V* characteristics.

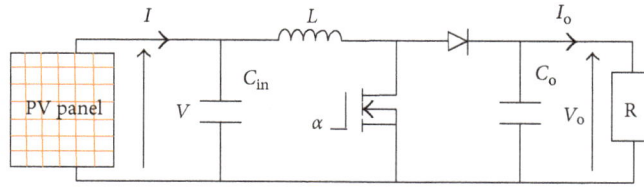

FIGURE 16: Boost converter.

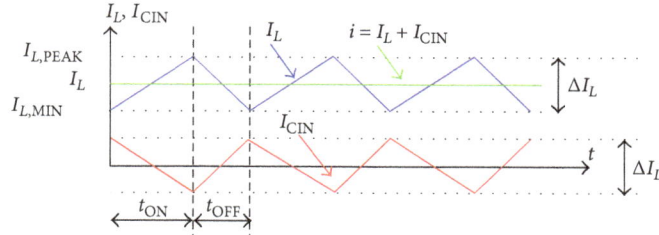

FIGURE 17: Current waveforms of the input capacitor and inductor in CCM.

cycle (α) generated by the MPPT controller is put between the panel and the load [23]. The interest of this addition is to remove the mismatch between the panel and the load, and then the PV panel can operate at MPP.

3.1. Boost Converter Design.
Figure 16 presents the circuit of the boost converter; this converter is used as an adapter between the source and the load [24].

The operation principle of this converter is described by the following equations [25]:

$$V_O = \frac{V}{(1 - \alpha)},$$

$$I_O = I(1 - \alpha). \tag{5}$$

By using (5), (6) is obtained, which is the relationship between the resistance seen by the PV panel (R_{eq}) and the load resistance (R). Hence, based on this equation, the MPPT controller can find the optimum α to remove the mismatch between the load and MPP. Therefore, The boost converter is required to get maximum power available from the panel.

$$R_{eq} = \frac{V}{I} = \frac{V_O(1 - \alpha)^2}{I_O} = R(1 - \alpha)^2. \tag{6}$$

3.1.1. Selection of the Inductor.
The choice of the inductor can directly influence the performance of the boost converter. Moreover, the selection of the inductance is a trade-off between its cost, its size, and the inductor current ripple. A higher inductance value results in a minor inductance current ripple; however, that results in a higher cost and larger inductor's size, which means a larger PCB surface.

By the way, the inductance value can be given as follows: During T_{ON} state,

$$V = L\frac{\Delta I_L}{T_{ON}} \Rightarrow L = \frac{V\alpha}{\Delta I_L F}, \tag{7}$$

where ΔI_L can be computed as below, and r is the inductor current ripple ratio, which is optimal in the range [0.3, 0.5] [26]:

$$\Delta I_L = r \times I. \tag{8}$$

Therefore, the optimum inductor value can be computed by using

$$L \geq \frac{V\alpha}{rIF}. \tag{9}$$

Based on Figure 17, in order to guarantee the performance of boost converter in the continuous conduction mode, the following equation must be verified:

$$I \geq \frac{\Delta I_L}{2} \Rightarrow \Delta I_L \leq 2I. \tag{10}$$

Therefore,

$$L \geq \frac{\alpha(1 - \alpha)^2 R}{2F}. \tag{11}$$

3.1.2. Selection of the Output Capacitor.
The choice of the output capacitor is made by using output voltage ripple as follows:

During T_{ON},

$$I_O = C_O\frac{\Delta V_O}{T_{ON}} \Rightarrow C_O = \frac{\alpha I_O}{\Delta V_O F}. \tag{12}$$

Therefore, the output capacitor value can be calculated as below, where the desired ΔV_O equals to 2% of output voltage [24]:

$$C_O \geq \frac{\alpha}{0.02 \times F \times R}. \tag{13}$$

3.1.3. Selection of the Input Capacitor.
An input capacitor is used to decrease the input voltage ripple and to deliver an alternative current to the inductor. The input voltage ripple

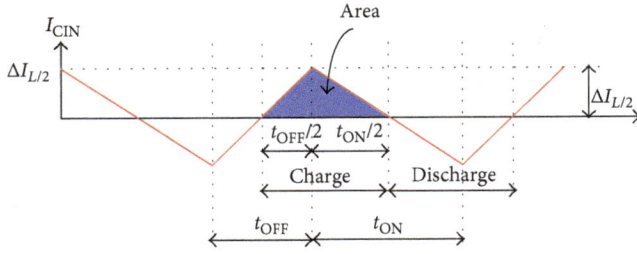

FIGURE 18: Current waveforms of the input capacitor in CCM.

matches to the charge voltage during the charge phase of the capacitor, and during this phase, I_{Cin} is greater than zero, so this phase is illustrated by the blue area in Figure 18; therefore, this area is used to calculate the input capacitor as follows:

$$I_{Cin} = C_{in} \frac{\Delta V}{\Delta t} \Rightarrow \Delta V = \frac{I_{Cin} \Delta t}{C_{in}}. \qquad (14)$$

Based on Figure 18, and by using (14):

$$\Delta V = \frac{\Delta I_L}{8FC_{in}} \Rightarrow C_{in} = \frac{\Delta I_L}{8F\Delta V} = \frac{V\alpha}{8F^2 L\Delta V}. \qquad (15)$$

Therefore, the input capacitor can be calculated by (16), where the desired ΔV equals to 1% of input voltage [27]:

$$C_{in} \geq \frac{\alpha}{8 \times F^2 \times L \times 0.01}. \qquad (16)$$

The design of the used boost is presented in Table 2.

3.2. Problem with the Conventional INC Algorithm. A good MPPT algorithm balances between the tracking speed and steady-state performance. In accordance with these requirements, the INC algorithm can be used even if it can fail in some cases [20] and in this study, it will be modified in order to improve its performance. INC algorithm is founded in the fact that slope of P-V characteristic is zero at the MPP [28]. Therefore, this algorithm can be modeled as follows:

$$\frac{dP}{dV} = 0 \quad \text{at MPP,}$$

$$\frac{dP}{dV} > 0 \quad \text{left to MPP,} \qquad (17)$$

$$\frac{dP}{dV} < 0 \quad \text{right to MPP.}$$

Since

$$\frac{dP}{dV} = \frac{d(IV)}{dV} = V\frac{dI}{dV} + I; \qquad (18)$$

then,

$$\frac{dI}{dV} = -\frac{I}{V} \quad \text{at MPP,} \qquad (19)$$

$$\frac{dI}{dV} > -\frac{I}{V} \quad \text{left to MPP,} \qquad (20)$$

TABLE 2: Design of the boost converter.

Parameters	Values
L	1.2 mH
C_{in}	75 μF
C_O	75 μF
F	10 kHz
R	50 Ω
α_{MPP}	0.69

$$\frac{dI}{dV} < -\frac{I}{V} \quad \text{right to MPP.} \qquad (21)$$

The flowchart of the INC algorithm is presented in Figure 19 [28]. This algorithm measures the current and voltage of the panel. If (21) is met, the duty cycle is increased, and vice versa if (20) is met. Then, there is nothing to do if (19) is met. Therefore, theoretically, if MPP is reached, there is no more perturbation of α; consequently, steady-state oscillations are decreased, and that is the main advantage of INC algorithm.

However, the conventional INC algorithm fails to make a good decision when the irradiance is suddenly increased [20]. As presented in Figure 20, once the solar irradiance is at 500 W/m^2 and the PV system operates at load_2, the INC technique controls the PV system in order to reach the MPP (point B). When the irradiance is increased to 1000 W/m^2, load_2 will lead the system to point G in I-V characteristic, which matches to point C in P-V characteristic. The INC technique calculates the slope between point C and point B which is positive. Therefore, the INC algorithm will decrease the duty cycle and consequently, the PV panel voltage will be increased. But since the MPP of 1000 W/m^2 is at point A, and the slope between point A and C is negative, then the PV panel voltage should have been decreased in order to reach point A, instead of increase voltage and recede from point A as made by the conventional INC algorithm. In addition, as presented in Figure 6, generally when the solar irradiance increases, the MPP moves to the right and consequently, the same problem will occur.

Conversely, this weakness does not happen if the solar irradiation is decreased. Because as shown in Figure 20, the slope is positive between point A and D, and also between point B and D.

3.3. Modified INC Algorithm. Based on the above analysis, it is noted that when the solar irradiance increases, both the voltage and the current are increased. Therefore, the sudden increase in solar irradiation can be detected, by checking if the MPP was reached and both the voltage and current are increased. Therefore, a permitted error is accepted (22) to detect that the MPP is reached.

$$\left| \frac{dI}{dV} + \frac{I}{V} \right| < 0.07. \qquad (22)$$

The proposed algorithm is presented in Figure 21. So as shown, the addition is the check if the MPP was reached by using (22), then set Var to one. After that, when (22) is

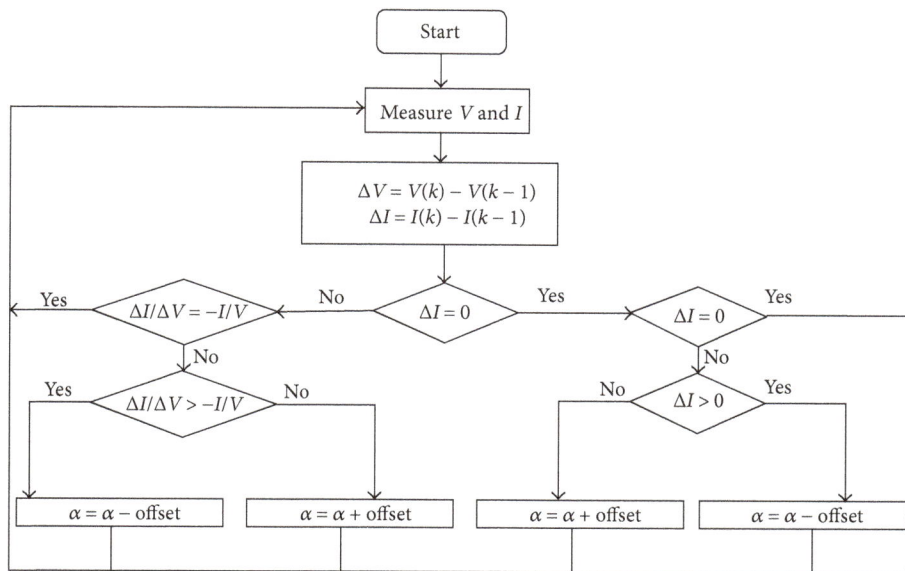

FIGURE 19: Flowchart of INC algorithm.

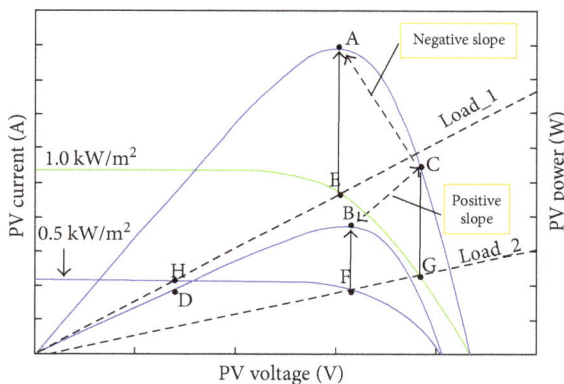

FIGURE 20: P-V and I-V curves for solar irradiation $500\,W/m^2$ and $1000\,W/m^2$.

not met and Var is one, the proposed technique checks if both voltage and current are increased; in this case, the duty cycle is increased instead of decreased as made by the conventional algorithm. Hence, the INC algorithm is modified to overcome the incorrect decision made by the conventional algorithm when the irradiance is increased.

4. Results and Discussions

The test was made for the conventional and the proposed techniques. At first, the solar irradiance is suddenly increased from $500\,W/m^2$ to $1000\,W/m^2$ at $t = 0.11\,s$, and then it is decreased from $1000\,W/m^2$ to $500\,W/m^2$ at $t = 0.23\,s$. Figure 22 shows the test result of the conventional technique, and Figure 23 shows the test result of the proposed technique. Therefore, as shown in these figures, the steady-state oscillations are minimized by using the proposed algorithm and admit an error equals to 0.07. Contrary to

the conventional algorithm, the power oscillates between (28.5–29.8 W), and this can generate a loss in PV energy.

In addition, as presented in Figure 22, by using the conventional algorithm, when solar irradiation is suddenly increased, the power diverges with a value greater than 62 W and after that, the conventional INC reverses the direction and the power diverges with a value lower than 56 W; and like that, the system takes a long time to converge around MPP that is due to the wrong decision made by the conventional technique. Also, even if the MPP is reached, the power oscillates between (59–61 W). On the other side, as presented in Figure 23, the proposed technique detects the fast increase of irradiance and makes a correct decision in duty cycle. As a result, the power converges to the new MPP from the first step and it is maintained at it (60 W). In addition, it only needs 0.001 s to reach the MPP. Hence, by using the proposed INC, the power converges faster compared with the conventional algorithm which needs more time to reach the MPP.

Table 3 summarizes a comparison of the proposed technique to other improved INC techniques proposed in the scientific literature in term of the oscillation level, tracking efficiency, the response time during sudden increase in irradiation, and if the technique makes an incorrect decision under sudden increase of irradiation. As presented, the proposed technique shows a very fast tracking speed, a higher efficiency, and neglected oscillations around the MPP compared to other techniques. Thus, only the proposed algorithm and that proposed in [29] make a correct decision under sudden increase of irradiation, contrary to the conventional technique and those proposed in [30, 31] which make an incorrect decision.

5. Conclusion

In this paper, PV panel's parameters are found using Math-Works tool (PV array); hence, by using these parameters, a

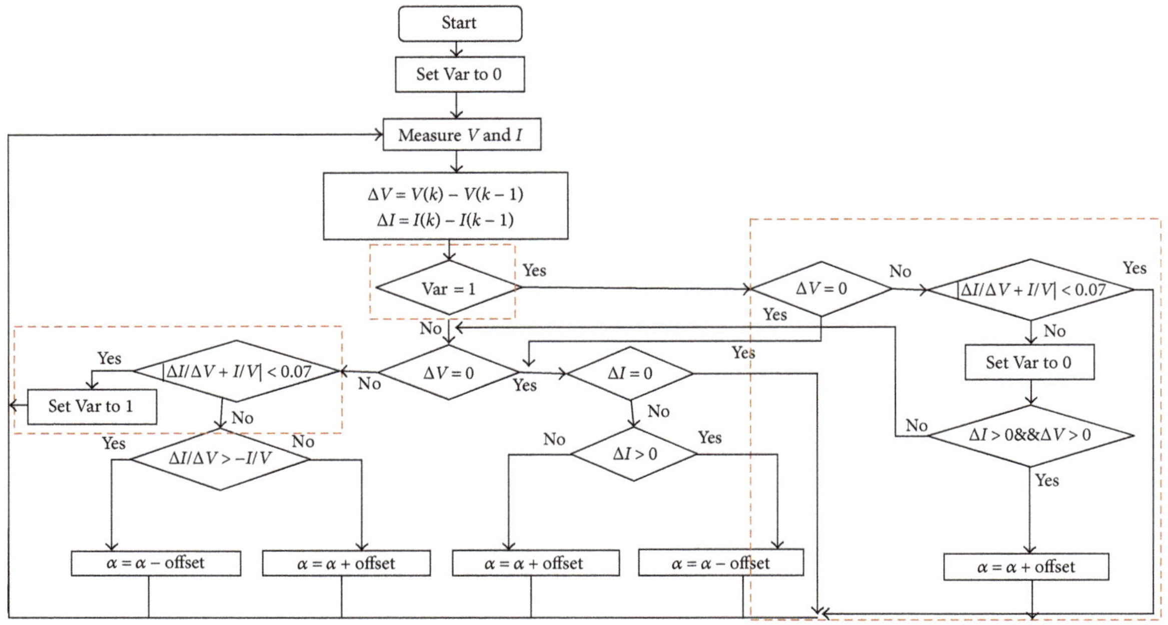

FIGURE 21: Flowchart of the proposed INC algorithm.

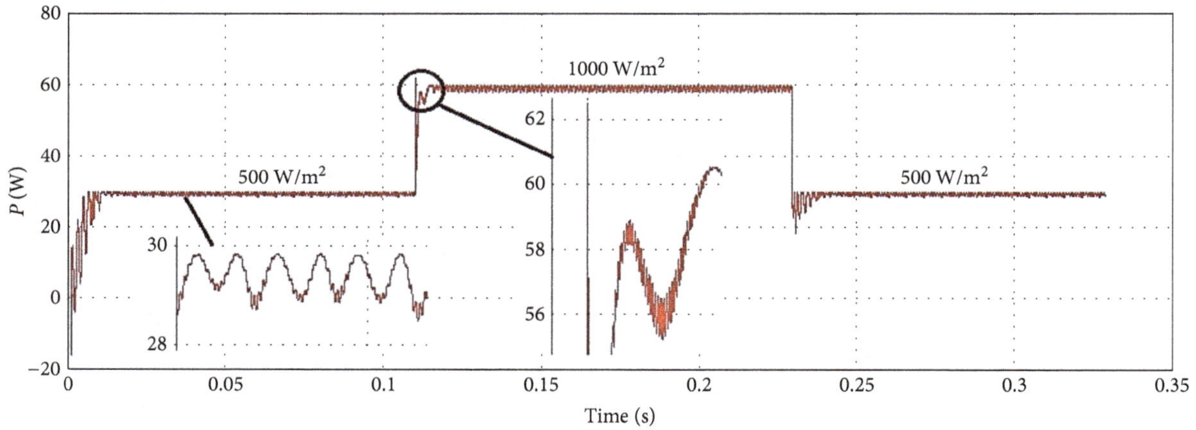

FIGURE 22: Test result of the INC algorithm.

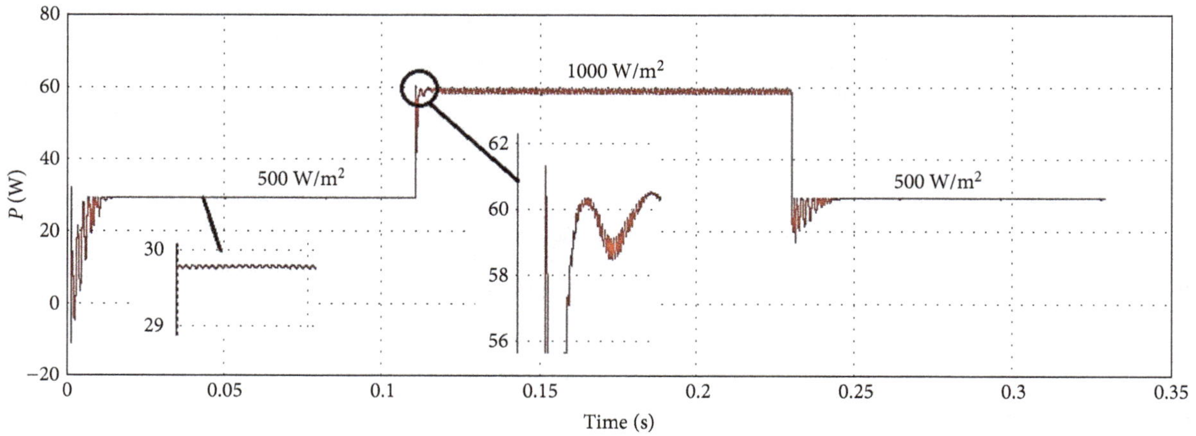

FIGURE 23: Test result of the modified INC algorithm.

TABLE 3: Comparison of the proposed algorithm with other improved incremental conductance algorithms proposed in scientific literature.

Technique	Oscillation level	Efficiency (%)	Response time during sudden increase in irradiation	Incorrect decision under sudden increase of irradiation
Conventional	2.5 W	96	Slow	Yes
[29]	1 W	96.40	Fast	No
[30]	1.5 W	98.5	Fast	Yes
[31]	1 W	97.5	Medium	Yes
Proposed	Neglected	98.8	Very fast	No

PV panel and a PV array are modeled, and the results show that the model is in accordance with experimental data of the used panel (MSX-60). In addition, a modified INC algorithm which can overcome the confusion faced by the conventional INC technique is proposed in this paper. As a result, the tests show that the modified technique detects the fast increase of irradiation and makes a correct decision, contrary to the conventional technique. Moreover, by using the modified algorithm, steady-state oscillations are almost neglected. Hence, the loss of energy is minimized; consequently, the efficiency is equal to 98.8% instead of 96% obtained by the conventional technique.

As a perspective, the modified INC algorithm can be more improved and then implemented in an embedded hardware device.

Nomenclature

a: Diode's ideality factor
I: Output current of the panel (A)
I_s: Diode saturation current (A)
I_{ph}:   Panel photocurrent (A)
G: Solar irradiation (W/m^2)
K: Boltzmann constant (J·K^{-1})
q: Electron charge (C)
R: The load (Ω)
R_{eq}:   The resistance seen by the panel (Ω)
R_s: Series resistance (Ω)
R_{sh}:   Shunt resistance (Ω)
T: Junction temperature (K)
V: Output voltage of the panel (V)
V_O: Output voltage of the boost converter (V)
I_O: Output current of the boost converter (A)
F: Switching frequency (Hz)
ΔV: Input voltage ripple of the boost converter (V)
ΔV_O: Output voltage ripple of the boost converter (V)
ΔI_L: Inductor current ripple (A).

Greek Letters

α: Duty cycle.

Abbreviations

CCM: Continuous conduction mode
FSCC: Fractional short-circuit current
FOCV: Fractional open-circuit voltage
INC: Incremental conductance
MPP: Maximum power point

MPPT: Maximum power point tracking
P&O: Perturb and observe
PV: Photovoltaic
STC: Standard test conditions.

Conflicts of Interest

The authors declare that they have no conflicts of interest.

References

[1] H. Patel and V. Agarwal, "MATLAB-based modeling to study the effects of partial shading on PV array characteristics," *IEEE Transactions on Energy Conversion*, vol. 23, no. 1, pp. 302–310, 2008.

[2] S. Motahhir, A. Chalh, A. Ghzizal, S. Sebti, and A. Derouich, "Modeling of photovoltaic panel by using proteus," *Journal of Engineering Science and Technology Review*, vol. 10, no. 2, pp. 8–13, 2017.

[3] S. Motahhir, A. El Ghzizal, S. Sebti, and A. Derouich, "Proposal and implementation of a novel perturb and observe algorithm using embedded software," in *2015 3rd International Renewable and Sustainable Energy Conference (IRSEC)*, pp. 1–5, Marrakech, Morocco, December 2015.

[4] H. S. Rauschenbach, *Solar Cell Array Design Handbook*, Van Nostrand Reinhold, 1980.

[5] N. Barth, R. Jovanovic, S. Ahzi, and M. A. Khaleel, "PV panel single and double diode models: optimization of the parameters and temperature dependence," *Solar Energy Materials and Solar Cells*, vol. 148, pp. 87–98, 2016.

[6] K. Nishioka, N. Sakitani, Y. Uraoka, and T. Fuyuki, "Analysis of multicrystalline silicon solar cells by modified 3-diode equivalent circuit model taking leakage current through periphery into consideration," *Solar Energy Materials and Solar Cells*, vol. 91, no. 13, pp. 1222–1227, 2007.

[7] N. Yıldıran and E. Tacer, "Identification of photovoltaic cell single diode discrete model parameters based on datasheet values," *Solar Energy*, vol. 127, pp. 175–183, 2016.

[8] K. Ishaque, Z. Salam, S. Mekhilef, and A. Shamsudin, "Parameter extraction of solar photovoltaic modules using penalty-based differential evolution," *Applied Energy*, vol. 99, pp. 297–308, 2012.

[9] K. Ishaque and Z. Salam, "An improved modeling method to determine the model parameters of photovoltaic (PV) modules using differential evolution (DE)," *Solar Energy*, vol. 85, no. 9, pp. 2349–2359, 2011.

[10] M. F. AlHajri, K. M. El-Naggar, M. R. AlRashidi, and A. K. Al-Othman, "Optimal extraction of solar cell parameters using pattern search," *Renewable Energy*, vol. 44, pp. 238–245, 2012.

[11] Mathworks, *PV Array*, 2015, Mai 2017, https://fr.mathworks.com/help/physmod/sps/powersys/ref/pvarray.html.

[12] C. Carrero, J. Amador, and S. Arnaltes, "A single procedure for helping PV designers to select silicon PV modules and evaluate the loss resistances," *Renewable Energy*, vol. 32, no. 15, pp. 2579–2589, 2007.

[13] T. Radjai, L. Rahmani, S. Mekhilef, and J. P. Gaubert, "Implementation of a modified incremental conductance MPPT algorithm with direct control based on a fuzzy duty cycle change estimator using dSPACE," *Solar Energy*, vol. 110, pp. 325–337, 2014.

[14] J. Ahmed and Z. Salam, "A modified P&O maximum power point tracking method with reduced steady-state oscillation and improved tracking efficiency," *IEEE Transactions on Sustainable Energy*, vol. 7, no. 4, pp. 1506–1515, 2016.

[15] A. Amir, J. Selvaraj, and N. A. Rahim, "Study of the MPP tracking algorithms: focusing the numerical method techniques," *Renewable and Sustainable Energy Reviews*, vol. 62, pp. 350–371, 2016.

[16] A. Gupta, Y. K. Chauhan, and R. K. Pachauri, "A comparative investigation of maximum power point tracking methods for solar PV system," *Solar Energy*, vol. 136, pp. 236–253, 2016.

[17] D. Verma, S. Nema, A. M. Shandilya, and S. K. Dash, "Maximum power point tracking (MPPT) techniques: recapitulation in solar photovoltaic systems," *Renewable and Sustainable Energy Reviews*, vol. 54, pp. 1018–1034, 2016.

[18] J. P. Ram, T. S. Babu, and N. Rajasekar, "A comprehensive review on solar PV maximum power point tracking techniques," *Renewable and Sustainable Energy Reviews*, vol. 67, pp. 826–847, 2017.

[19] S. Motahhir, A. El Ghzizal, S. Sebti, and A. Derouich, "Shading effect to energy withdrawn from the photovoltaic panel and implementation of DMPPT using C language," *International Review of Automatic Control*, vol. 9, no. 2, pp. 88–94, 2016.

[20] K. S. Tey and S. Mekhilef, "Modified incremental conductance MPPT algorithm to mitigate inaccurate responses under fast-changing solar irradiation level," *Solar Energy*, vol. 101, pp. 333–342, 2014.

[21] Solarex MSX60 and MSX64 photovoltaic panel, datasheet1998, April 2017, https://www.solarelectricsupply.com/media/custom/upload/Solarex-MSX64.pdf.

[22] F. Bayrak, G. Ertürk, and H. F. Oztop, "Effects of partial shading on energy and exergy efficiencies for photovoltaic panels," *Journal of Cleaner Production*, vol. 164, pp. 58–69, 2017.

[23] S. Motahhir, A. El Ghzizal, S. Sebti, and A. Derouich, "MIL and SIL and PIL tests for MPPT algorithm," *Cogent Engineering*, vol. 4, p. 1378475, 2017.

[24] S. Sivakumar, M. J. Sathik, P. S. Manoj, and G. Sundararajan, "An assessment on performance of DC–DC converters for renewable energy applications," *Renewable and Sustainable Energy Reviews*, vol. 58, pp. 1475–1485, 2016.

[25] N. Mohan and T. M. Undeland, *Power Electronics: Converters, Applications, and Design*, John Wiley & Sons, 2007.

[26] S. Maniktala, *Switching Power Supplies A-Z*, Elsevier, 2012.

[27] M. H. Rashid, *Power Electronics: Circuits, Devices, and Applications*, Pearson Education India, 2009.

[28] A. Loukriz, M. Haddadi, and S. Messalti, "Simulation and experimental design of a new advanced variable step size incremental conductance MPPT algorithm for PV systems," *ISA Transactions*, vol. 62, pp. 30–38, 2016.

[29] A. Belkaid, I. Colak, and O. Isik, "Photovoltaic maximum power point tracking under fast varying of solar radiation," *Applied Energy*, vol. 179, pp. 523–530, 2016.

[30] M. A. G. De Brito, L. Galotto, L. P. Sampaio, G. de Azevedo e Melo, and C. A. Canesin, "Evaluation of the main MPPT techniques for photovoltaic applications," *IEEE Transactions on Industrial Electronics*, vol. 60, no. 3, pp. 1156–1167, 2013.

[31] P. C. Sekhar and S. Mishra, "Takagi–Sugeno fuzzy-based incremental conductance algorithm for maximum power point tracking of a photovoltaic generating system," *IET Renewable Power Generation*, vol. 8, no. 8, pp. 900–914, 2014.

4

The III–V Triple-Junction Solar Cell Characteristics and Optimization with a Fresnel Lens Concentrator

Yin Guo,[1,2] **Qibing Liang,**[1,2] **Bifen Shu** ⓘ,[1,2] **Jing Wang,**[1,2] **and Qingchuan Yang**[1,2]

[1]*Institute for Solar Energy Systems, Sun Yat-Sen University, Guangzhou, China*
[2]*Guangdong Provincial Key Laboratory of Photovoltaic Technologies, Guangzhou, China*

Correspondence should be addressed to Bifen Shu; shubifen@163.com

Academic Editor: Mahmoud M. El-Nahass

At present, the Fresnel lens are commonly used as the condenser in high-concentrating photovoltaic (HCPV) modules. It is ideally believed that the output power of a III–V triple-junction solar cell which is placed on the focal plane of a Fresnel lens is the largest, because the intensity of the sunlight on the focal plane is the largest. Actually, according to our work, the dispersion of sunlight through a Fresnel lens and the nonparallelism and divergence of the incident light will lead to changes in the spectrum and the homogeneity of illumination, and cause a drop of the solar cell output. In this paper, the influence of the dispersion and nonparallel incidence of the light on the output of a triple-junction solar cell at different positions near the focal plane were theoretically studied, combined with the light-tracing simulation method and triple-junction solar cell circuit network model. The results show that the III–V triple-junction solar cell has the highest output power in both sides of the focal plane positions. The output power can be increased by about 15% after being optimized. The simulation results were verified by the experiments.

1. Introduction

The triple-junction cell is currently one of the most common multijunction cells used in high-concentrating photovoltaic (HCPV) modules. It is composed of three p-n junctions in a series. Each subcell in it absorbs different wavelengths light [1–4]. HCPV modules based on Fresnel lens have now become a research hotspot [5, 6]. Work was carried out for Fresnel lens design [7, 8] focused on improving the illumination energy on the cell, by improving the efficiency of the optical lens. However, the output power of the solar cell is not only related to the illumination energy but also to the spectral distribution and the uniformity of the illumination. The three subcells in a typical triple-junction solar cell respond to three spectral bands: the short band, 300–700 nm; the medium band, 700–900 nm; and the long band, 900–1700 nm [9]. Owing to the dispersion, and considering that the refractive indices of different spectral bands are different, the illumination distributions of the three spectral

bands are also different and nonuniform on the focal plane of lens.

Steiner et al. [10] discovered that the output current of their new multijunction solar cell decreased slightly on the designed focal plane of the lens. However, no work was carried out for further research. In this paper, we used the method of ray-tracing simulation [11] and the solar cell circuit network model [12–18] to analyze the illumination distribution and the power performance of the solar cell at different positions of the optical axis. The illumination distribution of different spectral bands and the power output performance of triple-junction solar cells were obtained.

2. Mathematical Modelling

2.1. The Dispersion of Fresnel Lens. Because a lens has different refractive indices for various frequencies of light, the direction of transmission of a monochromatic light is different after sunlight enters the lens, so it disperses when

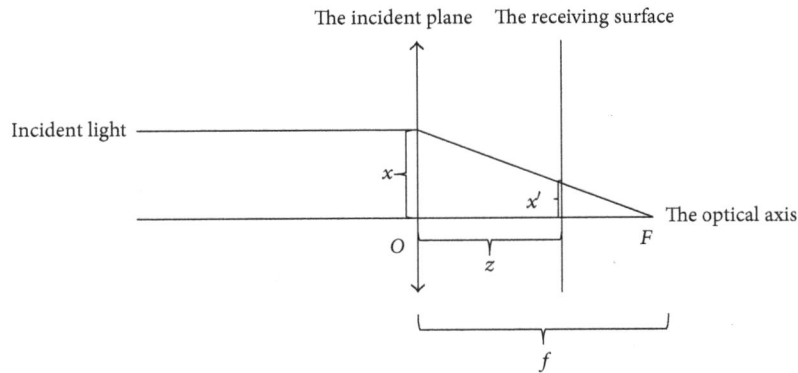

FIGURE 1: The light path of a normal incidence light passes through a Fresnel lens.

leaving the lens, and the colors are arranged in a certain order to form spectra. This is the dispersion phenomenon of the lens. Let us make a detailed analysis as shown below.

Figure 1 shows the normal incidence light passing through a Fresnel lens, where x is the distance between the incident light and the optical axis, x' is the half width of the light spot on the receiving surface, z is the distance between the receiving surface and Fresnel lens, f is the focal length, and F is the focal point.

By the geometric relation in Figure 1, there is

$$\frac{x'}{x} = \frac{f-z}{f}. \tag{1}$$

The calculation formula above can be written as

$$x' = Ax. \tag{2}$$

As you can see, given f and z, a certain x corresponding to a certain x', and x and x' have the same distribution form. For a parallel light incident, with different positions x_1, $x_2, \ldots x_n$ of incident light, there is

$$x'_n = Ax_n. \tag{3}$$

Figure 2 shows the schematic diagram of a parallel incident light path. When the vertical incidence of light is uniformly distributed at the incident plane, the energy distribution of the convergent surface on the receiving surface is also uniform without considering energy loss caused by the refraction of light passing through the lens.

If x is the maximum distance between the incident light and the optical axis, then x' is the maximum distance between the light spot of the receiving surface and the optical axis, so that the range of light for a given wavelength will be given at the given receiving surface. For different wavelengths of light, the focal length is different, so the light range on the receiving surface is also different.

Figure 3 is the light path diagram of a parallel polychromatic light. Here the definition of illumination width is half of the actual illumination range, represented by the subscript m, where X_m is the width of incident light illumination, $x'_{m\gamma}$ is the width of a certain wavelength of light on the receiving

surface, and f_γ is the focal length of a certain wavelength of light. There is

$$x'_{m\gamma} = \frac{\left| f_\gamma - z \right|}{f_\gamma} x_m. \tag{4}$$

When position Z is larger than f, it is still valid.

If the illumination distribution of incident light is uniform, it is also uniform for the light of a certain wavelength on the receiving surface, so that the illumination energy distribution of the receiving surface can be calculated. Set w_γ as the unit wavelength light radiation power on the incident surface of the unit area and w'_γ as the unit wavelength light radiation power on the per receiving surface of the unit area. Because of the conservation of energy, there is

$$w_\gamma d\gamma \cdot 4x_m^2 = w'_\gamma d\gamma \cdot 4x'^2_{m\gamma}. \tag{5}$$

Therefore,

$$w'_\gamma = \frac{f_\gamma^2}{\left(f_\gamma - z \right)^2} w_\gamma. \tag{6}$$

It can be seen that the illumination distribution of different wavelengths of light is generally different and the range of illumination is also different, so the illumination distribution on the receiving surface must be nonuniform. We can calculate the degree of nonuniformity of the illuminated distribution caused by the dispersion.

2.2. Concentrating Illumination of Nonparallel and Divergent Incident Light.
For a monochromatic light at a certain wavelength, when the ray is not parallel to the optical axis, the light does not gather at the focal point of the lens but converges to a point on the plane passing through the focus and perpendicular to the axis; this plane is called the focal plane, as shown in Figure 4. When the incident light is divergent, the light in different directions converges at different points on the focal plane which improves the quality of the illumination energy near the focal point. The greater the divergence angle of the light source is, the more scattered the light energy will be.

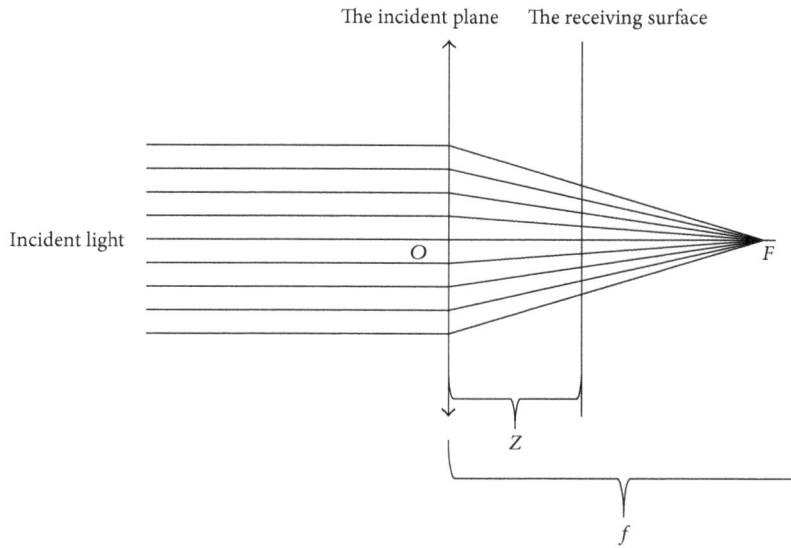

FIGURE 2: The schematic diagram of parallel incident light path.

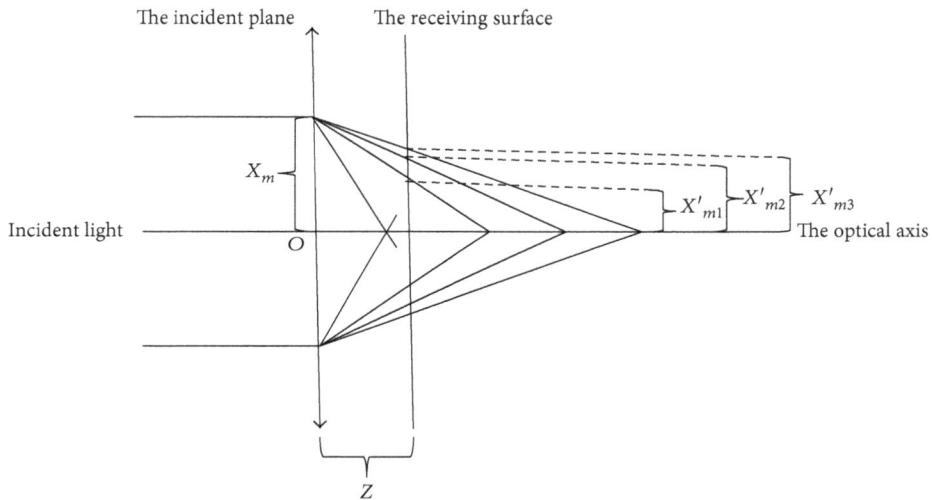

FIGURE 3: The light path diagram of parallel polychromatic light.

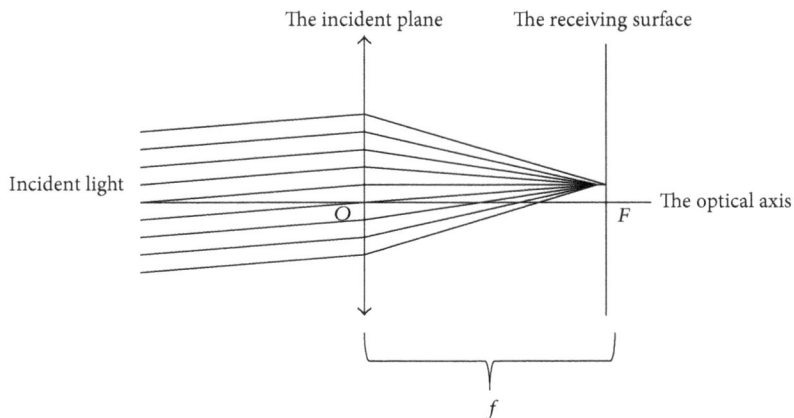

FIGURE 4: Schematic diagram of a nonparallel monochromatic incident light focusing through a lens.

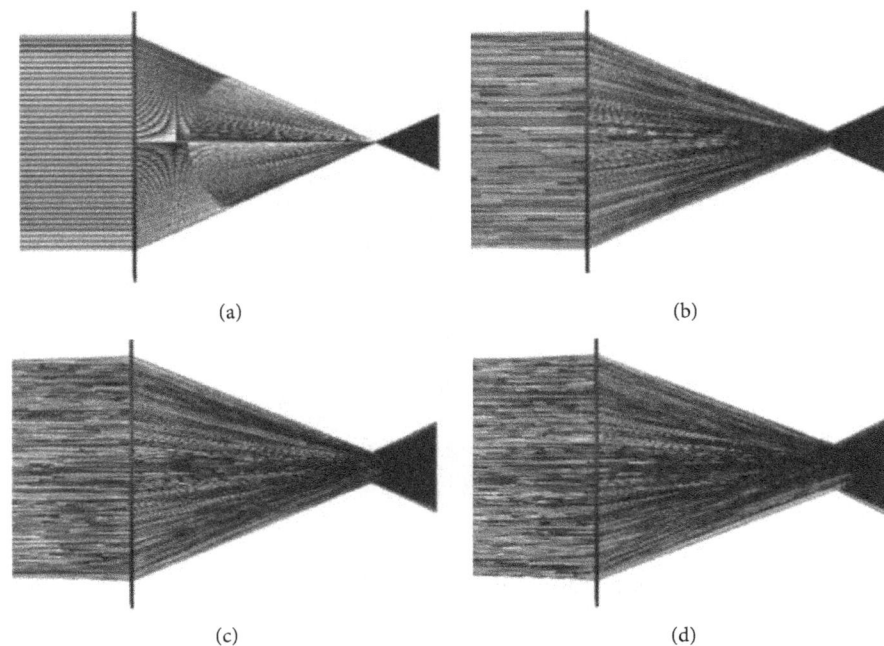

FIGURE 5: The diagram of the result of a 500 nm monochromatic incident light with different divergence angles concentrated by a lens. (a) Divergence angle at 0°. (b) Divergence angle at 1°. (c) Divergence angle at 2°. (d) Divergence angle at 3°.

Figures 5 and 6 are the results of our previous work [19]. Figure 5 shows the results of a 500 nm monochromatic incident light with different divergence angles concentrated by a lens. For parallel light, the light converges to the focal point, and if the solar cell is located at the focal point, the center of the cell will have great light intensity. When the light has a divergent angle, the light through the lens will not be concentrated entirely on the focal point. The larger the divergent angle is, the more scattered the spot near the focus will be, and the larger the spot is. The size of the spot is sensitive to the divergence angle of the light source, and the divergence angle, even at 1°, will have a very obvious effect on the spot size.

Figure 5 is the result of light for a single wavelength. The real situation is that the distribution of illumination on the solar cells is the result of the convergence of multiple wavelengths of light in the case of the spectral distribution of AM1.5d. Figure 6 is the distribution of the light intensity ratio on the surface of a solar cell at the focal point under the AM1.5d spectrum distribution, and the width of the cell is 10 mm. The light intensity ratio here is defined as the ratio of the concentrated light intensity to the standard light intensity.

Figure 6 shows that the parallel incident light (divergence angle is 0°) is concentrated highly on the center of the cell. When the divergence angle increases, the light intensity ratio decreases, and the maximum light intensity ratio of the cell center decreases. This indicates that the divergence of light will obviously weaken the nonuniformity of the illumination on the solar cell surface. Because the nonuniformity of illumination has a negative impact on the performance of the cell, the divergence angle of the light source has a greater impact on the cell performance.

2.3. Circuit Network Model for Triple-Junction Solar Cells under Nonuniform Illumination. The equivalent circuit of the conventional triple-junction cell is shown in Figure 7(a). The triple-junction cell is composed of three p-n junctions in the upper, middle, and lower parts, and each p-n junction is connected by the tunnel diode. The three junctions of the triple-junction cell are equivalent to three subcells, and each subcell is replaced by a dual-diode model. The equivalent circuit of a triple-junction cell can be used to solve the electrical characteristics of the cell under different concentration ratios and temperatures.

We know that triple-junction cells actually work under nonuniform illumination. In order to consider the influence of the heterogeneity of illumination on the cell performance, the cell can be divided into many smaller cells, with each smaller cell corresponding to different irradiation intensities at different locations. The basic idea of the circuit network model is to divide the cell into many smaller cells. The primary cell circuit is equivalent to a circuit formed by a parallel connection of many small cells. Each small cell has the electrical characteristics similar to the original cell. Each small cell is replaced by a dual diode model, respectively. The cross connection between smaller cells, the contact between the surface and the metal grid of the original cell, and the interconnection of the metal grid itself, are connected by resistances. The circuit network is shown in Figure 7(b) [19].

3. Results and Analysis

3.1. Concentrating Characteristics of Fresnel Lens. In order to study the photovoltaic characteristics of triple-junction cells under Fresnel lens concentrators, we first studied the concentrating characteristics of Fresnel lens for three bands of solar

(a)

(b)

(c)

(d)

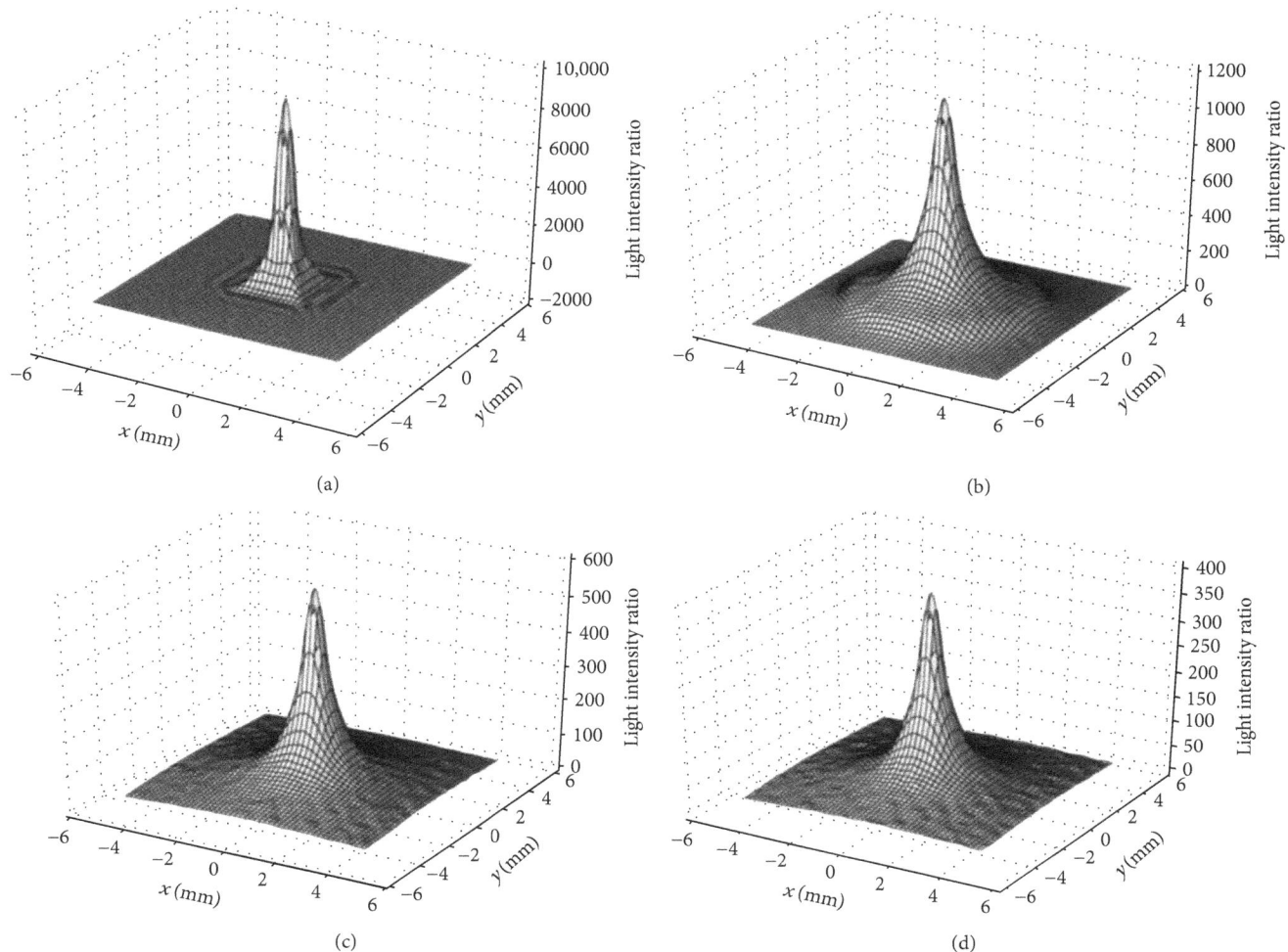

FIGURE 6: The distribution of the light intensity ratio on the surface of a solar cell at the focal point under the AM1.5d spectrum distribution; the width of the cell is 10 mm. (a) Divergence angle at 0°. (b) Divergence angle at 1°. (c) Divergence angle at 2°. (d) Divergence angle at 3°.

light (short: 300–700 nm, medium: 700–900 nm, and long: 900–1700). These three bands correspond to the three sub-cells of triple-junction cells.

Figure 8 shows the variation of the spot diameter of three bands with the position of the optical axis under a Fresnel lens for parallel incident light. For Z value, $Z = 0$ is the focal plane of the lens of full-band solar light. The negative value Z is expressed above the focal plane, close to the Fresnel lens, and the positive value Z is expressed below the focal plane and far away from the lens. The Fresnel lens material is PMMA.

It can be seen from Figure 8 that the diameter of the spot of the three bands below the Fresnel lens varies with the change of the optical axis position. For the shortwave band, there is a minimum spot diameter of 3.7 mm at 5.5 mm above the focal plane. The minimum spot diameter is the same as 0.8 mm of the middle and long wavelengths, but the location is slightly different. The middle band is 4.5 mm below the focal plane, and the long band is 5.5 mm below the focal plane. At the focal plane position of the lens, the diameter of the short-band light spot is 5.8 mm, and the diameter of the middle-long-band light spot is about 2 mm. The

difference of the diameter of the light spot in different bands shows that the degree of nonuniformity of light in different bands is different for a certain size of the cell.

Figure 9 shows that the light spots vary with the installation position of the triple-junction cell in the experimental test. The size of the III–V triple-junction cell is 5 mm*5 mm. The Fresnel lens is 100 mm*100 mm with a focal length of 110 mm, and the material of the lens is PMMA. The outdoor experiment is carried out on the roof of our Institute in Guangzhou. The DNI is about 700–800 W/m², and the atmospheric temperature is about 25°C.

From Figure 9 we can observe that when the distance between the triple-junction cell and the lens changes, the spot on the cell will change significantly. In the process of moving from 5 mm above the focal plane to 5 mm below the focal plane, no change can be observed to the overall size of the spot, almost covering the surface of the cell, while the spot color slowly changed from red and orange to blue and violet. We have known from Figure 8 that the refractive indices of different wave bands of sunlight are different, so the illumination distribution of different bands in the optical axis will be different. Figure 9 shows that when $Z = -5$ mm, the light spot

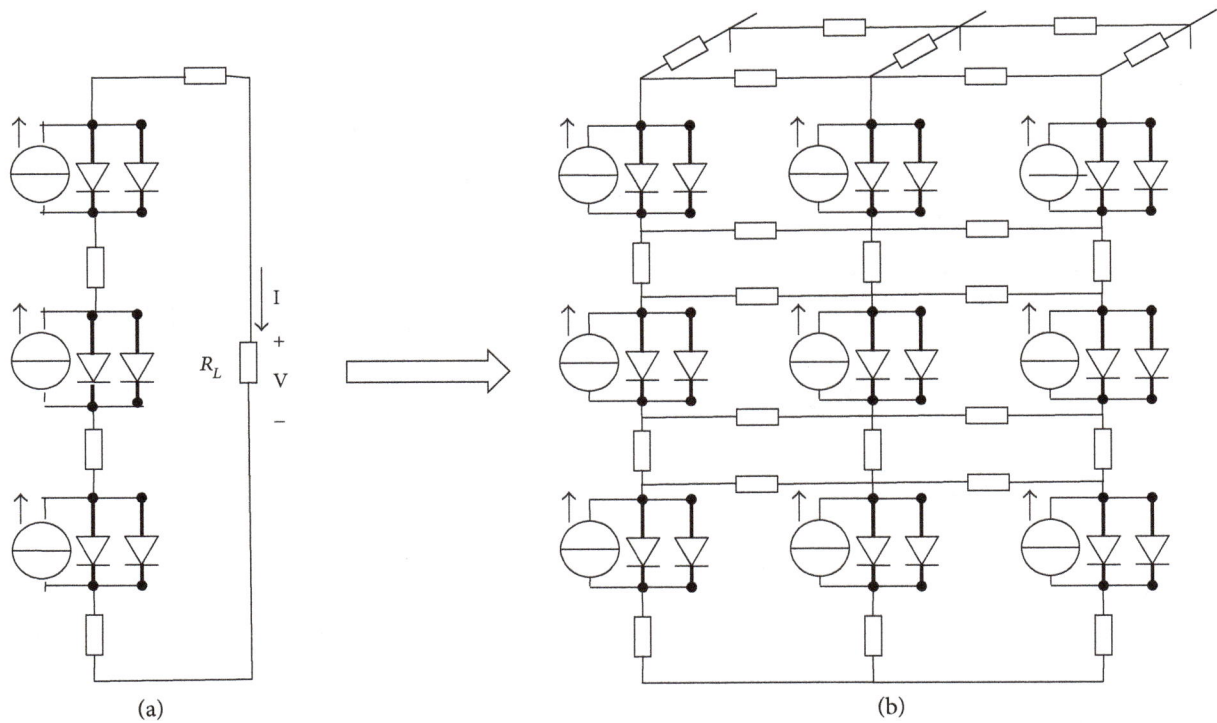

FIGURE 7: The circuit network model of triple-junction cells under nonuniform illumination.

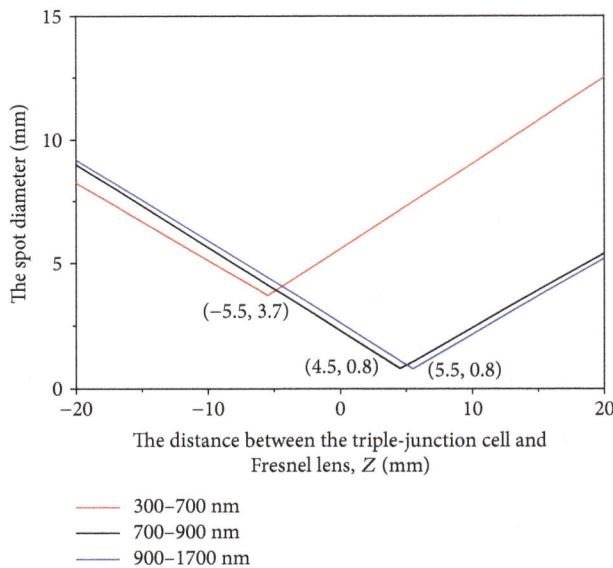

FIGURE 8: The variation of the spot diameter of three bands with the position of the optical axis under Fresnel lens.

of the short-wavelength band is small and the light spot of the long-wavelength band is larger, so the outer ring of the spot is observed to be orange red. When $Z = 5$ mm, the short-wavelength band light spot is large and the long-wavelength light spot is smaller, so the outer ring of the spot is blue purple. When $Z = 0$ mm (the focal plane), the three bands of light converge to be white light, and the light spot is relatively concentrated.

Because a triple-junction cell is composed of three sub-cells absorbing three different wave bands of light, the differences of illumination distribution on the triple-junction cell surface will inevitably affect the power output performance of triple-junction cells.

3.2. Triple-Junction Solar Cell Characteristics. Sunlight can be considered as parallel light. However, due to factors such as dust and the limitations of lens-processing technology, light will inevitably diverge after passing through a lens. So we simulated the photovoltaic characteristics of triple-junction cells under the illumination of light sources with different divergence angles.

Figure 10 is the change of the short circuit current, maximum-output power, fill factor, and efficiency of a 5 mm*5 mm triple-junction cell with the position of the optical axis when the divergence angles of a light source are 0°, 0.1°, 0.2°, 0.5°, 1°, 2°, and 4.4° respectively.

Seen from Figure 10, at the focus of the lens (focal plane) position, when the incident light is more parallel (divergence angle less than 0.2 degrees), the short circuit current of the cell is greatly reduced. This is because the differences of the intensity of the short-, medium-, and long-light bands is large, and it leads to the current mismatch of the three sub-cells. Moreover, the illumination distribution is very nonuniform, so the fill factor (FF) is also very low, causing the efficiency and the output power of the cell to be very low. When the divergence degree is greater than 0.2° (or equal to 0.2°) and less than 0.5°, the focal spot becomes larger, which is almost the same as the area of the cell. The nonuniform degree of illumination distribution on the cell decreases, so the short circuit current and fill factor are all high, which

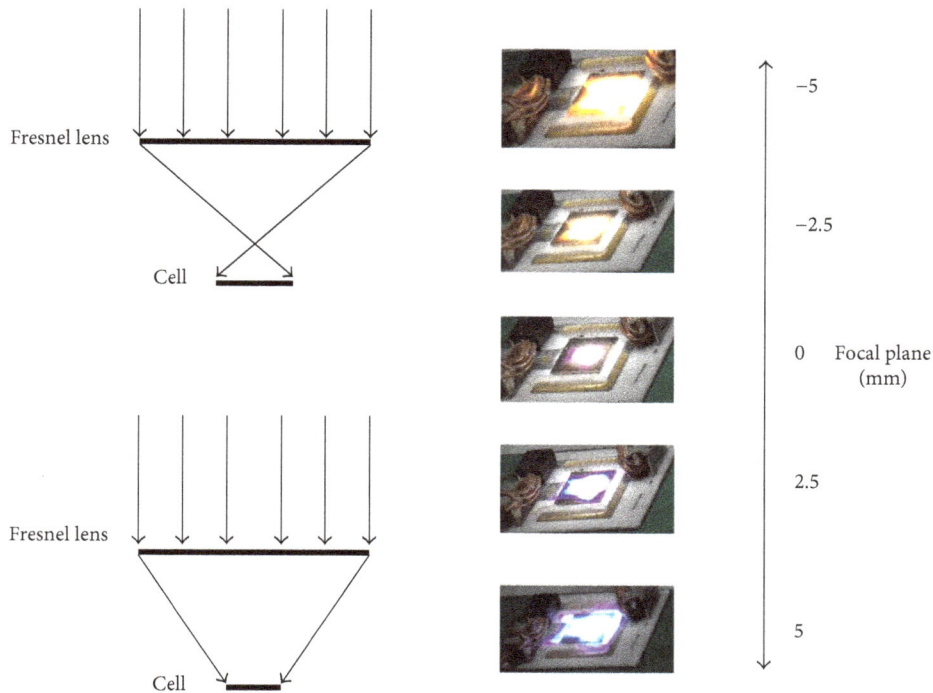

FIGURE 9: Light spots vary with the installation position of the triple-junction cell (along the direction of the optical axis).

makes the efficiency and output power larger. When the divergence is greater than 0.5 degrees, the spot is too scattered, and its area is larger than the area of the cell. Parts of the illumination were out of the cell. Although the illumination on the cell is more homogeneous, the energy decreases. So although the fill factor is larger, the short circuit current is too low, which makes the cell efficiency and output power smaller.

From the results of Figure 10, we also find that when the divergence degree is 0.2 degrees, the performance of the cell is the best if the cell is located at the upper and lower sides of the focal plane ($Z = -9$ mm, $Z = 11$ mm). We know that, in practical application, sunlight can be considered as parallel light. However, due to factors such as dust and limitations of lens-processing technology, light will inevitably diverge after passing through the lens. Normally, it is considered at about 0.2 degrees. Therefore, we can determine the position of the upper and lower sides of the focal plane as the optimal position of the cell performance. The maximum-output power of the cell (2.7 W) at this optimal position is about 17% higher than that at the focal plane (2.3 W).

In order to verify the results of the numerical simulation, we use the existing experimental platform to carry out the experiment. The experimental platform and device are shown in Figure 11. The test cell is a triple-junction cell with an area of 5 mm*5 mm.

We tested the short circuit current of the triple-junction solar cell in the different optical axis positions many times. We found that the short circuit current in different axial positions presents regular distribution.

Figure 12 shows the influence of the optical axis position on the short circuit current of the triple-junction cell. It uses repeated testing data 3 times, and tests 20 experimental points from 5 mm above the focal plane to 5 mm below the focal plane, with a 0.5 mm interval.

Neglecting the influence of some outdoor measurement errors, we can see from the diagram that the short circuit current of the triple-junction cell has a minimum value at the focal plane, and there is a maximum value at the both sides of the focal plane. Compared with the focus position, the short circuit current at both sides of the focus increases by about 10%.

Figure 13 shows the change in the maximum-output power of the triple-junction cell with the position of the optical axis. We can see from the diagram above, the maximum-output power also has a minimum value at the focal plane, and there is a maximum value at both sides of the focal plane. Compared with the focus position, the maximum-output power at both sides of the focus increases by about 15%.

Figure 14 shows the changes in the fill factor of the triple-junction cell with the position of the optical axis. It can be seen that the change regulation of the fill factor is similar to that of the maximum-output power. Compared with the focus position, the fill factor at both sides of the focus also increases by about 15%. It can be found from the results above that the experimental results are in agreement with the simulated results.

4. Conclusion

In this paper, we first establish theoretical models to simulate the concentrating characteristics of Fresnel lens, considering the effects of the lens dispersion and the nonparallel incidence of light. It is found that for different bands of light,

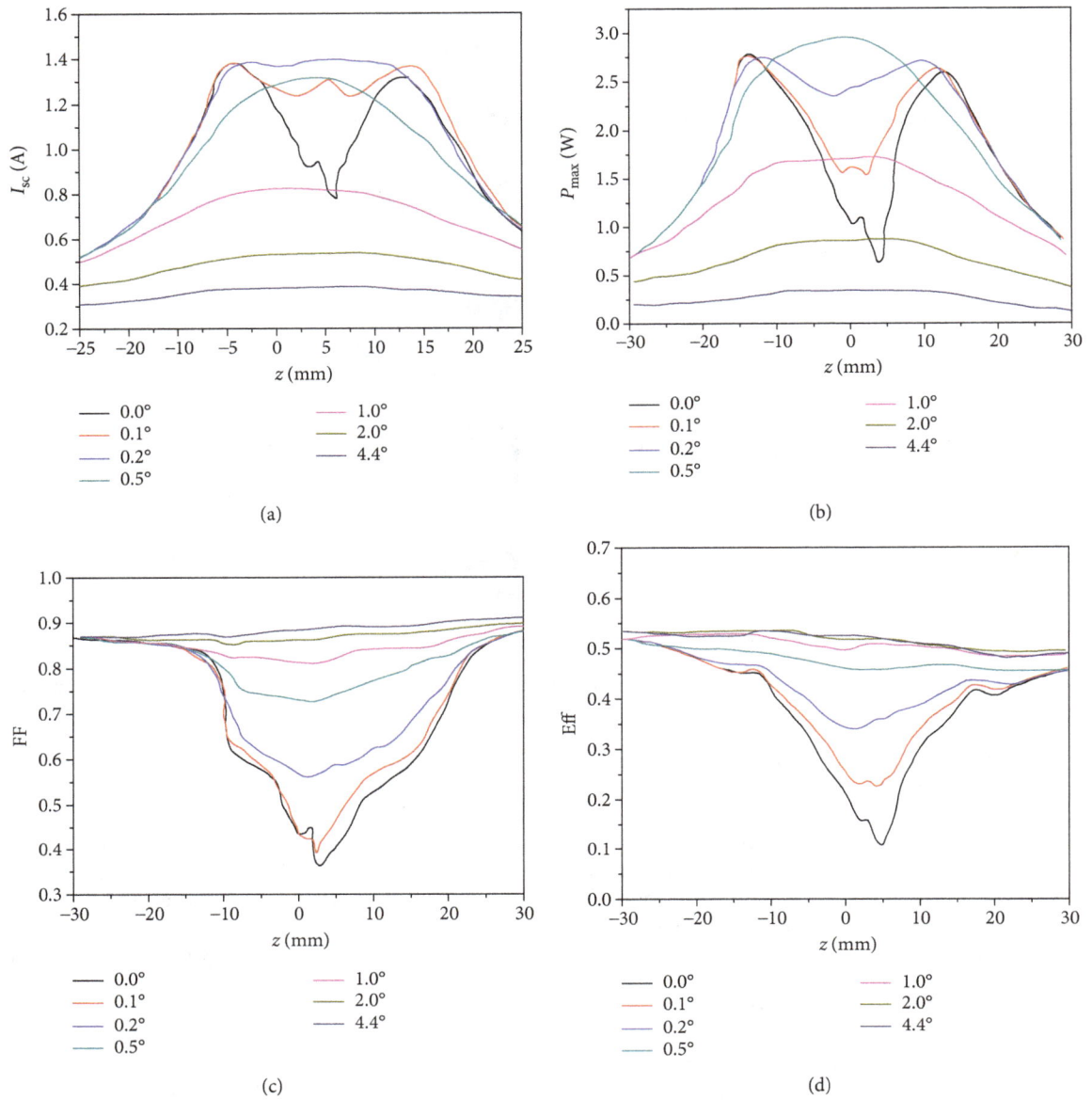

FIGURE 10: Variation of the cell characteristic parameters with the optical axis position under different divergence angles of light sources (cell size 5 mm*5 mm): (a) short circuit current; (b) maximum-output power; (c) fill factor; (d) cell efficiency.

FIGURE 11: Experimental platform and device.

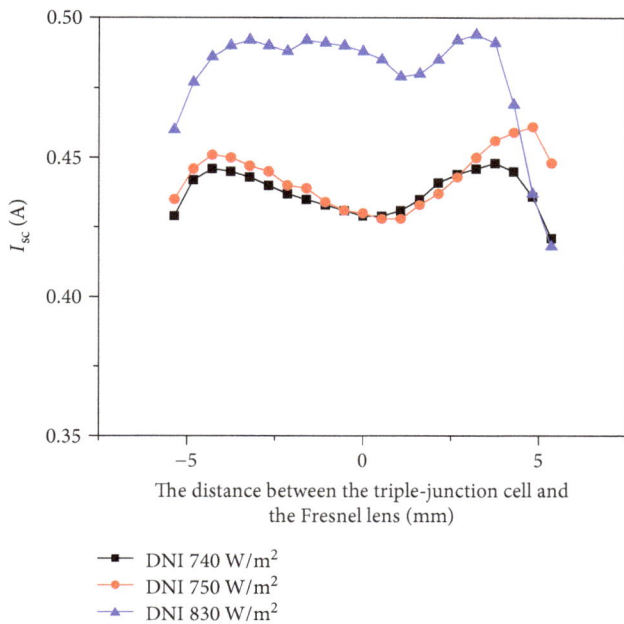

FIGURE 12: The short circuit current of the triple-junction cell along with the change of optical axis positions.

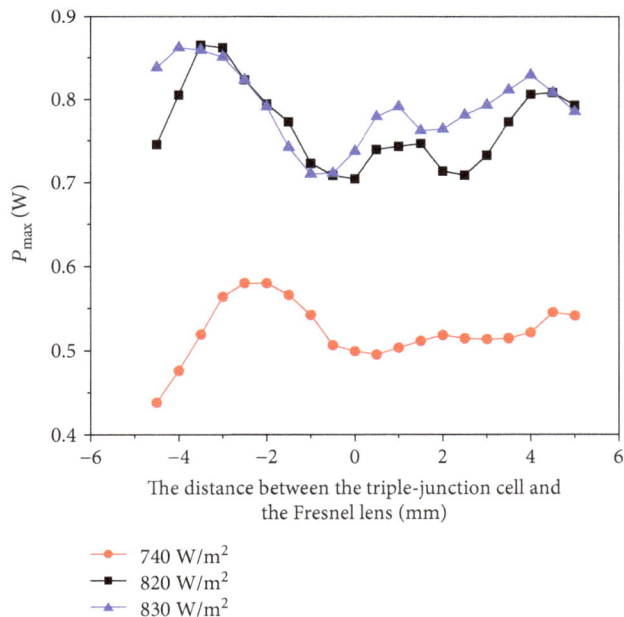

FIGURE 14: Changes in the fill factor of the triple-junction cell with the position of the optical axis.

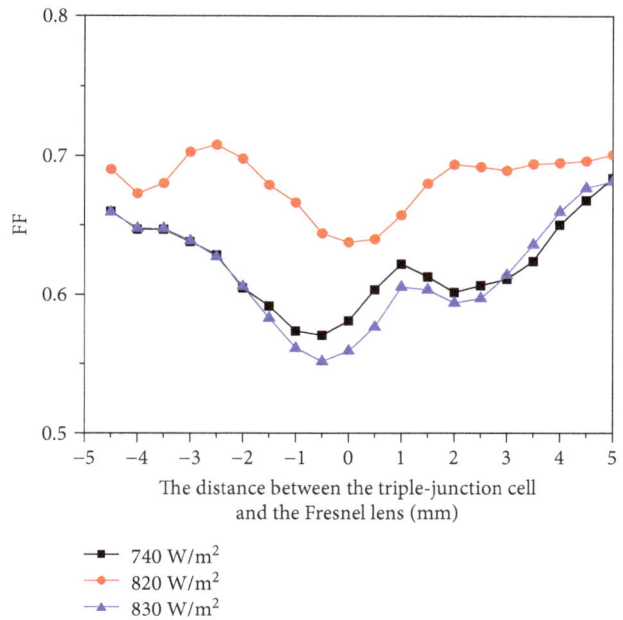

FIGURE 13: Changes in the maximum-output power of the triple-junction cell with the position of the optical axis.

due to the dispersion phenomenon of the light through the lens and the nonparallel incidence of light, the location of Fresnel's focus, the distribution of light, and the size of spot are different after the lens. Then, we simulated the photoelectric performance of the triple-junction cell. It is found that, at the focal plane of the lens, due to the difference of illumination distributions between three wave bands, there is current mismatch between the subcells of the triple-junction cell,

causing the decrease of the short circuit current. And, due to the greater inhomogeneity of the illumination at the focal plane of the lens, the fill factor of the cell decreases. Therefore, there is a minimum value of the short circuit current, maximum-output power, and fill factor of the triple-junction cell in the focal position of the Fresnel lens, and there is a maximum value on both sides of the focal point. The output power on both sides of the focal point is increased by more than 15% over that of the focal plane. The simulation results are verified by the experimental results.

Conflicts of Interest

The authors declare that they have no conflicts of interest.

Acknowledgments

The authors acknowledge the financial support of the National Natural Science Foundation of China (U1707603).

References

[1] H. Helmers, M. Schachtner, and A. W. Bett, "Influence of temperature and irradiance on triple-junction solar subcells," *Solar Energy Materials and Solar Cells*, vol. 116, pp. 144–152, 2013.

[2] W. Zhang, C. Chen, R. Jia et al., "Analysis of the interdigitated back contact solar cells: the n-type substrate lifetime and wafer thickness," *Chinese Physics B*, vol. 24, no. 10, article 108801, 2015.

[3] E. F. Fernández and F. Almonacid, "A new procedure for estimating the cell temperature of a high concentrator photovoltaic grid connected system based on atmospheric parameters," *Energy Conversion and Management*, vol. 103, pp. 1031–1039, 2015.

[4] F.-X. Chen, L.-S. Wang, and W.-Y. Xu, "Enhanced optical absorption by Ag nanoparticles in a thin film Si solar cell," *Chinese Physics B*, vol. 22, no. 4, article 045202, 2013.

[5] G. Zubi, J. L. Bernal-Agustín, and G. V. Fracastoro, "High concentration photovoltaic systems applying III–V cells," *Renewable and Sustainable Energy Reviews*, vol. 13, no. 9, pp. 2645–2652, 2009.

[6] N. F. Chen and Y. M. Bai, "Concentrating photovoltaic system," *Physics*, vol. 36, p. 862, 2007.

[7] G. H. Yang, M. Wei, B. Z. Chen, M. C. Dai, L. M. Guo, and Z. Y. Wang, "Design and research of equal-thickness slab Fresnel lens," *Journal Applied Optics*, vol. 898, p. 34, 2013.

[8] F. Languy, K. Fleury, C. Lenaerts et al., "Flat Fresnel doublets made of PMMA and PC: combining low cost production and very high concentration ratio for CPV," *Optics Express*, vol. 19, no. S3, p. A280, 2011.

[9] M. A. Green, K. Emery, Y. Hishikawa, W. Warta, and E. D. Dunlop, "Solar cell efficiency tables (version 45)," *Progress in Photovoltaics: Research and Applications*, vol. 23, no. 1, pp. 1–9, 2015.

[10] M. Steiner, A. Bösch, A. Dilger et al., "FLATCON® CPV module with 36.7% efficiency equipped with four-junction solar cells," *Progress in Photovoltaics: Research and Applications*, vol. 23, no. 10, pp. 1323–1329, 2015.

[11] Y. Ota and K. Nishioka, "Three-dimensional simulating of concentrator photovoltaic modules using ray trace and equivalent circuit simulators," *Solar Energy*, vol. 86, no. 1, pp. 476–481, 2012.

[12] M. Steiner, S. P. Philipps, M. Hermle, A. W. Bett, and F. Dimroth, "Validated front contact grid simulation for GaAs solar cells under concentrated sunlight," *Progress in Photovoltaics: Research and Applications*, vol. 19, no. 1, pp. 73–83, 2011.

[13] M. Steiner, W. Guter, G. Peharz, S. P. Philipps, F. Dimroth, and A. W. Bett, "A validated SPICE network simulation study on improving tunnel diodes by introducing lateral conduction layers," *Progress in Photovoltaics: Research and Applications*, vol. 20, no. 3, pp. 274–283, 2012.

[14] G. Segev, G. Mittelman, and A. Kribus, "Equivalent circuit models for triple-junction concentrator solar cells," *Solar Energy Materials and Solar Cells*, vol. 98, pp. 57–65, 2012.

[15] P. Rodrigo, E. F. Fernández, F. Almonacid, and P. J. Pérez-Higueras, "Models for the electrical characterization of high concentration photovoltaic cells and modules: a review," *Renewable and Sustainable Energy Reviews*, vol. 26, pp. 752–760, 2013.

[16] S.-G. Yi, W.-H. Zhang, B. Ai, J.-W. Song, and H. Shen, "Analysis of each branch current of serial solar cells by using an equivalent circuit model," *Chinese Physics B*, vol. 23, no. 2, p. 028801, 2014.

[17] X. J. Jia, B. Ai, X. X. Xu, J. M. Yang, Y. J. Deng, and H. Shen, "Two-dimensional device simulation and performance optimization of crystalline silicon selective-emitter solar cell," *Acta Physica Sinica*, vol. 63, article 068801, 2014.

[18] Q. B. Liang, B. F. Shu, L. J. Sun, Q. Z. Zhang, and M. B. Chen, "Optimization of light spot intensity and coverage to a triple-junction solar cell under non-uniform illumination," *Acta Physica Sinica*, vol. 63, article 168801, 2014.

[19] Q. B. Liang, "The optimization on electrical performance and thermal stress of high-concentration CPV modules," Ph. D. Dissertation (Sun Yat-sen University), Guangzhou, China, 2015.

5

Effect of Surface Structure on Electrical Performance of Industrial Diamond Wire Sawing Multicrystalline Si Solar Cells

Shaoliang Wang,[1] Xianfang Gou,[2,3] Su Zhou (ID),[2,4] Junlin Huang,[2,4] Qingsong Huang,[2] Jialiang Qiu,[2] Zheng Xu (ID),[1] and Honglie Shen[4]

[1]Institute of Optoelectronics Technology, Beijing Jiaotong University, Beijing 100044, China
[2]CECEP Solar Energy Technology (Zhenjiang) Co. Ltd., Zhenjiang 212132, China
[3]Beijing University of Technology, Beijing 100124, China
[4]Nanjing University of Aeronautics and Astronautics, Nanjing 211106, China

Correspondence should be addressed to Su Zhou; zhousu2003@hotmail.com and Zheng Xu; zhengxu@bjtu.edu.cn

Academic Editor: Germà Garcia-Belmonte

We report industrial fabrication of different kinds of nanostructured multicrystalline silicon solar cells via normal acid texturing, reactive ion etching (RIE), and metal-assisted chemical etching (MACE) processes on diamond wire sawing wafer. The effect of different surface structure on reflectivity, lifetime, and electrical performance was systematically studied in this paper. The difference between industrial acid, RIE, and MACE textured multicrystalline silicon solar cells to our knowledge has not been investigated previously. The resulting efficiency indicates that low reflectivity surface structure with the size of 0.2–0.8 μm via RIE and MACE process do not always lead to low lifetime compared with acid texturing process. Both RIE and MACE process is promising candidate for high efficiency processes for future industrial diamond wire sawing multicrystalline silicon solar cells.

1. Introduction

Recently, diamond wire sawing technique on multicrystalline silicon wafer grew up at a very high speed due to cost-effectiveness compared to traditional slurry sawing process [1, 2]. It was demonstrated that diamond wire sawing (DWS) has several superiorities over multiwire slurry sawing (MWSS), that is, higher slicing speed, less saw damage layer, better environmental friendship, and potential for cutting thinner wafers [3, 4]. However, diamond wire sawing brings a big texturing problem due to the surface with less saw damage layer and less start point for normal acid texturing [5]. For mass production, there are now three main types of industrial techniques for fabricating DWS mc-Si solar cells: acid texturing with additive [6], reactive ion etching (RIE) [7], and metal-catalyzed chemical etching (MACE) [8–10].

In this work, the comparison between different process and the effect of different surface structure on reflectivity, lifetime, and electrical performance was systematically studied. We also demonstrate that DWS mc-Si solar cells can

achieve average efficiency as high as 19.05% and 18.86% by RIE and MACE, respectively.

2. Experimental

The acid texturing, RIE, and MACE processes presented in this work are applied in the following solar cell fabrication process using industrial production machine:

Acid texturing for DWS mc silicon wafers by etching in HF/HNO$_3$/H$_2$O/additive solution (49 wt% HF: 65 wt% HNO$_3$: H$_2$O = 1:2:1.2 by volume, 10°C) and subsequent cleaning in KOH solution and HF/HCl solution via RENA Inline text machine. Acid texturing for MWSS mc silicon wafers as reference by etching in HF/HNO$_3$/H$_2$O solution (49 wt% HF: 65 wt% HNO$_3$: H$_2$O = 1:3:2 by volume, 10°C) and subsequent cleaning in KOH solution and HF/HCl solution via RENA Inline text machine. RIE for DWS mc silicon wafers by etching in HF/HNO$_3$/H$_2$O solution (49 wt% HF: 65 wt% HNO$_3$: H$_2$O = 1:3:2 by volume, 10°C) to remove saw damage layer, maskless RIE in SF$_6$/O$_2$/Cl$_2$ plasma with

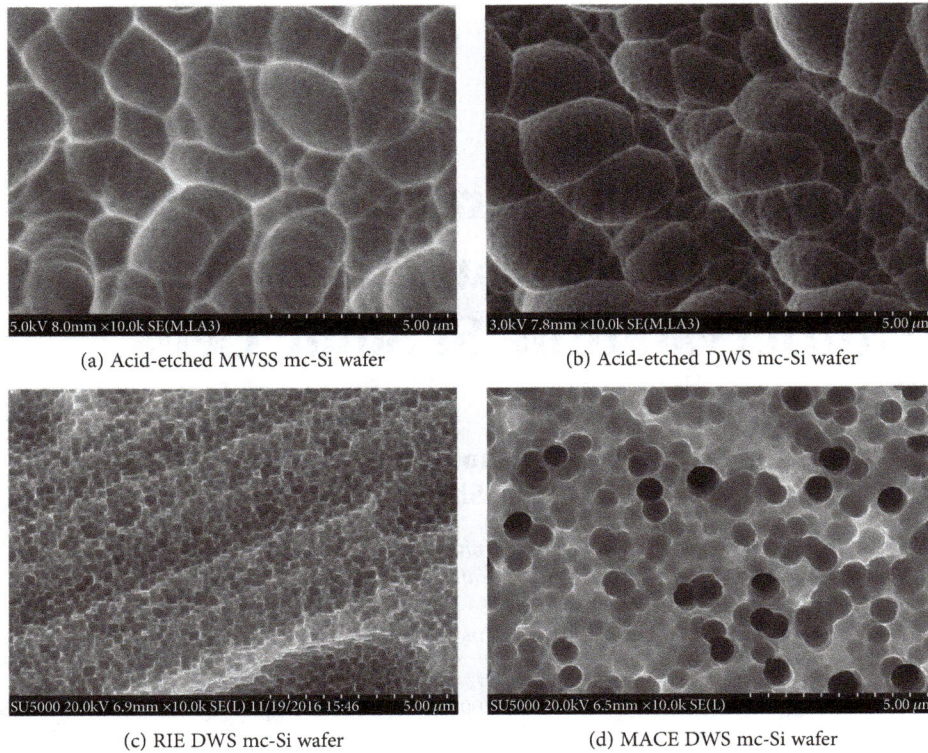

(a) Acid-etched MWSS mc-Si wafer

(b) Acid-etched DWS mc-Si wafer

(c) RIE DWS mc-Si wafer

(d) MACE DWS mc-Si wafer

FIGURE 1: SEM pictures of different surface structures.

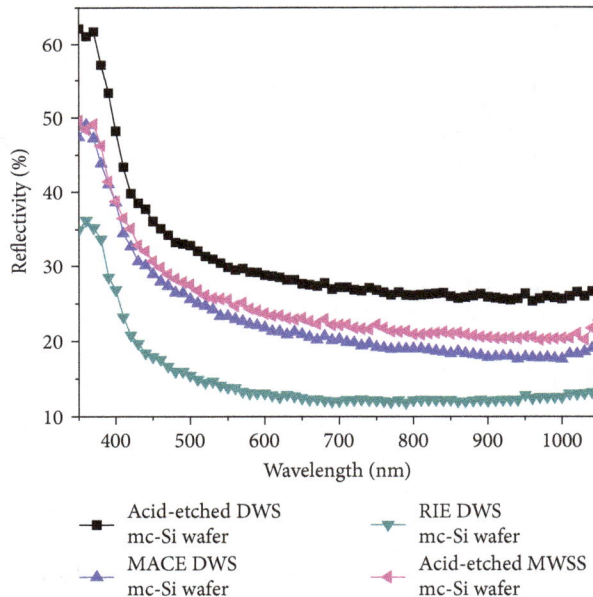

FIGURE 2: Reflectivity versus wavelength curves of wafers with different microstructures.

13.56 MHz radiofrequency using Belight system, and subsequent etching in $HF/NH_4F/H_2O_2$ solution. MACE for DWS mc silicon wafers by etching in $HF/HNO_3/H_2O/AgNO_3$ solution (49 wt% HF: 65 wt% HNO3: H2O = 1 : 2 : 2 by volume, 25°C), cleaning in hot HNO_3 solution and subsequent

etching in $HF/HNO_3/H_2O$ solution. The etch depth of all texturing process are about 2.9 μm per each wafer side.

Emitter formation using a tube furnace from CT Systems with liquid $POCl_3$ as dopant source at a temperature of 835°C and atmospheric pressure for 60 min in O2 and N2 ambient followed by edge isolation and removal of phosphor-silicate glass (PSG) using Rena tool. Plasma enhanced chemical vapor deposition (PECVD) of 80 nm hydrogenated amorphous silicon nitride (SiNx:H) antireflective coating at 400°C using a CT tool. Screen printing of Al rear contact, Ag rear contact, and Ag front contact with standard paste, which was fired using a Despatch infrared fast-firing furnace, with a peak temperature set point of 925°C and a belt speed of 6000 mm/min.

Samples used for lifetime measurement were deposited by SiN on both side and fired at the same temperature as the cell. The microstructure, reflection, lifetime of minority carrier, and IV curves of the resulted mc-Si wafers or solar cells were measured by SEM (Hitachi, S4800, Japan), reflectometer (Radiation Technology D8, China), IV measurement system (Halm, Germany), and WT2000 (Sinton WCT120, USA), respectively.

3. Results and Discussion

3.1. Microstructure of mc-Si Wafers with Different Texture Process. Figure 1 shows the SEM pictures of different surface structures including acid etching on MWSS wafer and acid etching, RIE, and MACE on DWS wafer. Figures 1(a) and 1(b) compare the surface morphologies of acid etching on

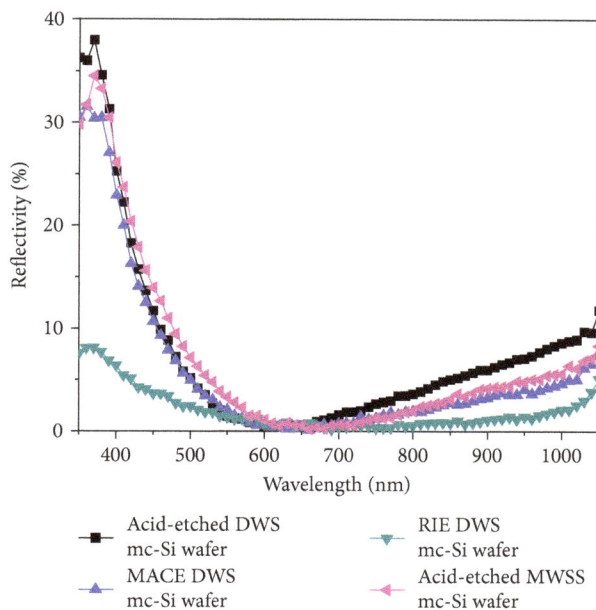

FIGURE 3: Reflectivity versus wavelength curves of SiN-deposited wafers with different microstructures.

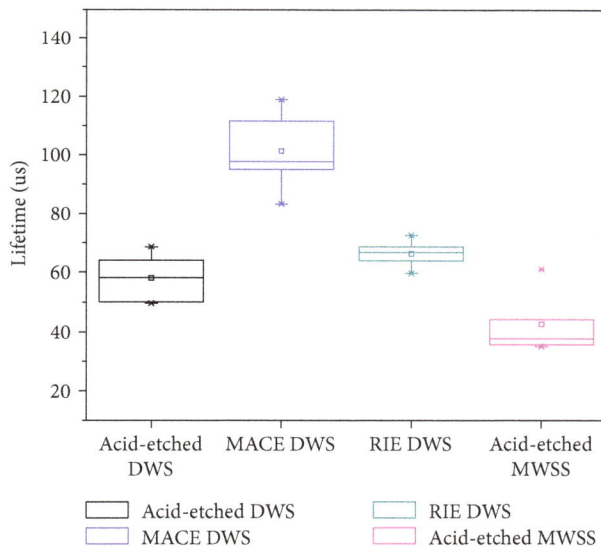

FIGURE 4: Minority carrier lifetime of wafers with different surface structures.

MWSS and DWS wafer. Obviously, the size of oval pits is quite similar and about 2–5 μm. A uniform nanostructure fabricated by RIE was successfully integrated into the former etching pit on DWS wafer in Figure 1(c). The diameter of the nanostructure is in the range of 200–400 nm. Figure 1(d) shows that uniform small round etching pits with the size of 500–800 nm can be obtained by MACE process on DWS wafer.

3.2. Reflectance of mc-Si Wafers and SiN-Deposited Wafer with Different Texture Process. Figure 2 shows the reflectivity

R versus the wavelength (350–1050 nm) curves of wafers with different microstructures. The average reflectivity R_{ave} of the acid etching DWS wafer (31.0%) is 5.7% higher than that of the MWSS wafer (25.3%), while the R_{ave} of RIE DWS mc-Si wafer (15.1%) and MACE DWS mc-Si wafer (23.2%) is 15.9% and 7.8% less than that of the acid-etched DWS mc-Si wafer (31.0%). The reflectivity difference between acid-etched DWS and MWSS may be attributed to the deeper etching pit in the MWSS wafer than those of the DWS wafer because a thin amorphous silicon layer only found in the DWS wafer will prevent effective etching of the surface [8, 11]. The size of surface structure should be responsible for the significant reflectivity reduction of RIE and MACE DWS mc-Si wafer compared to acid-etched ones.

Figure 3 shows the reflectivity R versus the wavelength (350–1050 nm) curves of SiN-deposited wafers with different microstructures. The average reflectivity R_{ave} of the acid-etching DWS wafer (7.7%) is 0.8% higher than that of the MWSS wafer (6.9%), while the R_{ave} of RIE DWS mc-Si wafer (2.1%) and MACE DWS mc-Si wafer (5.8%) is 5.6% and 1.9% less than that of the acid-etched DWS mc-Si wafer (7.7%).

3.3. Minority Carrier Lifetime of mc-Si Wafers with Different Texture Process. Figure 4 shows the minority carrier lifetime of wafers with different microstructures. The average lifetime of the acid-etching DWS mc-Si wafer (58.1 us) is 15.5 us higher than that of the MWSS mc-Si wafer (42.6 us), while the R_{ave} of RIE DWS mc-Si wafer (66.2 us) and MACE DWS mc-Si wafer (101.3 us) is 8.1 us and 43.2 us higher than that of the acid-etched DWS mc-Si wafer (58.1 us). The increasement of lifetime of acid-etched DWS mc-Si wafer compared to that of MWSS wafer was due to the thinner saw damage layer of DWS mc-Si wafer. It is important that even the size of surface structure decrease to several hundred nanometers, good surface passivation can still be obtained on DWS mc-Si wafer with proper process. It can be inferred that within the size of several hundred nanometers, the surface area is not the key factor to affect surface passivation. Especially for MACE process, the HF/HNO$_3$/H$_2$O treatment after Ag-removal acts like "rounding" process in texturing for HIT solar cells, providing possibility to deposit high-quality passivation film [12]. Both RIE and MACE processes have the potential to optimize surface recombination velocity and reflectance simultaneously.

3.4. Electrical Parameters of mc-Si Solar Cells with Different Surface Structure. Table 1 shows electrical performance of mc-Si solar cells with different surface structure. Even with some additive to improve the reflectance of acid-etched DWS mc-Si wafer, higher reflectivity leads to lower short-circuit current density (36.25 mA/cm^2) and efficiency (18.51%) compared to that of acid-etched MWSS mc-Si wafer (36.60 mA/cm^2 and 18.60%). Both RIE and MACE processes show significant current improvement compared to acid-etching process on DWS mc-Si solar cells due to the lower reflectance. Comparing RIE DWS with MACE DWS mc-Si solar cells, the RIE process shows higher J_{sc} and MACE process shows higher V_{oc} which is in accord with the reflectivity and lifetime results. The average

TABLE 1: Electrical performance of mc-Si solar cells with different surface structure.

	V_{oc}/mV	J_{sc}/(mA/cm^2)	FF/%	η/%
Acid-etched MWSS	635.1 ± 1.5	36.60 ± 0.10	80.00 ± 0.16	18.60 ± 0.09
Acid-etched DWS	636.2 ± 1.3	36.25 ± 0.07	80.25 ± 0.15	18.51 ± 0.08
RIE DWS	635.6 ± 1.9	37.48 ± 0.08	79.96 ± 0.13	19.05 ± 0.10
MACE DWS	636.9 ± 2.8	36.89 ± 0.13	80.28 ± 0.12	18.86 ± 0.16

efficiency and open-circuit voltage of RIE (19.05%) and MACE (18.86%) also demonstrate their potential to achieve higher performance and the possibility to be mainstream texturing process for DWS mc-Si solar cells.

4. Conclusion

In summary, 19.05% and 18.86% efficiency DWS mc-Si solar cells with uniform nanotexture have been fabricated through RIE and MACE processes on an industrial production line. The results confirmed that both RIE and MACE process technique can provide an effective texture for DWS mc-Si solar cells. Because of the significant improvement of light-trapping ability on RIE and surface passivation ability on MACE DWS mc-Si solar cells, the J_{sc} of RIE DWS mc-Si solar cell is 1.23 mA/cm^2 higher and the V_{oc} of MACE DWS mc-Si solar cell is 0.7 mV higher than that of an acid-etched one. Those results demonstrate the biggest obstacle for industrial wide spread of DWS mc-Si wafer has been removed by the application of RIE or MACE process and this will benefit for high efficiency and low cost of the PV industry.

Conflicts of Interest

The authors declare that there is no conflict of interest regarding the publication of this paper.

Acknowledgments

The authors express their thanks to the National Natural Science Foundation of China under Grant no. 61575019; the Fundamental Research Funds for the Central Universities with Grant no. 2017RC034, no. 2017RC015, and no. 2017JBZ105; the Doctoral Program of Higher Education no. 20130009130001; and Jiangsu Planned Projects for Postdoctoral Research Funds (1701072B).

References

[1] N. Watanabe, Y. Kondo, D. Ide, T. Matsuki, H. Takato, and I. Sakata, "Characterization of polycrystalline silicon wafers for solar cells sliced with novel fixed-abrasive wire," *Progress in Photovoltaics: Research and Applications*, vol. 18, no. 7, pp. 485–490, 2010.

[2] H. Wu, "Wire sawing technology: a state-of-the-art review," *Precision Engineering*, vol. 43, pp. 1–9, 2016.

[3] C. Yang, H. Wu, S. Melkote, and S. Danyluk, "Comparative analysis of fracture strength of slurry and diamond wire sawn multicrystalline silicon solar wafers," *Advanced Engineering Materials*, vol. 15, no. 5, pp. 358–365, 2013.

[4] A. Kumar, S. Kaminski, S. N. Melkote, and C. Arcona, "Effect of wear of diamond wire on surface morphology, roughness and subsurface damage of silicon wafers," *Wear*, vol. 364-365, pp. 163–168, 2016.

[5] B. Meinel, T. Koschwitz, C. Blocks, and J. Acker, "Comparison of diamond wire cut and silicon carbide slurry processed silicon wafer surfaces after acidic texturisation," *Materials Science in Semiconductor Processing*, vol. 26, pp. 93–100, 2014.

[6] M. Lippold, F. Buchholz, C. Gondek, F. Honeit, E. Wefringhaus, and E. Kroke, "Texturing of SiC-slurry and diamond wire sawn silicon wafers by HF–HNO$_3$–H$_2$SO$_4$ mixtures," *Solar Energy Materials and Solar Cells*, vol. 127, pp. 104–110, 2014.

[7] J. Yoo, G. Yu, and J. Yi, "Large-area multicrystalline silicon solar cell fabrication using reactive ion etching (RIE)," *Solar Energy Materials and Solar Cells*, vol. 95, no. 1, pp. 2–6, 2011.

[8] F. Cao, K. Chen, J. Zhang et al., "Next-generation multicrystalline silicon solar cells: diamond-wire sawing, nanotexture and high efficiency," *Solar Energy Materials and Solar Cells*, vol. 141, pp. 132–138, 2015.

[9] Z. G. Huang, X. X. Lin, Y. Zeng et al., "One-step-MACE nano/microstructures for high-efficient large-size multicrystalline Si solar cells," *Solar Energy Materials and Solar Cells*, vol. 143, pp. 302–310, 2015.

[10] Y. Jiang, H. Shen, T. Pu et al., "High efficiency multi-crystalline silicon solar cell with inverted pyramid nanostructure," *Solar Energy*, vol. 142, pp. 91–96, 2017.

[11] A. Bidiville, K. Wasmer, R. Kraft, and C. Ballif, "Diamond wire-sawn silicon wafers - from the lab to the cell production," in *24th European Photovoltaic Solar Energy Conference*, pp. 1400–1405, Hamburg, Germany, 2009.

[12] G. Li, Y. Zhou, and F. Liu, "Influence of textured c-Si surface morphology on the interfacial properties of heterojunction silicon solar cells," *Journal of Non-Crystalline Solids*, vol. 358, no. 17, pp. 2223–2226, 2012.

6

Distribution Network Voltage Control Based on Coordinated Optimization of PV and Air-Conditioning

Qi Wang⑩,[1,2] Bin Lu,[1,2] Xiaobo Dou⑩,[3] Yichao Sun,[1,2] and Jin Liu[1,2]

[1]*School of Electrical & Automation Engineering, Nanjing Normal University, Nanjing 210042, China*
[2]*Jiangsu Province Gas-electricity Integrated Energy Engineering Laboratory, Nanjing 210046, China*
[3]*School of Electrical Engineering, Southeast University, Nanjing 210096, China*

Correspondence should be addressed to Qi Wang; wangqi@njnu.edu.cn

Academic Editor: Mahmoud M. El-Nahass

This paper proposes a coordinated optimal control strategy of PV generators and air-conditioning loads, in order to handle the possible voltage beyond limit issues resulting from high penetration of PV generators in the distribution networks. This strategy is achieved via coordinately considering the node voltage sensitivity and the adjustment-compensation bid model, to improve the economy of the whole system. As a result, the shortage of the PVs' reactive power control capability is compensated by the adjustable air-conditioning loads, so that the waste of the photovoltaic power can be reduced or even avoided. The case study using an IEEE standard 33-node system, which is further updated with the installation of 5 PV generators and 5 air-conditioning loads, validates the correctness and effectiveness of the proposed control strategy.

1. Introduction

As the problem of energy shortage and environmental pollution is more critical, access of a distributed generator (DG) such as PV and wind power to the distribution network has become an important trend of distribution network development. Distributed PV access to the distribution network can realize the local balance of energy, cut the investment, and reduce the loss of long-distance power transmission. At the same time, making use of clean and renewable PV power can replace traditional fossil energy and improve energy structure. Due to the randomness and volatility of the PV power, a large number of distributed PV access also bring power quality problems, such as voltage fluctuation and flicker and voltage off-limit to the power grid [1], which endanger the safe and stable operation of the power grid.

The problem of voltage off-limit has become one of the main factors restricting the scale and permeability of distributed PV power in the distributed network [2]. When the PV output is high and the load is light, it may easily cause the problem of voltage off-limit [3]. Traditional solutions are generally focused on the supply side, and the main solutions are adjusting the reactive power of the PV inverter [4], installing reactive-load compensation equipment such as a reactor [5], and calling energy storage equipment to absorb the extra PV power [6, 7]. If necessary, the measure of removing part of active power, namely, "waste of PV power," must be taken [8].

During the scheme above, due to the limit of the power factor, adjusting the reactive power of the inverter has limited effect on voltage adjustment, and the installing of reactive compensation equipment and energy storage equipment brings additional investment. The "waste of PV power" reduces the utilization of renewable energy. If the voltage can be ensured under the limit and the economy can be improved at the same time, it will promote the scale and permeability of distributed PV in the distribution network.

With the deeper research of smart grid and demand-side response, more and more researchers begin to pay attention to the great potential of flexible loads such as air-conditioning in power system regulation [9]. At present, relevant researches mainly focus on the establishment of an air-conditioning load model [10], the prediction of air-conditioning load [11], and air-conditioning load participating in peak load regulation and frequency modulation [10, 12]. If the regulation ability of an air-conditioning load can be used, though increasing air-conditioning power to consume the excessive PV power, it will be able to make up for the inadequacy of photovoltaic reactive power regulation ability and at the same time reduce the waste of PV power and the investment of energy storage and reactive compensation equipment, which provides a new method to solve the problem of voltage off-limit.

This paper considers the regulation potential of the air-conditioning load which coordinates with the voltage adjustment ability of PV itself and puts forward a voltage control strategy for the distribution network of coordinated optimization of photovoltaic and air-conditioning loads. In the distribution network, when voltage exceeds the limit due to the excess photovoltaic output, based on the voltage sensitivity and the pricing model of photovoltaic and air conditioner, the PV reactive power, air-conditioning load and PV active power regulation is allocated, so that the voltage limit control can be realized at the lowest amount of compensation. At the same time, this strategy maximizes the amount of photovoltaic power generation. On the other hand, the air conditioner user gets the compensation during the regulation, and after the adjustment, it can also maintain indoor temperature at a relatively low power. This strategy takes into account the benefits of the photovoltaic, air conditioner, and power grid side, which has high economic efficiency.

2. The Pricing Model of PV and Air-Conditioning

2.1. Adjustment-Compensation Bid Model of PVs.
When using the reactive voltage regulation capability of PV, the maintenance cost should be taken into consideration:

$$c_{\text{pv,q}} = c_{\text{q}}, \tag{1}$$

where $c_{\text{pv,q}}$ is the compensation price for the reactive power adjustment of PV and c_{q} is the maintenance cost of PV adjusting reactive power.

When using the active voltage regulation capability of PV, the maintenance cost and the loss of power should be taken into consideration:

$$c_{\text{pv,p},i} = c + c_{\text{p}} + k_{\text{pv}} P_{\text{pv}}, \tag{2}$$

where $c_{\text{pv,p}}$ is the compensation price for the active power adjustment of PV; c is the selling price of electricity, c_{p} is the maintenance cost of PV adjusting active power; k_{pv} is the user preference coefficient, which is determined by each user; and P_{pv} is the amount of active power adjustment.

2.2. Adjustment-Compensation Bid Model of Air-Conditioning Loads.
In this paper, the adjustment-compensation bid model of air-conditioning loads is formulated by the

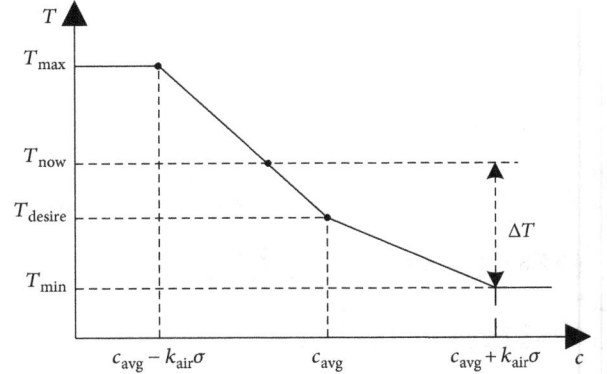

FIGURE 1: The air-conditioning curve of cost and temperature.

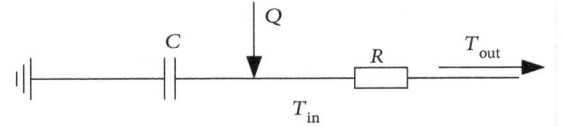

FIGURE 2: Equivalent thermal parameter model of air-conditioning.

following steps: first, users define their temperature-compensation model based on their choices between comfort and earning. Then the adjustment-compensation model is formulated by combining the temperature-compensation model and the thermodynamic model of the building where the air-conditioning is located.

2.2.1. Temperature-Compensation Model.
The temperature-compensation model is shown in Figure 1, where T_{max}, T_{min}, and T_{desire} are user-specified maximum, minimum, and desired indoor temperatures, respectively; c_{avg} is the average compensation price of a certain price history; σ is the standard variation of the compensation prices during the given history; k_{air} is the user preference coefficient, which is determined by each user; and the ΔT part in Figure 1 shows the adjustable range [13].

The curve in Figure 1 can be expressed as the following piecewise function:

$$T = \begin{cases} T_{\text{max}}, & c_{\text{air}} \leq c_{\text{avg}} - k_{\text{air}}\sigma, \\ k_1 c_{\text{air}} + b_1, & c_{\text{avg}} - k_{\text{air}}\sigma < c_{\text{air}} \leq c_{\text{avg}}, \\ k_2 c_{\text{air}} + b_2, & c_{\text{avg}} < c_{\text{air}} \leq c_{\text{avg}} + k_{\text{air}}\sigma, \\ T_{\text{min}}, & c_{\text{air}} > c_{\text{avg}} + k_{\text{air}}\sigma, \end{cases} \tag{3}$$

where c_{air} is the compensation price for air-conditioning and k_1, k_2, b_1, and b_2 are coefficients.

2.2.2. Thermodynamic Model of the Building Where the Air-Conditioning Is Located.
In this paper, the equivalent thermal parameter (ETP) model is used in the thermodynamic model of the building where the air-conditioning is located, which is shown in Figure 2 [14].

The ETP model has the following expression:

$$T_{\text{in},t+1} = \text{coe} \cdot T_{\text{in},t} + (1 - \text{coe})T_{\text{out},t} + \text{COP} \cdot R \cdot (1 - \text{coe}) \cdot P_{\text{air}}, \tag{4}$$

$$\text{coe} = e^{-\Delta t/RC}, \tag{5}$$

where C is the equivalent heat capacity, R is the equivalent heat resistance, $T_{\text{out},t}$ is the outside temperature at t time, $T_{\text{in},t}$ is the indoor temperature at t time, P_{air} is the average power from t time to $t + 1$ time, COP is the energy efficiency, and Δt is the time interval.

By combining (3) with (4) and (5), the adjustment-compensation model is finally formulated with the following expression:

$$P_{\text{air}} = \begin{cases} P_{\min}, & c_{\text{air}} \leq c_{\text{avg}} - k_{\text{air}}\sigma, \\[2mm] \dfrac{k_1 c_{\text{air}} + b_1 - e^{-\Delta t/RC} \cdot T_{\text{now}} - \left(1 - e^{-\Delta t/RC}\right)T_{\text{out},t}}{R(1 - e^{-\Delta t/RC})\text{COP}}, & c_{\text{avg}} - k_{\text{air}}\sigma < c_{\text{air}} \leq c_{\text{avg}}, \\[2mm] \dfrac{k_2 c_{\text{air}} + b_2 - e^{-\Delta t/RC} \cdot T_{\text{now}} - \left(1 - e^{-\Delta t/RC}\right)T_{\text{out},t}}{R(1 - e^{-\Delta t/RC})\text{COP}}, & c_{\text{avg}} < c_{\text{air}} \leq c_{\text{avg}} + k_{\text{air}}\sigma, \\[2mm] P_{\max}, & c_{\text{air}} > c_{\text{avg}} + k_{\text{air}}\sigma, \end{cases} \tag{6}$$

where $P_{\min}$ is the minimum air-conditioning power, which is equal to the larger value between 0 and air-conditioning power when $T = T_{\max}$, and $P_{\max}$ is the maximum air-conditioning power, which is equal to the smaller value between the rated power and the power when $T = T_{\min}$.

3. Coordinate Optimization Control Strategy of PV and Air-Conditioning Load

3.1. Overall Framework. Figure 3 shows the coordinative optimization control strategy of photovoltaic and air-conditioning load. When the distribution network exists at the voltage limit, voltage-centralized controllers make a power flow calculation online in the distribution network. Based on the Jacobi matrix, this paper deduces the active and reactive power sensitivities of all nodes related to PV or air-conditioning. Meanwhile, each PV and air-conditioning submits its own adjustment quantity-compensation price model. After sensitivities and adjustment, a quantity-compensation price model is built; the target is to minimize the total compensation. The objective function is established by considering constraints which relate to voltage overlimit elimination, power limit at each node, and users' comfort index. The objective function is resolved to optimize the adjustment of each PV and air-conditioning. Each PV and air-conditioning gains access to the regulation information. According to the adjustment, each PV and air conditioner adjusts its own output to solve the voltage overlimit problem.

3.2. Objective Function. For nodes where PV and air-conditioning are installed, the target is to determine the minimum compensation payment of regulation in the distribution network, namely,

$$F\left(P_{\text{pv},1}, \ldots, P_{\text{pv},N_{\text{pv}}}, Q_{\text{pv},1}, \ldots, Q_{\text{pv},N_{\text{pv}}}, P_{\text{air},1}, \ldots, P_{\text{air},N_{\text{air}}}\right)$$
$$= \min\left[\left(\sum_{i=1}^{N_{\text{pv}}}\left(c_{\text{pv},p,i}P_{\text{pv},i} + c_{\text{pv},q,i}Q_{\text{pv},i}\right) + \sum_{j=1}^{N_{\text{air}}}c_{\text{air},j}P_{\text{air},j}\right)\Delta t\right], \tag{7}$$

where N_{pv} and N_{air} are the quantities of PV and air-conditioning, respectively, which are involved in the regulation; $c_{\text{pv},p,i}$ and $c_{\text{pv},q,i}$ are the compensation price of active and reactive regulations, respectively, of the ith PV; and $P_{\text{pv},i}$ and $Q_{\text{pv},i}$ are the active and reactive regulations, respectively, of the ith PV. $P_{\text{air},j}$ is the power regulation of the jth air-conditioning, and $c_{\text{air},j}$ is the compensation price of the ith air-conditioning.

3.3. Constraints. On the basis of the objective function, the constraints which need to be satisfied are as follows.

3.3.1. Constraint of the Elimination of Voltage Violation. The elimination of voltage violation can be determined as follows:

$$\left|\Delta U_n - \left(\sum_{i=1}^{N_p}j_{p,i,n}P_i + \sum_{k=1}^{N_q}j_{q,k,n}Q_k\right)\right| \leq \alpha, \quad n = (1, 2, \ldots, N), \tag{8}$$

where ΔU_n is the difference between voltage at the nth node and 1.0 p.u.; N is the number of nodes in the distribution network; N_p and N_q are the nodes involved in active and reactive power regulations, respectively; $j_{p,i,n}$ is the active sensitivity of the ith node related to the nth node; $j_{q,k,n}$ is the reactive sensitivity of the kth node related to the nth node; P_i is the active regulation of the ith node; Q_k is the reactive regulation of the kth node; and α is the maximum voltage deviation allowed during voltage controlling, which can be

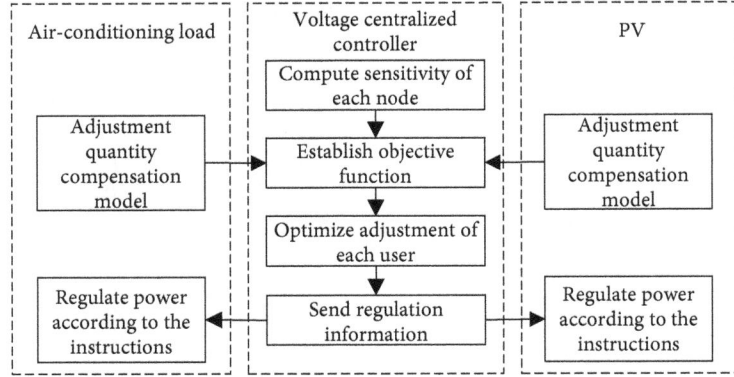

FIGURE 3: Coordinated optimal control structure of PV and air-conditioning.

determined according to the actual voltage level and voltage regulation requirements. This constraint means that the voltage is within the range of $1 - \alpha$ p.u. to $1 + \alpha$ p.u. after regulation.

3.3.2. Upper Bound of Power Regulation. For each PV generation, we have

$$P_{\mathrm{pv},i} \leq P_{\mathrm{pv,max}},$$

$$Q_{\mathrm{pv},i} \leq Q_{\mathrm{pv,max}},$$

$$P_{\mathrm{pv},i}^2 + Q_{\mathrm{pv},i}^2 \leq S_{\mathrm{pv},i}^2, \qquad (9)$$

$$\cos \varphi_i \leq 0.98.$$

Namely, solar power should not only meet the upper bound of the active and reactive power output but also meet the constraint of the power factor.

For each air-conditioning load, we have

$$P_{\mathrm{air},i} \leq P_{\mathrm{air,max}}, \qquad (10)$$

which means that air-conditioning only needs to meet the upper bound of the active power.

3.3.3. Constraint of Users' Comfort. For each air-conditioning load, it should also be that in the regulation process, the indoor temperature does not exceed upper and lower temperature limits that are defined by the users. Accordingly, we have

$$T_{\min} \leq T_{\mathrm{in},t+1} \leq T_{\max}. \qquad (11)$$

4. Case Study

In order to verify the effectiveness of the proposed strategy, this section makes a simulation analysis under the IEEE 33-node distribution system, as shown in Figure 4.

The 33-node test system was built in the DigSILENT/ PowerFactory simulation platform; 5 nodes are chosen from the 33 nodes to install photovoltaic panels and another 5 nodes are placed with air-conditioning loads. The user type and voltage sensitivity of each node are shown in Table 1. Assume that the outdoor temperature is 34°C and the indoor temperature is 26°C. The maximum temperature, minimum

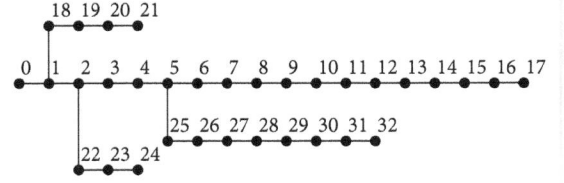

FIGURE 4: IEEE 33-node test system.

TABLE 1: The user type and voltage sensitivity of each node.

Node number	User type	Installed capacity	Voltage sensitivity (p.u./MW (Mvar))
7	PV	1 MVA	Active: 0.014041 Reactive: 0.015123
11	PV	2 MVA	Active: 0.027051 Reactive: 0.025023
17	PV	2 MVA	Active: 0.053279 Reactive: 0.054883
19	PV	1.5 MVA	Active: 0.000424 Reactive: 0.000355
32	PV	2 MVA	Active: 0.010038 Reactive: 0.009773
8	Air-conditioning	1 MW	Active: 0.019006
13	Air-conditioning	1 MW	Active: 0.037459
15	Air-conditioning	1 MW	Active: 0.044487
20	Air-conditioning	1 MW	Active: 0.000424
31	Air-conditioning	1 MW	Active: 0.010052

temperature, and expected temperature designated by air-conditioning users are 26°C, 18°C, and 22°C, respectively. The air-conditioning load of each node is initially operated with a minimum power to maintain the room temperature not more than 26°C. The air-conditioning user benefit coefficient $k_{\mathrm{air}} = 2$, the PV panel user benefit coefficient $k_{\mathrm{pv}} = 0.001$, and the maximum of permissible voltage deviation $\alpha = 0.03$ p.u. The operating and maintaining costs of the PV active power and reactive power adjustment are \$0.0008 per kW·h and \$0.0014 per kvar·h, respectively [15].

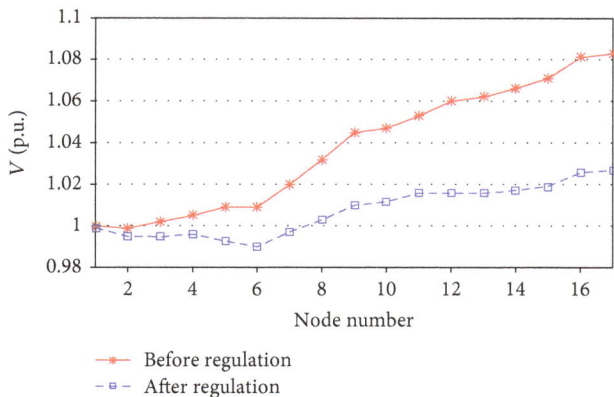

FIGURE 5: The voltage of each node in the main feeder.

TABLE 2: The amount of adjustment of each user.

Node number	User type	Active power regulation (kW)	Reactive power regulation (kvar)
7	PV	0	101.5295
11	PV	0	304.5885
17	PV	0	203.0590
19	PV	0	0
32	PV	0	304.5885
8	Air-conditioning	0	—
13	Air-conditioning	0	—
15	Air-conditioning	565.15	—
20	Air-conditioning	0	—
31	Air-conditioning	0	—

TABLE 3: The compensation price for participants in the regulation.

Node number	Compensation price
7	$0.0014/kvar·h
11	$0.0014/kvar·h
17	$0.0014/kvar·h
32	$0.0014/kvar·h
15	$0.0487/kW·h

Taking some node voltage of the 33-node main feeder is overlimited as a consequence of the excess of photovoltaic power at some time. Since the main feeder is a chain structure, making reactive power compensation at the end of the feeder is the most effective [16]. Therefore, considering voltage sensitivities of the nodes where the PV and air conditioners are located at the number 17 node at the end of the main feeder, this section analyzes the optimal results. Through online power flow calculation, the voltage sensitivities mentioned above can be achieved. The calculation result is shown in Table 1.

4.1. Result of Control of Voltage beyond Limits. We adopt the control strategy proposed in this paper and use the voltage regulating equipment at each node to adjust the main feeder voltage. A comparison of the raw and the adjusted voltage is shown in Figure 4.

Due to the excess of photovoltaic power, without adopting the proposed strategy, voltage goes beyond the limit at nodes 11–17 in Figure 5. The voltage limit violation is the most serious at the number 17 node, where the voltage has approached 1.083 p.u. That is why we need to take drastic measures to ensure that the voltage at each node of the distribution network meets the operational requirements. In the strategy adopted, the voltage at each node is regulated within an allowable range of 0.97 p.u. to 1.03 p.u., where voltage overlimit constraints can be eliminated.

The amount of power adjustment of each user is shown in Table 2. Voltage sensitivity of the number 7, number 11, number 17, and number 32 nodes is high. And the PV reactive power output of these four nodes reached the upper limit of the reactive power output limited with the power factor constraints. For the sake of lower voltage sensitivity, the 19th node has pool regulation effects and will not be involved in the voltage regulation. The active power sensitivity of the 15th node is the most visible. The remaining power is assumed by increasing the output of the air-conditioning at the 15th node to meet the regulatory requirement. As a result, the rest of the air-conditioning was loaded only with the minimum power to maintain the indoor temperature to meet user comfort constraints, being unnecessary in participating in the regulation. Besides, PV panels can maintain the

current generation power instead of curtailing PV power to control voltage.

In summary, the proposed strategy can not only solve the problem of distribution network voltage beyond the limit but also avoid the curtailment of PV power.

4.2. Economic Analysis. The compensation price for participants in the regulation is shown in Table 3.

The duration time of regulation is 5 min. We figure out that the cost of the compensation is $2.3538. Making full use of the reactive power regulation, the cost of which is low, minimizes the use of PV power curtailment which is expensive. Thus, this strategy can reach the minimum cost of compensation paid by the grid. Minimizing the use of PV power curtailment achieves the PV power accommodation maximization.

When the regulation ends, the air-conditioning load of the number 15 node involved in this regulation turns off and the room temperature recovers naturally. According to (4) and (5), it takes 10 min to recover from room temperature to 26°C. In order to highlight the effect of the control strategy, the following 2 control schemes are used for comparison. In case 1, we adopt the strategy proposed in this paper. In case 2, without using the proposed strategy, we maintain the room temperature at 26°C. Considering the payment of electricity and the variation of temperature, we compare case 1 with case 2. The electricity price is $0.0783/kW·h which is an experienced value based on the electricity price in China.

As is shown in Table 4, even if the air-conditioning is turned off when the regulation ends, the temperature in the room will not rise quickly in a short period of time. It is because the buildings can store heat. The consumption is

TABLE 4: Economic comparison of air-conditioning in different participation cases.

Contrast terms	Under regulation	Without regulation
Temperature at the beginning	26°C	26°C
Temperature at the end	25.77°C	26°C
Temperature 10 min later after regulation	26.01°C	26°C
Total electricity consumption	72.20 kW·h	73.52 kW·h
Compensation gained	$2.33	$0
Actual electricity payment	$3.37	$5.86

very low for the duration. Therefore, the total electricity consumption is maintained regardless of the increase in temporary consumption. Besides, users participating in the regulation can gain compensation from the grid company, so the users actually can reduce the electricity payment. It can be seen that the regulation and control proposed will not cause any damage to users' economic interests.

To summarize, the proposed strategy can effectively solve the voltage beyond the limit and has high economic efficiency.

5. Conclusions

This paper presents a distribution network voltage control strategy based on the coordinated optimization of PV and air-conditioning. The result of the case study shows that the strategy can optimize and allocate the adjustment amount of each PV and air-conditioning according to the sensitivity and the adjustment-compensation bid models. Users who are located in highly sensitive nodes and have lower compensation price participate in the arrangement preferentially, so that the grid can take the optimal control effect with the lowest payment. The adjustment-compensation bid models are submitted by PV and air-conditioning users themselves, which ensures the interest of each user. In addition, air-conditioning users can keep the indoor temperature with lower power after the arrangement, and the waste of the photovoltaic power can be reduced or even avoided. The investment of reactive compensation equipment can be saved to a certain degree as well. As a result, the benefits of all the PV generators, the air-conditioning loads, and the grid are guaranteed.

The presented strategy can be promoted to the coordinated optimization of different kinds of loads and power sources to solve problems like voltage drop caused by the loss of active power. This makes contributions to restraining the fluctuation of renewable power and improving the penetration of distributed generations.

Considering that the output of PV and air-conditioning is time-variant, models reflecting the impact of response characteristics accurately and strategies based on the output characteristics of different loads and generators deserve further research.

Conflicts of Interest

The authors declare that there is no conflict of interest regarding the publication of this paper.

References

[1] C. H. Wei, A. I. Xin, W. U. Tao, and L. Hui, "Influence of grid-connected photovoltaic system on power network," *Electric Power Automation Equipment*, vol. 33, no. 2, pp. 26–32, 2013.

[2] R. Yan and T. K. Saha, "Investigation of voltage stability for residential customers due to high photovoltaic penetrations," *IEEE Transactions on Power Systems*, vol. 27, no. 2, pp. 651–662, 2012.

[3] R. A. Shayani and M. A. G. de Oliveira, "Photovoltaic generation penetration limits in radial distribution systems," *IEEE Transactions on Power Systems*, vol. 26, no. 3, pp. 1625–1631, 2011.

[4] Y. Wang, F. Wen, B. Zhao, and X. Zhang, "Analysis and countermeasures of voltage violation problems caused by high-density distributed photovoltaics," *Proceedings of the CSEE*, vol. 36, pp. 1200–1206, 2016.

[5] Y. Che, W. Li, X. Li, J. Zhou, S. Li, and X. Xi, "An improved coordinated control strategy for PV system integration with VSC-MVDC technology," *Energies*, vol. 10, no. 10, p. 1670, 2017.

[6] N. Jayasekara, P. Wolfs, and M. A. S. Masoum, "An optimal management strategy for distributed storages in distribution networks with high penetrations of PV," *Electric Power Systems Research*, vol. 116, pp. 147–157, 2014.

[7] W. Zhang, Z. Liu, and L. Shen, "Flexible grid-connection of photovoltaic power generation system with energy storage system for fluctuation smoothing," *Electric PowerAutomationEquipment*, vol. 33, no. 5, article 106G111, 2013.

[8] Q. Li and J. Zhang, "Solutions of voltage beyond limits in distribution network with distributed photovoltaic generators," *Automation of Electric Power Systems*, vol. 39, no. 22, pp. 117–123, 2015.

[9] N. Mahdavi, J. H. Braslavsky, M. M. Seron, and S. R. West, "Model predictive control of distributed air-conditioning loads to compensate fluctuations in solar power," *IEEE Transactions on Smart Grid*, vol. 8, no. 6, pp. 3055–3065, 2017.

[10] D. Yang, X. Zhang, and B. Zhou, "Modeling and control of air conditioning loads for consuming distributed energy sources," *Energies*, vol. 10, no. 10, p. 1630, 2017.

[11] X. Xu, G. Huang, H. Liu, L. Chen, and Q. Liu, "The study of the dynamic load forecasting model about air-conditioning system based on the terminal user load," *Energy and Buildings*, vol. 94, pp. 263–268, 2015.

[12] K. Ma, C. Yuan, J. Yang, Z. Liu, and X. Guan, "Switched control strategies of aggregated commercial HVAC systems for demand response in smart grids," *Energies*, vol. 10, no. 7, p. 953, 2017.

[13] Y. Yao and P. Zhang, "Transactive control of air conditioning loads for mitigating microgrid tie-line power fluctuations," in *2017 IEEE Power & Energy Society General Meeting*, pp. 1–5, Chicago, IL, USA, July 2017.

[14] N. Lu, "An evaluation of the HVAC load potential for provid-

ing load balancing service," *IEEE Transactions on Smart Grid*, vol. 3, no. 3, pp. 1263–1270, 2012.

[15] G. Ma, G. Xu, R. Ju, and T. Wu, "Study on optimal configuration of the grid-connected wind–solar–battery hybrid power system," *International Journal of Sustainable Energy*, vol. 36, no. 7, pp. 668–681, 2017.

[16] X. Xu, Y. Huang, C. Liu, W. Wang, and Y. L. Wang, "Influence of distributed photovoltaic generation on voltage in distribution network and solution of voltage beyond limits," *Power System Technology*, vol. 10, no. 34, pp. 140–146, 2010.

7

Optimal Design Method of a Hybrid CSP-PV Plant Based on Genetic Algorithm Considering the Operation Strategy

Rongrong Zhai ⓘ, Ying Chen, Hongtao Liu, Hao Wu, and Yongping Yang

School of Energy, Power and Mechanical Engineering, North China Electric Power University, Beijing 102206, China

Correspondence should be addressed to Rongrong Zhai; zhairongrong01@163.com

Guest Editor: Mohammad O. Hamdan

Solar energy is the most abundant renewable energy and it has a great potential for development. There are two ways to transfer solar energy to electricity: photovoltaic power generation (PV) and concentrated solar power (CSP). CSP-PV hybrid system can be fully integrated with the advantages of the two systems to achieve low cost, stable output, and manageable to generate electricity. In this paper, the operation strategy of the CSP-PV system is proposed for parabolic trough CSP system and PV system which are now commercially operated. Genetic algorithm is used to optimize the design of the system and calculate PV-installed capacity, battery capacity, and storage capacity of CSP system, making the system to achieve the lowest cost of electricity generation. The results show that the introduction of the CSP system makes it possible to ensure the stability of the output power of hybrid system when the battery capacity is small, which greatly improves the annual utilization time of the PV and reduces solar abandonment. When the system is optimized by operation characteristics of Spring Equinox, the lowest LCOE is 0.0627 $/kWh, the rated capacity of PV and CSP system are 222.462 MW and 30 MW, respectively, and the capacity of heat storage and battery are 356.562 MWh and 14.687 MWh. When the system is optimized by the operation characteristics of the whole year, the lowest LCOE is 0.0555 $/kWh, the rated capacity of PV and CSP system are 242.954 MW and 30 MW, respectively, and the capacity of heat storage and battery are 136.059 MWh and 8.977 MWh. The comparison shows that the power generation curves of the hybrid system are similar in the two optimization-based methods—Spring Equinox based and annual based, but LCOE is lower when optimized by the annual operation characteristic, and the annual utilization rate of the system is higher when optimized by Spring Equinox based.

1. Introduction

Climate change and the scarcity of natural resources make the world to look for a cleaner and more efficient way to use energy to meet the growing energy needs. At present, renewable energy has a great potential and developed rapidly, and it will occupy an important share of future energy structure [1]. However, the major shortcomings of renewable energy are that their variability and intermittency can cause frequent imbalances and serious problems with the grid. The researchers have suggested that different strategies can be used to improve the safety and quality of energy supply, such as the use of more flexible thermal power plants [2], the introduction of appropriate energy storage equipment, and the use of multicomplementary strategy [3].

Solar energy is one of the most abundant renewable resources; solar radiation to reach the Earth's surface is 1800 times as the world's primary energy consumption [4]. There are two ways to convert solar energy into electricity: solar photovoltaic power generation and concentrated solar power [5]. The development of solar photovoltaic power generation is fast, and the ratio of installed solar photovoltaic will reach 16% of the global energy consumption [6]. The power generation costs of photovoltaic systems is relatively lower because of the low price of PV modules, and it can also be achieved grid parity in the absence of any market incentives [6]. However, the high price of photovoltaic energy storage system obstacles the further large-scale application of photovoltaic systems. Many scholars have focused on the concentrated solar power in the recent years; due to that, solar thermal system can be combined with the thermal storage system, so that the solar power plant can meet the requirements of grid operation and the peak load after the sun can still be deployed [7]. But due to the slow development of technology, the cost of solar thermal system is decreased lower than the solar photovoltaic system [8]. In this context, PV system and CSP system were initially considered to be competitors, but they are complementary in fact. The combination of the two technologies is increasingly concerned. PV-CSP hybrid system is a viable way to generate power, and it can meet the local electricity demand and cost less than a single concentrated solar power [9]. The combination of solar photovoltaic power and solar thermal power can improve the capacity factor of the system and can be dispatched to meet the load demand of peak period [10]. The photovoltaic power is rich and cheap to meet the power load during the day; the peak load in the night will be met by solar thermal power system with storage. Then the stability and schedulability of the system can be ensured at low cost.

At present, a CSP-PV hybrid power system is being built in Ottana of Italy, the system is comprised of a 600 kW linear Fresnel concentrated solar power with 15 MWh heat storage and a 400 kW photovoltaic system with 430 kWh battery. Cocco et al. contrasted the two hybrid ways of the system, which are partially integrated and full-integrated, and found that the annual power generation and annual operating hours of the system on full-integrated were higher [11]. Cau et al. optimized the system's operational strategy by meteorological condition to maximize the annual power generation of the system while meeting the energy conservation and the minimum climbing time [12].

The CSP-PV hybrid system on the Atacama Desert in Chile coupled 20 MW photovoltaic solar system and solar tower power system with 300 MWh storage; the cost of power generation is lower than CSP-alone power system but higher than photovoltaic power system. The levelized cost of electricity (LCOE) in 2014 is 14.69 US cent/kWh and 13.88 cent/kWh, respectively, based on Bluemap and Roadmap [9]. Capacity factor of typical intermittent energy is about 20%–40%, this system can reach about 90% [12]. Green et al. proposed an operation with priority on the basis of this system; the priority of the output power is 50 MW, 100 MW, and 130 MW [13]. Hlusiak et al. found that the collector has the greatest impact on the total cost, followed by coal prices; impacts of photovoltaic capacity and thermal storage system on the cost of power generation is relatively small; the cost of the CSP-PV hybrid system is 13% cheaper than the stand-alone concentrated solar power system with molten salt heat storage [14]. Bootello et al. divided power generation into three grades: the energy consumption of the tracking system, the energy consumption of the auxiliary equipment of the power plant and power generation connected to grid, and put forward the operation mode of the hybrid power station [15]. Larchet found that CSP-PV hybrid system with coal-fired backup unit has the lowest cost of electricity generation and project capital output. LCOE of this system is 42% and 52% lower than PV-alone system and PV-diesel hybrid system, respectively. CSP-PV hybrid system increased the investment but decrease the emission compared

to coal-fired power system [16]. The combination of solar thermal and photovoltaic can provide stable energy and increase the capacity factor of the solar thermal power system. It not only can meet the basic requirement of the energy system but also can provide a low-risk investment option [17].

The research on CSP-PV hybrid system are mainly focused on the operation strategy and technical-economic analysis. But the research on the optimal configuration of CSP-PV hybrid system considering the operation strategy is few. In this study, the operation strategy of CSP-PV system is proposed for parabolic trough CSP system and PV system which are now commercially operated. Genetic algorithm is innovatively used to optimize the design of system and calculate PV installed capacity, battery capacity, and storage capacity of CSP system, making the system to achieve the lowest cost of electricity generation. The operation strategy proposed in this paper provides a new idea for the design and operation of CSP-PV hybrid power generation system. The optimization method can be used in the preliminary design of power station.

2. System Description

CSP-PV hybrid power system is composed of concentrated solar power system and photovoltaic system as shown in Figure 1. The upper dashed line box is the photovoltaic power generation subsystem, and the lower dashed box is the concentrated solar power subsystem. The PV power generation system includes PV array, inverter subsystem, and electronic storage system. The PV array consists of a number of subarrays, each of which consists of 20 PV modules with a rated power of 250 W. Each PV subarray is connected to the inverter to ensure that the DC will be converted into AC. At the same time, each inverter is equipped with a maximum power point tracking (MPPT) device to ensure that the photovoltaic subarray can run at the maximum power point. The PV modules are placed south and have a certain tilt angle, which maximizes the annual generation of photovoltaic power generation systems. The PV power generation technology is mature and the system is simple and has flexible layout and low operating costs. However, solar energy resources are fluctuating and intermittent, which makes the output power of photovoltaic power unstable, and it will have great impact on the power grid to a certain extent. To improve the stability of photovoltaic power output, the batteries are configured.

The CSP system consists of trough collector subsystem, two tanks thermal storage subsystem, and power block subsystem. Heat transfer oil is used as heat transfer fluid, and molten salt is used as the heat storage medium. The inlet temperature of the heat transfer oil is 295°C, and the outlet temperature is 395°C. The feed water is heat by the heat transfer oil to become superheated steam and work in the steam turbine. When solar energy is sufficient, one part of the energy of the collector system is sent into the thermal cycle of power generation, another is storage by molten salt for power cycle when the solar energy is poor. When the stored energy is used out, the gas backup can be started to meet the required load. The concentrated solar power system

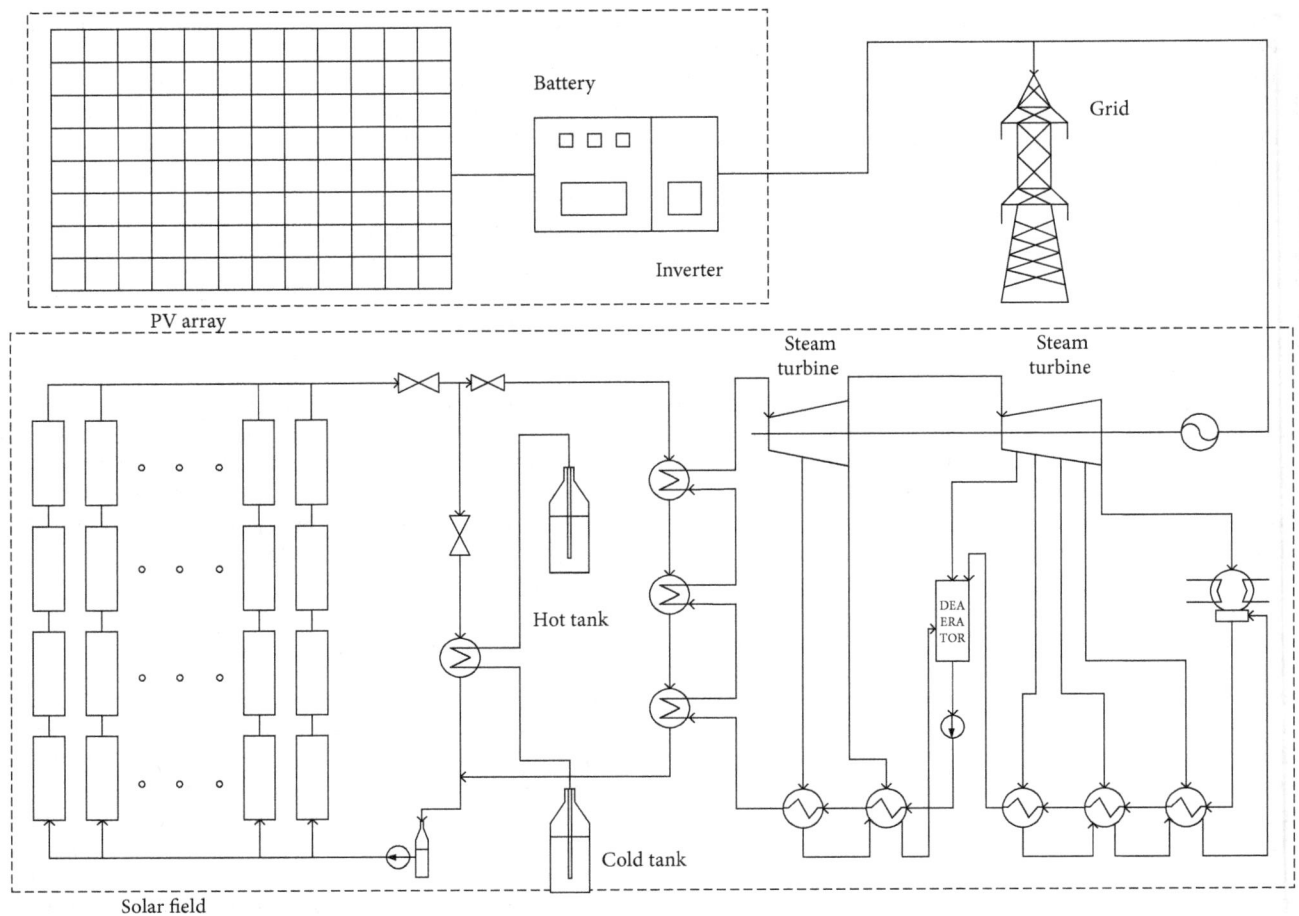

FIGURE 1: CSP-PV hybrid power system.

adopts the energy conversion form of light-heat-electricity, in which the thermal delay of the system and the heat storage system makes the output power of the concentrated solar power system stable, reduce fluctuation of solar energy, and improve the manageability of renewable energy. Therefore, it is expected that a low-cost manageable output solar power generation system will be established by coupling a low-cost photovoltaic power generation system with a concentrated solar power system that can be used for peaking.

The main parameters that affect the energy and economic performance of the system are the collector field area, the storage capacity, the Rankine cycle power, the PV-installed capacity, and the capacity of the battery. In this paper, based on the 30 MW CSP system, the storage capacity, the PV-installed capacity, and the capacity of the battery are selected as the optimization objects to make the power generation cost of the system minimal.

The main parameters of power block in the design condition are shown in Table 1.

3. Model Establishment

Performance analysis of CSP-PV hybrid power system is realized in software Matlab [18]. The simulation model has been simplified. The collector field area, the heat storage

TABLE 1: Main parameters of power block.

	Pressure (MPa)	Temperature (°C)	Mass flow (kg/s)
Main steam	9.80	369.41	55.8
The first extraction	4.00	257.59	6.8
The second extraction	1.70	204.32	4.4
The third extraction	0.60	250.55	1.9
The fourth extraction	0.25	163.75	1.8
The fifth extraction	0.12	104.78	1.4
The sixth extraction	0.06	85.93	2.6
Exhaust steam	0.008	41.51	36.9

capacity, the Rankine cycle power, the PV-installed capacity, and the capacity of the battery are the main design parameters that affect the system performance. The design of the system is optimized in order to find the system structure with the lowest cost of power generation under

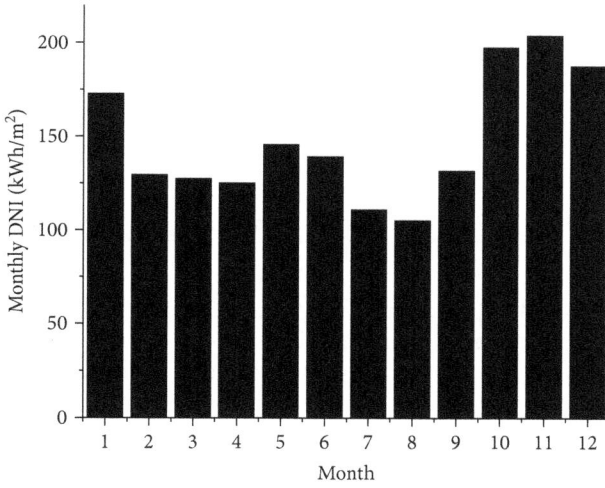

FIGURE 2: Monthly DNI of Lhasa.

TABLE 2: Main parameters of PV module.

PV module		Other assumptions	
Solar cell technology	Polycrystalline	Derating factor f_{PV}	0.8
Nominal power $P_{PV,REF}$	250 W	$U_{L,NOCT}$	9.5
Nominal efficiency $\eta_{PV,NOM}$	14.9%	U_L	$5.7 + 3.8\,V_{wind}$
Nominal operating cell temperature T_{NOCT}	46°C	Transmittance-absorptance coeff. $(\tau\alpha)$	0.8
Active panel area A_{MOD}	1.675 m^2	Inverter nominal efficiency	97.8%
Temperature coeff. of power γ	−0.41%/K	Nominal global irradiance GI$_{NOCT}$	800 W/m^2

the constraints of energy balance and storage energy. The CSP-PV hybrid system model includes the photovoltaic subsystem model and the solar thermal subsystem model. The input parameters of the model are the meteorological data of a certain location.

3.1. Solar Energy Resources. In this paper, the database of a typical meteorological year is from SAM software [19] and Lhasa (91.13°E 29.67°N) is selected. In this paper, the database of a typical meteorological year is from NREL [20]. The meteorological data include direct normal irradiation (DNI), global horizontal irradiance (GHI), ambient temperature, and wind speed.

Annual DNI and GHI of Lhasa are 1777 kWh/m^2 and 1818 kWh/m^2, respectively. In Lhasa, the annual variation of GHI is small and the radiation intensity is large, but DNI varies greatly with season changes and the radiation intensity of autumn and winter is higher than that of summer. The variation of solar radiation intensity makes the performance of the system change greatly. Monthly DNI of Lhasa is shown in Figure 2.

3.2. PV Subsystem Model. The PV system model consists of photovoltaic panels which output is rated at 250 W. PV panels are placed at a fixed angle and faced south. According to the equation proposed by Duffle, the effect of PV module temperature change on the power generation performance of the system is considered.

The main parameters of PV panels are shown in Table 2. The operating temperature (T_C) of PV panels is determined by the rated operating temperature of PV panels (equation (1)).

$$T_C = T_A + (T_{NOCT} - T_{A,NOCT}) \frac{GI}{GI_{NOCT}} \frac{U_{L,NOCT}}{U_L} \left[1 - \frac{\eta_{PV}}{\tau\alpha}\right], \tag{1}$$

where T_A is the ambient temperature; the ambient temperatures for nominal operating cell temperature ($T_{A,NOCT}$) is 20 °C; the solar radiation (GI$_{NOCT}$) is 800 W/

m^2; U_L and $U_{L,NOCT}$ is the actual and rated heat transfer factor; η_{PV} is the actual PV panel efficiency, which can be calculated by equation (2). $\tau\alpha$ is the transfer absorption factor [21]. The efficiency of PV panels

$$\eta_{PV} = \eta_{PV,NOM}[1 + \gamma(T_C - T_{C,REF})], \tag{2}$$

where $\eta_{PV,NOM}$ is the nominal efficiency; γ is the temperature factor; $T_{C,REF}$ is the PV module temperature under standard test conditions (25°C).

$$P_{PV} = n_{MOD} A_{MOD} GI \eta_{PV} \eta_{INV} f_{PV}. \tag{3}$$

The output power of PV panels can be calculated using equation (3). Where n_{MOD} is the number of PV subarray; A_{MOD} is the active area of each PV module, and η_{INV} is the inverter efficiency. A derating factor f_{PV} is finally considered to account for soiling of the panels, wiring losses, shading, snow cover, aging, and other secondary losses.

The use of batteries can make up for difference between PV power generation and energy demand. The energy available for the battery can be described as the charge state "SOC$_B$", which is the ratio of the stored energy to the rated storage capacity. The battery model can be calculated by the following formula. Where η_{BC} and η_{BD} are the battery efficiencies during charge and discharge processes, respectively. Battery efficiency depends on several operating parameters such as current, SOC, charging and discharging power, and battery lifetime. To simplify the model, a constant efficiency of 94% is assumed as declared by the manufacturer while a nominal Depth-of-discharge (DOD) of 80% is considered (with a minimum SOC of 10% and a maximum SOC of 90%).

$$SOC_B(t) = SOC_B(t-1) + \frac{(P_{BC}\eta_{BC} - (P_{BD}/\eta_{BD}))\Delta t}{E_B}. \tag{4}$$

TABLE 3: Main parameters of CSP module.

Design condition		Optical coefficient	0.98
DNI	800 W/m^2	Reflectance efficiency	0.93
Ambient temperature	3°C	Cleanliness factor	0.95
Trough collector		Tracking factor	0.99
Focal length	1.84 m	Dust cover coefficient	0.98
Width	5 m	End shadow coefficient	0.97
Unit length	8 m	Transmissivity factor	0.96
Oil inlet/outlet temp.	285/390°C	Coating absorption factor	0.95
		Other factor	0.96

TABLE 4: Physical property of solar salt.

Melting point	220°C	Viscosity	1.776 cP
Ceiling temperature	600°C	Thermal conductivity	0.519 W/(m·K)
Surface tension	109.2 mN/m	Thermal capacity	1495 J/(kg·K)
Density	1837 kg/m^3	Fusion heat	161 kJ/kg

The PV model is established based on the literature [22]. Rated output power is 30000 kW. The annual power generation of Matlab model is 130690 MWh. The plant electricity rate is defined as the electricity consumed by the power plant itself divided by the gross electricity produced by the power plant. The plant electricity rate is assumed as 2.43% according to statistics of Chinese power generation, then the actual power generation is 127514 MWh. The annual power generation of SAM is 122179 MWh, and the deviation of annual power generation is 4.18%. The reason for the higher power generation in Matlab is that the model ignores few losses of the actual power generation process.

3.3. CSP Subsystem Model.

The simulation of the CSP generation system is based on the parameters of the 30 MW SEGS VI CSP plant.

The solar input is determined by equation (5):

$$Q_{\text{solar}} = \text{DNI} \cdot A_{\text{net}} \cdot \text{KIA} \cdot f_{\text{opt}} \cdot f_{\text{ref}} \cdot f_{\text{trac}} \cdot f_{\text{end}} \cdot f_{\text{clean}} \cdot f_{\text{dust}} \cdot f_{\text{tran}} \cdot f_{\text{abs}}, \tag{5}$$

where A_{net} is the net aperture area of parabolic trough collector, KIA is incident angle correction, f_{opt} is the optical coefficient, f_{ref} is the reflectance efficiency, f_{trac} is the tracking factor, f_{end} is the coefficient to correct end loss effects, f_{clean} is the factor to correct for actual mirror cleanliness, f_{dust} is the factor to correct for dust cover, f_{tran} is the transmissivity factor, and f_{abs} is the coating absorption factor.

The main parameters of the CSP station are shown in Table 3 [22]. The hourly operating characteristics analysis of CSP system is based on the meteorological conditions, especially solar radiation and solar location. The heat loss of the collector field is taken into account. The Equinox day of Lhasa, annual average DNI, and annual average temperature is selected as the system design point.

LS-2 collector tube is used as solar collectors, and the row space is 15 m. The heat loss of the collector is affected by DNI, flow of heat transfer oil, ambient temperature, and wind speed. Usually, the impact of wind speed on heat loss can be ignored.

The available heat input depends on the solar heat input, the thermal losses of the receivers, and the field piping, as shown in equations (6) and (7) [23], where T_i and T_O are the inlet and outlet temperature of the heat transfer oil. The heat pipe is composed of a loop, and each circuit is connected through a pipe. The heat loss on the pipe is obtained from the empirical formula (8) [23], where ΔT is the temperature difference between the average temperature of the collector field and the ambient temperature.

$$Q_{\text{avail}} = Q_{\text{solar}} - Q_{\text{col,loss}} - Q_{\text{pipe}}, \tag{6}$$

$$Q_{\text{col,loss}} = \frac{a_0(T_0 - T_i) + (a_1/2)\left(T_0^2 - T_i^2\right) + (a_2/3)\left(T_0^3 - T_i^3\right) + (a_3/4)\left(T_0^4 - T_i^4\right)}{T_0 - T_i} + \frac{\text{DNI}\left[b_0(T_0 - T_i) + (b_1/3)\left(T_0^3 - T_i^3\right)\right]}{T_0 - T_i}, \tag{7}$$

$$Q_{\text{pipe}} = 0.01693\Delta T - 0.0001683\Delta T^2 + 6.78 \times 10^{-7}\Delta T^3. \tag{8}$$

The heat storage medium of CSP system is binary molten salt, which composes of 60% NaNO$_3$ and 40% KNO$_3$, and the physical properties are shown in Table 4 [24]. The available energy of the heat storage is obtained by the energy balance of the collector field, the power block, and the heat loss. In this paper, the heat loss of the thermal storage tank is assumed to be 2%. The thermal oil VP-1 is used for heat transfer fluid, which has a wide optimum use range of 12° to 400°C.

The solar thermal energy absorbed by the collectors is stored in the thermal storage system as thermal energy. When the power output from CSP subsystem is needed, the molten salt releases heat and water is heated and turned into steam, making the turbine work. The turbine efficiency is variable and related to the steam mass flow in the off-design condition. Variation of turbine efficiency under off-design conditions is considered in power block model. The reduction rate in turbine efficiency can be calculated using equation (9) [25], then turbine efficiency can be calculated using equation (10). The generator efficiency can be calculated using equation (11) [26].

$$\text{Reduction rate} = 0.191 - 0.409\frac{m}{m_{\text{ref}}} + 0.218\left(\frac{m}{m_{\text{ref}}}\right)^2, \quad (9)$$

$$\eta = (1 - \text{Reduction rate})\eta_{\text{ref}}, \quad (10)$$

$$\eta_{\text{generation}} = 0.9 + 0.258\text{load} - 0.3\text{load}^2 + 0.12\text{load}^3. \quad (11)$$

The CSP model is established based on the literature [19] and is compared to SAM software simulation results. The annual power generation of the model is 66584.75 MWh, the plant electricity rate is assumed as 5%; then, the actual power generation is 63255.5 MWh. The annual power generation of SAM is 61167.6 MWh, and the deviation of annual power generation is 3.3%. The main reason is that this model is simplified that some of the error is ignored, making the model of the annual power generation higher.

4. Methodology

4.1. Operation Strategy. Operation strategies of the system can be divided into two modes: prioritize the PV (mode 1) and minimize the turbine shutdown (mode 2). Using mode 1, the PV and the battery is prioritized over the CSP. That is to say that if the PV capacity is large enough to cover the load, it does and the CSP is shut down. The same can be said for the battery, if the battery discharge capacity is large enough to cover the load, it does and the CSP is shut down for the period that the battery discharges. This mode of operation will make the operation and maintenance costs of the system increase and the operating life decrease. However, the CSP must be shut down only if it remains offline for a sufficient amount of time. A turbine hot start can take as long as 1–2 hours. Therefore, in mode 2, the CSP plant will be shut down if it should be shut down for more than 2 hours.

The operation strategy was proposed by combining two modes and was shown in Figure 3. Where $W_{\text{min,turb}}$ is the minimum output of turbine, W_{set} is the rated output of the hybrid system, W_{CSP} and W_{PV} are the real output of CSP system and PV system, respectively, sto and BESS are the energy stored in the storage tank of CSP system and in the battery of PV system, respectively.

There are three operation modes according to the operation strategy.

(1) Energy of PV panels can meet energy demand and PV system runs alone. Exceeded energy is stored in a battery and the collected energy of CSP system is all stored in a storage tank

(2) Energy of PV panels cannot meet energy demand, and PV panels and battery are both used for energy generation. The collected energy of CSP system is all stored in a storage tank

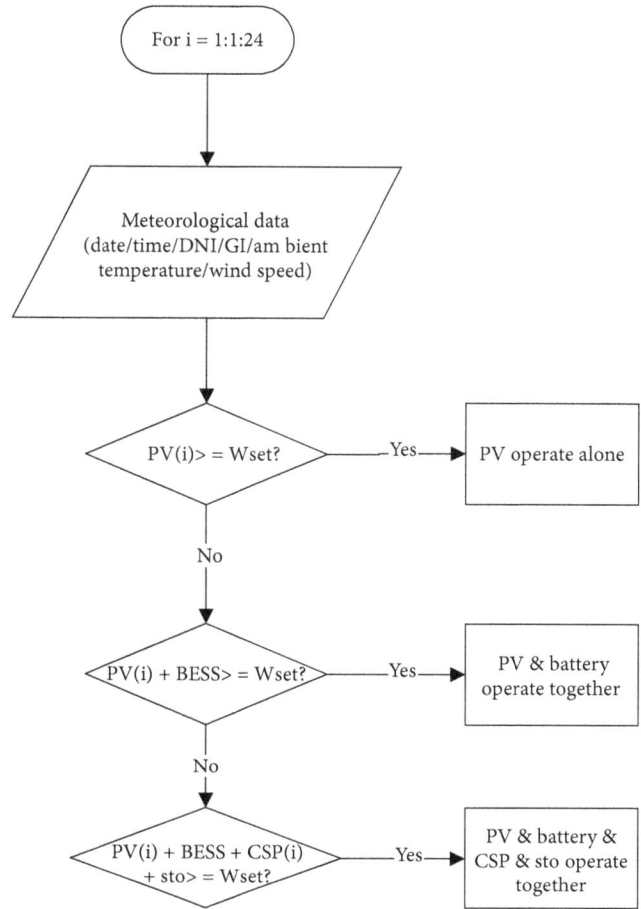

FIGURE 3: Operation strategy of CSP-PV hybrid power system.

(3) Energy of PV panels and battery cannot meet energy demand, and PV panels, battery, CSP system, and storage tank are all used for energy generation

4.2. Method of Optimization. Method of optimization adopted in this study is genetic algorithm. The classical optimization algorithm such as linear program and dynamic programming is easy to fall into the local optimal, and it is difficult to solve the global optimal problem. Genetic algorithm is a good way to overcome this shortcoming and is a global optimization algorithm with good convergence. Many scholars use genetic algorithms to optimize the system. The genetic algorithm is used to optimize the heat recovery system of the rotary kiln, and the mathematical relationship between the design parameters and the temperature and heat transfer rate of the heat recovery switch is deduced. The total heat transfer area and the total power consumption is reduced [27]. Gentils et al. optimized the support structure for offshore wind power, along the outer diameter and cross-sectional thickness of the support structure selected as the design variable. Optimizing makes the quality of the support structure reduced by 19.8% [28]. Li et al. used the reservoir as a decision variable to minimize the variance of the power output and maximize the annual power generation which is a goal by the NSGA-II [29].

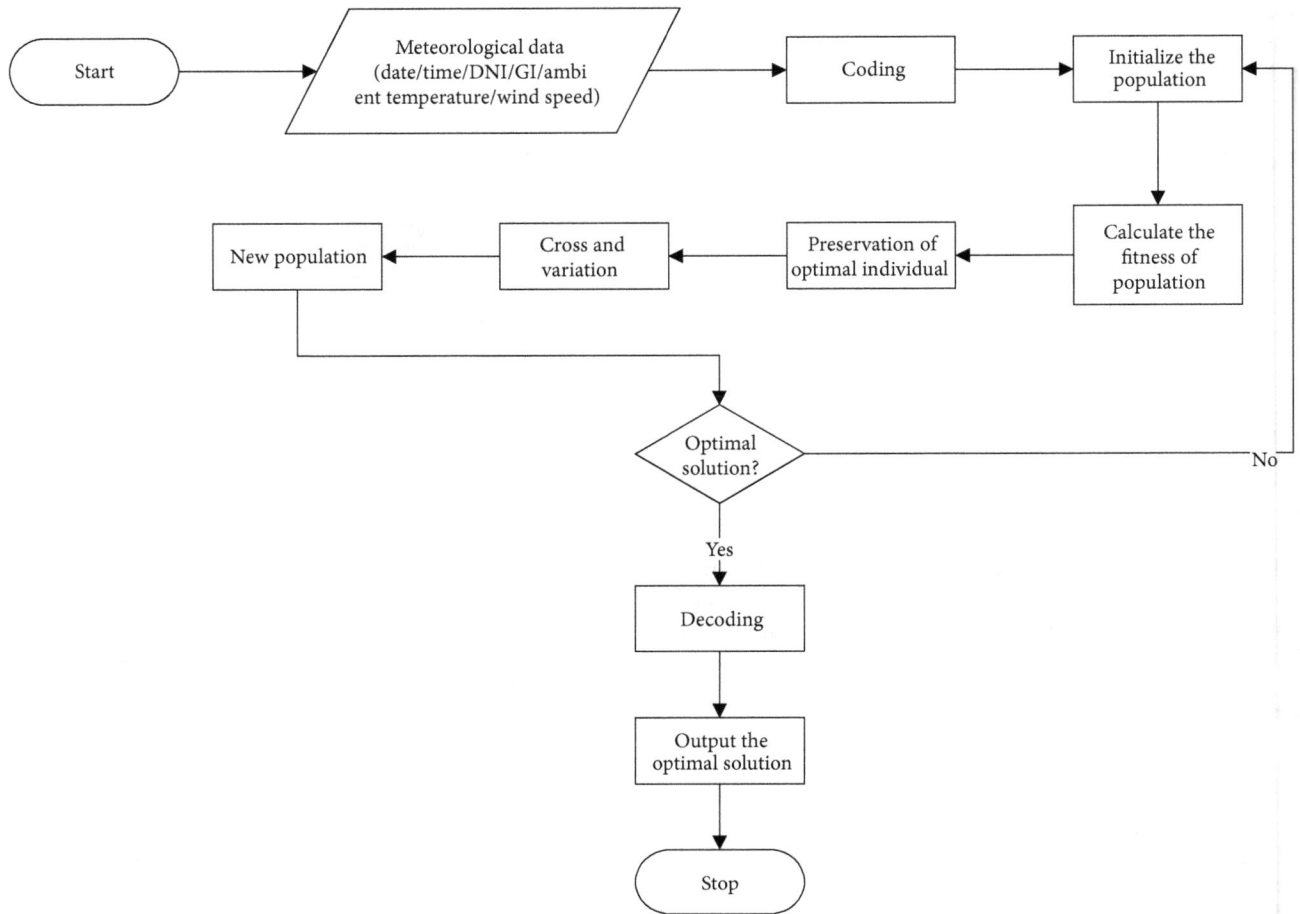

FIGURE 4: Process of genetic algorithm.

Genetic algorithm is an optimization method that simulated biological evolution. It is based on the principle of biological evolution and strategy of group optimization through iterative calculation of selection, replication, crossover, mutation, insertion, and migration. It is suitable for solving complex optimization problems [30]. The process of genetic algorithm is shown in Figure 4.

The process of genetic algorithm is as follows:

(a) Randomly generate the initial population of the determined length

(b) Calculate the fitness value of the population iterations and produce the next generation of groups through replication, crossover, and mutation

(c) The best individual in each generation as the result of the implementation of the algorithm

(d) After a given genetic algebra, compare all the results of the implementation to obtain the optimal solution as an optimal process

The fitness function chosen in this paper is LCOE (levelized cost of energy). LCOE is calculated by equation (12)

[31]. Where IC_{CSP} and IC_{PV} are the initial investment of CSP power station and PV station, respectively, AC_{CSP} and AC_{PV} are the annual cost of CSP power station and PV station, respectively (operation and maintain cost is included), E_{CSP} and E_{PV} are the output power of CSP system and PV system in the first year, d_{CSP} and d_{PV} are annual decay rate of power generation, i is the interest rate, and N is the lifetime of the system.

$$\text{LCOE} = \frac{IC_{CSP} + \sum_{n=1}^{N}(AC_{CSP}/(1+i)^n) + IC_{PV} + \sum_{n=1}^{N}(AC_{PV}/(1+i)^n)}{\sum_{n=1}^{N}((E_{CSP}(1-d_{CSP})^n + E_{PV}(1-d_{PV})^n)/(1+i)^n)},$$

(12)

where IC_{CSP} and IC_{PV} can be divided into two parts: direct cost and indirect cost. The direct component is the investment associated with the power block (C_{PB}), solar field (C_{SF}), piping (C_{PIP}), storage system (C_{TES}), salt purchase (C_{SALT}), and balance of plant ($C_{CSP,BoP}$). The indirect component covers all the remaining costs for the upfront investment that are not directly related to the equipment. These costs include the purchase of land (C_{LAND}) and the engineering, procurement, and construction costs (C_{EPC},

TABLE 5: Cost estimation of the CSP plant and PV system.

CSP direct cost		CSP annual cost	
Solar field cost (C_{SF})	240 \$/m^2 of collector area	O&M annual cost	1.5%
Piping cost (C_{PIP})	36 \$/m^2 of collector area	Insurance annual cost	0.5%
Tank cost (C_{TES})	750 \$/m^3 of storage volume	PV annual cost	
BoP cost ($C_{CSP,BoP}$)	300 \$/kW$_e$	O&M annual cost	1.5%
PV direct cost		Insurance annual cost	0.25%
Panel cost (C_{PV})	1200 \$/kW$_p$	CSP/PV indirect cost	
Inverter cost (C_{INV})	240 \$/kW	Land cost (C_{LAND})	12 \$/m^2
BoP cost ($C_{pv,BoP}$)	240 \$/kW	Engineering cost (C_{EPC})	20%
Battery cost (C_B)	1200 \$/kWh	Other assumption	
PV degradation rate	0.6%	Annual interest rate	7%
Operating lifetime	25 years	CSP degradation rate	0.2%

TABLE 6: Optimization design variables and range.

Variable	Range	Unit
Rated power of photovoltaic plant	0–100	MW
Capacity of heat storage	0–2000	MWh
Capacity of battery bank	0–2000	MWh

calculated as a percentage of the direct costs). The equations for the direct and indirect initial costs of the CSP section are therefore given by

$$IC_{CSP} = IC_{CSP,DIR} + IC_{CSP,IND},$$

$$IC_{PV} = IC_{PV,DIR} + IC_{PV,IND},$$

$$IC_{CSP,DIR} = [(C_{SF} + C_{PIP})A_{SF} + (C_{TES} + C_{SALT}\rho_{SALT})V_{TES}] + (C_{PB} + C_{CSP,BoP})P_{ORC,nom},$$

$$IC_{CSP,IND} = C_{LAND}A_{LAND,CSP} + IC_{CSP,DIR}C_{EPC},$$

$$IC_{PV,DIR} = [(C_{PV} + C_{INV} + C_{PV,BoP})P_{PV,nom} + C_B E_B],$$

$$IC_{PV,IND} = C_{LAND}A_{LAND,PV} + IC_{PV,DIR}C_{EPC}. \tag{13}$$

The main economic data is shown in Table 5 [21]. The floor area of CSP power station and PV power station are estimated by literature [32]. The floor area of the PV station whose output is over 20 MW and track by the fixed axis is 7.5 acres/MW, and the floor area of the CSP station is 10 acres/MW.

5. Case Study

The model is established in Matlab and it can simulate the operating performance of CSP-PV hybrid system. The output power of the CSP power station selected in this study is 30 MW. Input parameters are heat storage capacity, nominal power of PV station, and battery capacity, and output

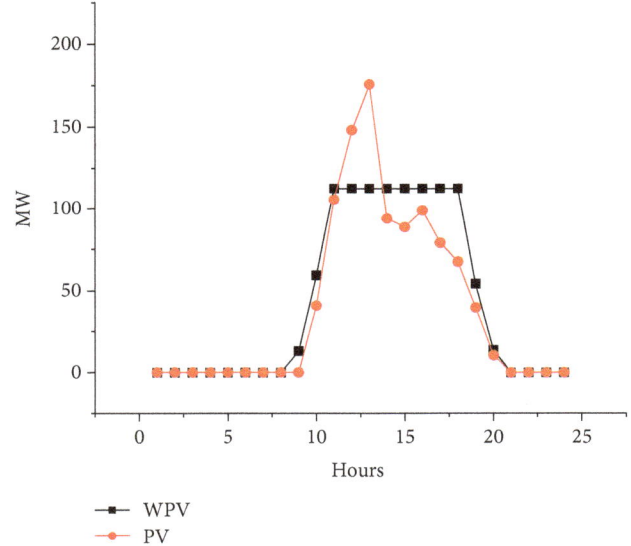

FIGURE 5: Comparison between the power generation capacity of PV system and the actual power generation on Equinox Day.

parameters are the output power of a hybrid system and LCOE. Then the model can simulate the daily, monthly, and annual performance of a hybrid system. The output power is 95% of the sum of the nominal power of the CSP and PV station. There is a 5% unmet load demand [21].

The objective function of optimization is LCOE and variable and their range are shown in Table 6.

5.1. Optimization Based on a Typical Day. Optimization is based on the operation characteristics of Equinox Day. According to the results of genetic algorithm, when the installed capacity of the CSP power system is 30 MW, the LCOE of the CSP-PV hybrid system reaches the lowest which is 0.0660 \$/kWh under the condition that the rated power capacity of PV is 222.462 MW, the battery capacity is 14.687 MWh, and the heat storage capacity is 356.562 MWh.

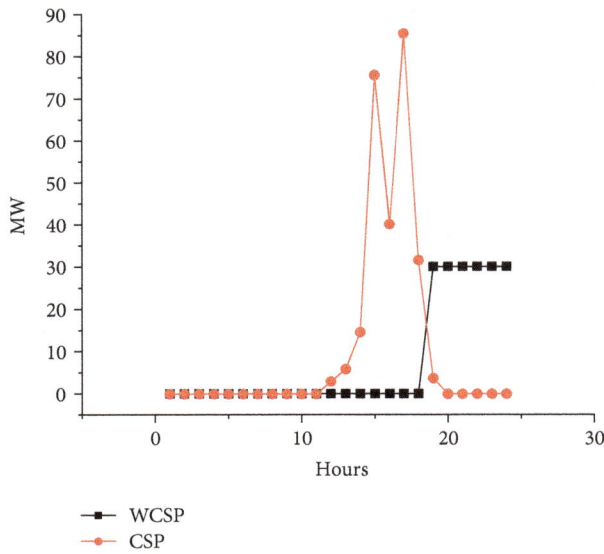

FIGURE 6: Comparison between the power generation capacity of CSP system and the actual power generation on Equinox Day.

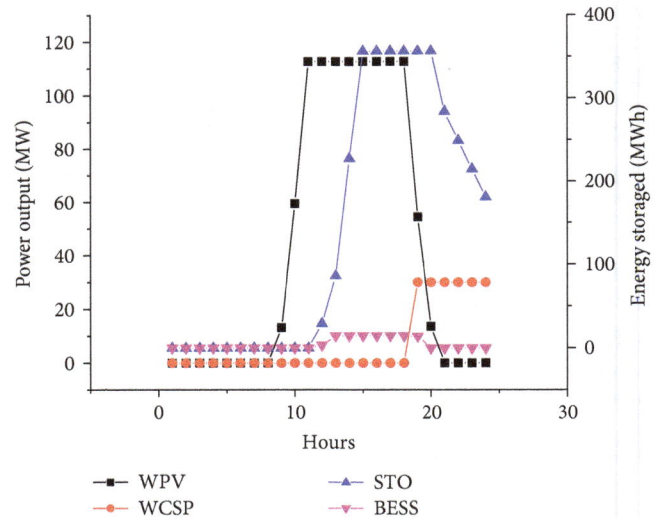

FIGURE 7: The system output power of Equinox Day of Equinox Day Optimization.

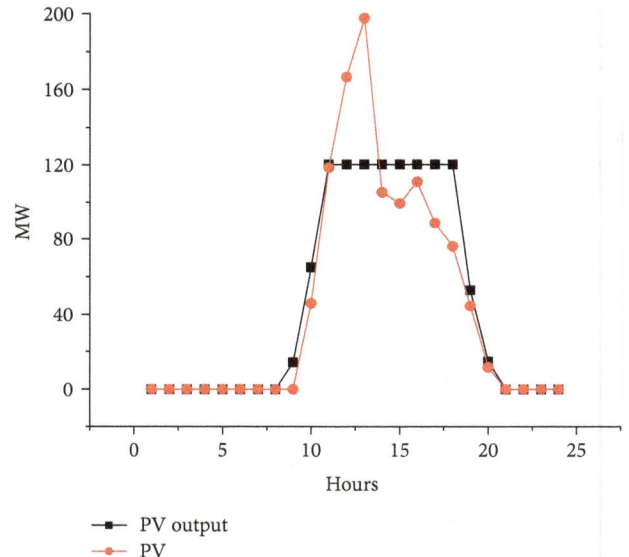

FIGURE 8: Comparison between the power generation capacity of PV system and the actual power generation on Equinox Day.

Figures 5 and 6 show the comparison between the power generation capacity of PV and concentrated solar power (PV/CSP) and the actual power generation (WPV/WCSP) in Equinox Day. The "PV" represents the power generated by the PV panels, and "WPV" represents the power output from the PV system after dispatch. The "CSP" represents the concentrated solar power energy, and it is stored in the TES. The "WCSP" is the actual power output from a solar thermal Rankine cycle. This is added in the manuscript. After the sun rises, the output power of the PV system increases as the irradiation intensity increases and reaches the maximum at 11 o'clock. The use of the battery makes the output power of the PV power plant does not change with the solar radiation changes, and the power generation curve tends to be gentle which makes the impact of PV system to reduce the power grid small and ensure that the system can still maintain the rated output power when the solar radiation is reduced. The CSP system generates power after 19 o'clock and maintains the rated power output. The use of the thermal storage system allows CSP system to continue to generate electricity after the sun goes down.

The system output power of Equinox Day is shown in Figure 7. After the sun rises, the output power of the PV system increases with the increase of solar radiation. The solar radiation intensity is the largest between 11:00 and 18:00, and the operating mode of the system is the power generated by the PV system alone; exceeded energy from the PV system is stored in the battery and the energy of CSP system is all stored in the heat storage tank. At 19 o'clock, the solar radiation is insufficient, the system operating mode changes from the PV alone into a PV system, the battery system and CSP system generate power at the same time, and the excess heat is stored in the thermal storage system. When at 20 o'clock, the sun is completely down without solar radiation, PV and CSP system has no energy source, then the system operating mode was changed into heat storage that generates power

alone. In the system, the power of CSP is used to make up the power generation of the PV system, and the output power of CSP system varies with the PV output power. The more stable PV output power curve is makes the response of CSP system better. The use of heat storage in the CSP system allows the CSP system to continue to generate electricity after the sun goes down and to compensate for the lack of power generation in the PV system without solar radiation.

According to the calculation results, annual power generation of CSP-PV hybrid system is 368722 MWh. Where annual power generation of PV system is 288291 MWh, annual utilization hours are 3261 h which is higher than PV-alone system. Annual power generation of CSP system

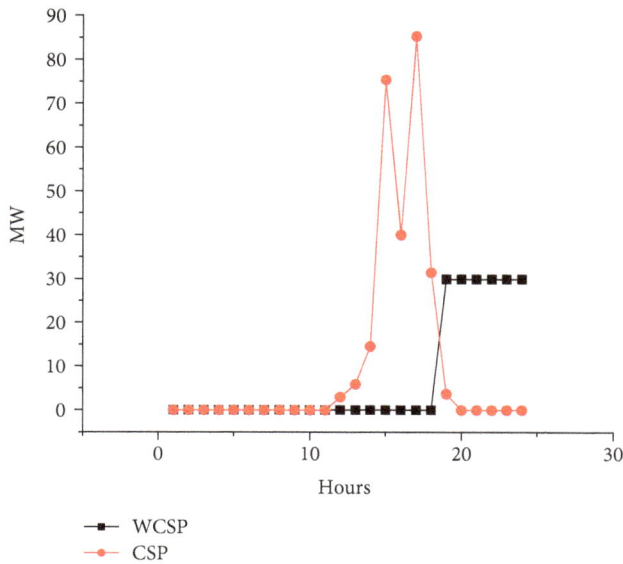

FIGURE 9: Comparison between the power generation capacity of CSP system and the actual power generation on Equinox Day.

TABLE 7: Comparison between annual and day optimization.

	Optimized by day operation	Optimized by annual operation
Annual output power (MWh)	368722	375695
LCOE ($/kWh)	0.0660	0.0555
CSP		
Capacity of heat storage (MWh)	356.562	136.059
Annual output power of CSP system (MWh)	80431	66659
Annual utilization hours of CSP system (h)	2681	2222
PV		
Installed capacity of PV system (MW)	222.462 MW	242.954
Capacity of battery (MWh)	14.687	8.977
Annual output power of PV system (MWh)	288291	309036
Annual utilization hours of PV system (h)	3261	3201

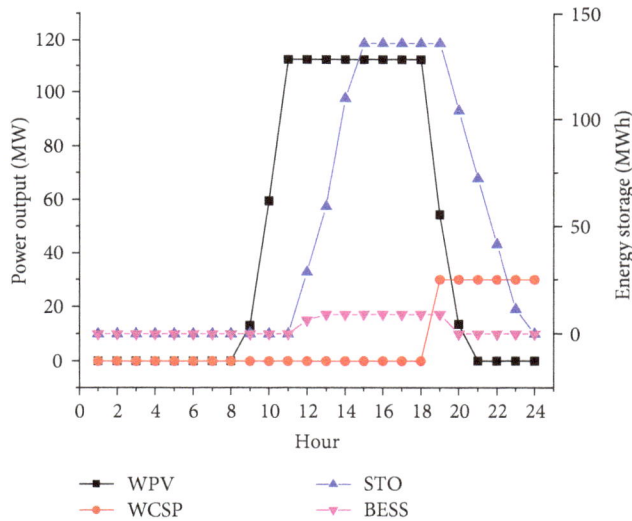

FIGURE 10: Output power of the system on Equinox Day of whole year optimization.

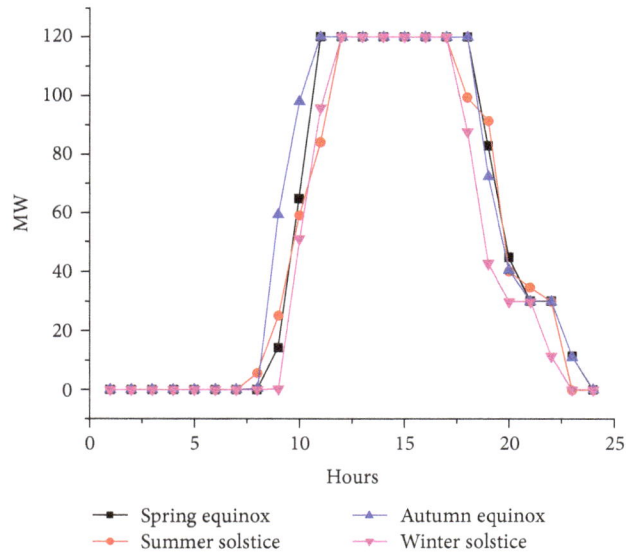

FIGURE 11: Comparative analysis of four typical days of whole year optimization.

is 80431 MWh; annual utilization hours are 2681 h which is lower than the 280 MW solar parabolic thermal power system in Solana; the reason is that the storage hour of Solana is 6 hours and the solar radiation is better. The optimized CSP-PV system combines the advantages of the PV power generation and CSP power generation system, which makes the annual operation hours of the PV system increase and the stability of the PV system operation improve, but the LCOE of the system is only 0.0660 $/KWh, which is lower than the CSP power generation alone.

5.2. *Optimization Based on the Whole Year.* GHI of Lhasa changes slowly with the seasons, and the DNI in summer

and autumn is higher. To optimize the annual operating characteristics of the system based on the operation strategy proposed above. According to the results of genetic algorithm, when the installed capacity of the CSP power system is 30 MW, the LCOE of the CSP-PV hybrid system reaches the lowest which is 0.0555 $/kWh under the condition that the rated power capacity of PV is 242.954 MW, the battery capacity is 8.977 MWh, and the heat storage capacity is 136.059 MWh. The LCOE is lower compared to

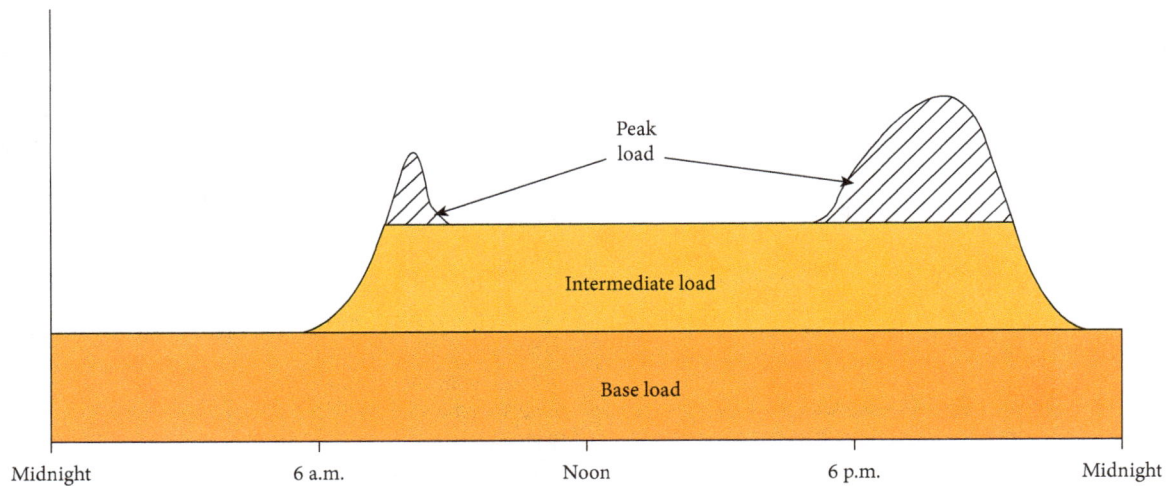

FIGURE 12: Electricity load curve.

TABLE 8: Impact of rated output power on CSP-PV system.

	85%	90%	95%	100%
Annual output power (MWh)	328957	399050	375695	384810
LCOE ($/kWh)	0.0592	0.0561	0.0555	0.0530
CSP				
Capacity of heat storage (MWh)	145.613	254.971	136.059	166.577
Annual output power of CSP system (MWh)	69925	80329	66659	74099
Annual utilization hours of CSP system (h)	2331	2678	2222	2470
PV				
Installed capacity of PV system (MW)	78.423	99.805	242.954	97.555
Capacity of battery (MWh)	10.061	11.274	8.977	5.752
Annual output power of PV system (MWh)	259032	318721	309036	310711
Annual utilization hours of PV system (h)	3303	3193	3201	3185

the CSP power generation alone and lower compared to the PV power generation with battery which is to ensure stability of the power output of the system.

It can be found in Figures 8 and 9 from the comparative analysis of the power generation capacity and actual power generation that the output power of PV increases with the increase of solar radiation and reaches the rated output power at 11 o'clock when the system output power is no longer changing with the increase of solar radiation. After 14 o'clock, the solar radiation drops, the PV system still maintains the rated output power, and the application of the battery makes the output power fluctuation of the PV system lower. CSP system does not participate in the system power generation in the daytime and generates electricity after 19:00. Heat storage tank makes the CSP system ensure stable power generation when the sun goes down or no sun irradiation.

The output power of the system in Equinox Day is shown in Figure 10. After the sun rises, the storage tank begins to store energy and the battery stores the excess energy of the PV system. The operation mode of the system

that PV operates alone from 11:00 to 18:00 and the exceeded energy is stored in a battery and heat storage tank. When at 19:00, the energy of the PV system is not enough to meet the energy needs, and the system operating mode is converted into the PV, battery, and CSP systems that operate at the same time. After 20:00, the sun has been down the mountain and the battery and the PV system cannot provide energy, then the power is provided by the heat storage tank.

According to the calculation results, annual power generation of the CSP-PV hybrid system is 375695 MWh. Where annual power generation of PV system is 309036 MWh, annual utilization hours are 3201 h which is higher than the PV-alone system. Annual power generation of the CSP system is 66659 MWh; annual utilization hours are 2222 h. It can be found that the use of CSP-PV hybrid system, which CSP system is to make up for the output power fluctuations of PV systems, can be combined with the advantages of both systems to increase annual operating hours of PV systems to reduce the rate of discards and decrease LCOE of the system but also can guarantee stable output.

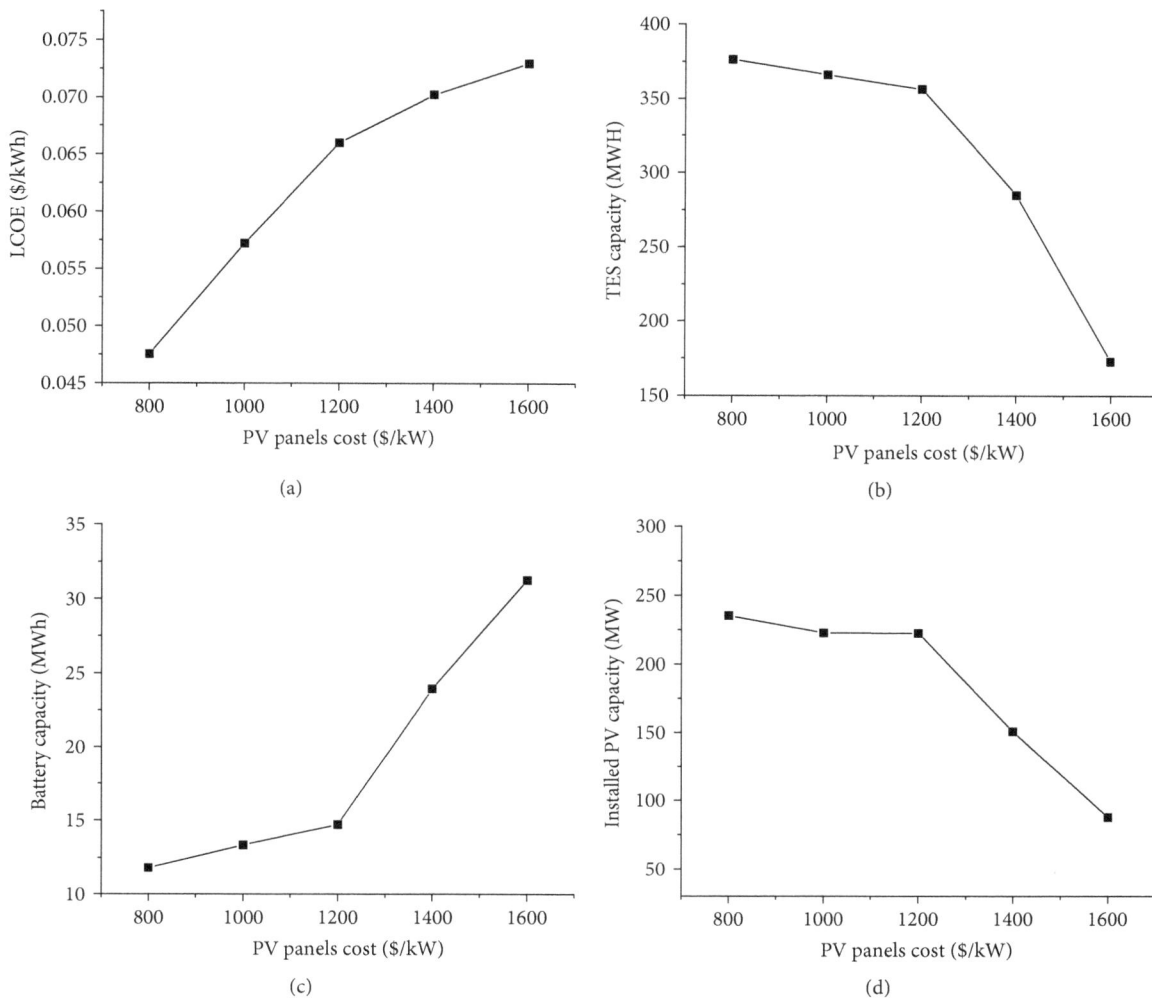

FIGURE 13: Influence of PV panels cost.

5.3. *Discussion.* The results of annual and day optimization are shown in Table 7. According to the comparison that can be found, the annual power generation of the system which is optimized by the annual operation characteristic is higher and the LCOE is lower. Annual utilization hours of PV and CSP system is lower due to the lower capacity of battery and heat storage tank which makes solar energy abandon more.

The comparative results of optimization of day and year which includes four typical days: Spring Equinox, Summer Solstice, Autumnal Equinox, and Winter Solstice are shown in Figure 11. It can be seen that the power generation curves of four typical days are basically the same. However, the power generation of Equinox Day is the worst because the solar radiation is poor, and Autumn Day is the best. Therefore, the system optimized with Equinox Day requires more energy storage system and more power generation, but LCOE is also higher in the meantime. The use of optimized CSP-PV system is to generate power at full load in the peak time. This system can

be used for intermittent energy power generation compared to the load curve shown in Figure 12.

The estimated gross to net conversion factor, which means the ratio of on-grid electricity to the generated electricity, influences the results greatly. The estimated gross to net conversion factor is set to be from 85% to 100%, and the results are shown in Table 8.

As the estimated gross to net conversion factor increases, the annual output power and LCOE of the system are improved. When the annual operating characteristics are optimized, with the increase of the output power, LCOE of the system is reduced. The main reason is that the battery capacity is reduced. The impact of battery capacity on the system is great. The decrease in battery capacity has reduced the number of utilization hours of PV, but it has little effect.

5.4. *Sensitivity Analysis.* The influence of PV panels cost and the thermal storage cost based on the Spring Equinox are analyzed. The PV panels cost is varied from 800 \$/kW to

(a)

(b)

(c)

(d)

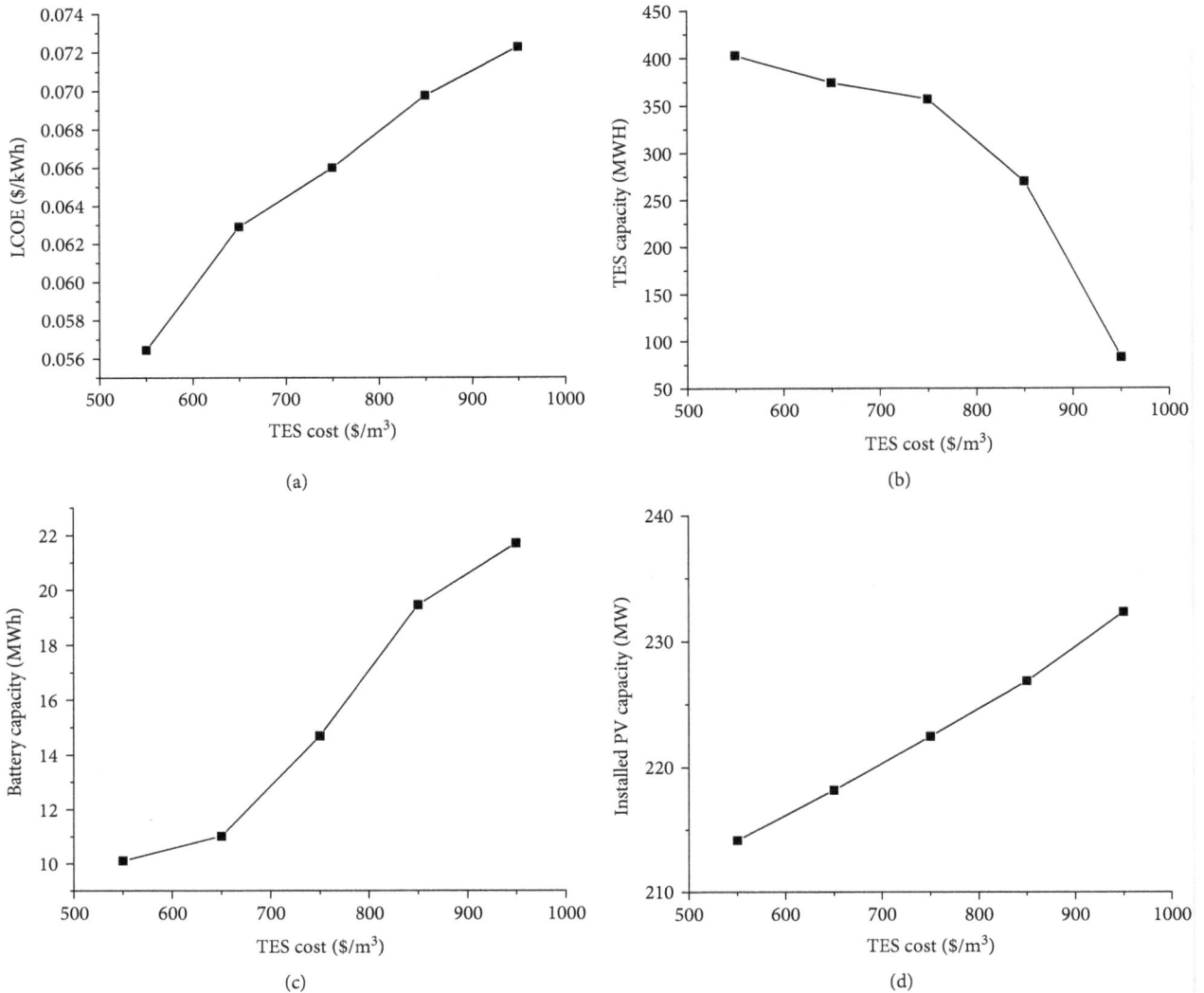

FIGURE 14: Influence of TES cost.

1600 $/kW, and the thermal storage cost varied from 550 $/m^3 to 950 $/m^3. The results are shown in Figures 13 and 14.

6. Conclusion

In this paper, the operation strategy of CSP-PV hybrid system is proposed. Based on this operation strategy, the ratio of PV and CSP in the system is optimized by genetic algorithm, which makes LCOE of the system to a minimum. Through the calculation of this paper, it is found that the increase of PV capacity in the CSP-PV system can reduce the power generation cost of the system; the cost of the battery is high and when the use of batteries will greatly improve the power generation cost of the system; the introduction of the CSP system makes it easy to ensure the stability of the output power of the system in the case of small battery capacity, which greatly improves the annual utilization of the PV and reduces the number of solar discard, but to a certain extent, the utilization efficiency of the CSP system is reduced; the integration of PV and CSP system not only can reduce

the power generation cost of CSP system but also can ensure the stability of the output of PV system.

When the output power is set to 95% of the sum of the PV-rated output power and the CSP-rated output power, the system is optimized by the operation characteristics of Spring Equinox and the result is as follows: the lowest LCOE is 0.0660 $/kWh, the capacity of PV and CSP system are 222.462 MW and 30 MW, respectively, and the capacity of heat storage and battery are 356.562 MWh and 14.687 MWh. According to the calculation results, annual power generation of the CSP-PV hybrid system is 368722 MW. Where the annual power generation of PV system is 288291 MW, annual utilization hours are 3261 h, and annual power generation of CSP system is 80431 MW, annual utilization hours are 2681 h.

When the system is optimized by the operation characteristics of the whole year, the result is that the lowest LCOE is 0.0555 $/kWh, the capacity of PV and CSP system are 242.954 MW and 30 MW, respectively, and the capacity of heat storage and battery are 136.059 MWh and 8.977 MWh.

According to the calculation results, annual power generation of PV system is 309036 MW, annual utilization hours are 3201 h, and annual power generation of the CSP system is 66659 MW, annual utilization hours are 2222 h.

The comparison shows that the power generation curves of the hybrid system are similar in the two optimization methods, but LCOE is lower when optimized by the annual operation characteristic, and the annual utilization rate of the system is higher when optimized by Spring Equinox.

The use of optimized CSP-PV hybrid system, which CSP system is to make up for the output power fluctuations of PV systems, can be combined with the advantages of both systems to increase annual operating hours of PV systems to reduce the rate of discards and decrease LCOE of the system but also can guarantee stable output. The operation strategy proposed in this paper provides a new idea for the operation of CSP-PV hybrid power generation system. The optimization method can be used in the preliminary design of power station.

Nomenclature

PV:	Photovoltaic
CSP:	Concentrated solar power
LCOE:	Levelized cost of energy
DC:	Direct current
AC:	Alternating current
MPPT:	Maximum power point tracking
DNI:	Direct normal irradiation
GHI:	Global horizontal irradiance
T_C:	Operating temperature of PV panels
T_A:	Ambient temperature
U_L:	Actual heat transfer factor
$U_{L,NOCT}$:	Rated heat transfer factor
η_{PV}:	Actual PV panel efficiency
$\tau\alpha$:	Transfer absorption factor
γ:	Temperature factor
$T_{C,REF}$:	PV module temperature under standard test conditions
n_{MOD}:	Number of PV subarray
A_{MOD}:	Active area of each PV module
η_{INV}:	Inverter efficiency
f_{PV}:	Derating factor
η_{BC}:	Battery efficiencies during charge processes
η_{BD}:	Battery efficiencies during discharge processes
SOC:	State of charge
DOD:	Depth-of-discharge
T_i:	Inlet temperature of the heat transfer oil
T_O:	Outlet temperature of the heat transfer oil
ΔT:	Temperature difference between the average temperature of the collector field and the ambient temperature
$W_{min,turb}$:	Minimum output of turbine
W_{set}:	Rated output of the hybrid system
W_{CSP}:	Real output of CSP system
W_{PV}:	Real output of PV system
BESS:	Battery Energy Storage System
IC_{CSP}:	Initial investment of CSP power station
IC_{PV}:	Initial investment of PV power station
AC_{CSP}:	Annual cost of CSP power station
AC_{PV}:	Annual cost of PV power station
E_{CSP}:	Output power of CSP system
E_{PV}:	Output power of PV system
d_{CSP}:	Annual decay rate of CSP power generation
d_{PV}:	Annual decay rate of PV power generation
i:	Interest rate
N:	Lifetime of system
O&M:	Operation and maintain.

Conflicts of Interest

The authors declare that they have no conflicts of interest.

Acknowledgments

The research work is supported by the National Major Fundamental Research Program of China (No. 2015CB251505) and the Fundamental Research Funds for the Central Universities (2018ZD04, 2016YQ04).

References

[1] Agency, International Energy, *World Energy Outlook 2013*, IEA, Paris, 2013.

[2] J. Spelling, *Solar Power Technologies - Solar Fundamentals*, KTH, Stockholm, 2012.

[3] H. Ibrahim, A. Ilinca, and J. Perron, "Energy storage systems—characteristics and comparisons," *Renewable and Sustainable Energy Reviews*, vol. 12, no. 5, pp. 1221–1250, 2008.

[4] M. Huber, D. Dimkova, and T. Hamacher, "Integration of wind and solar power in Europe: assessment of flexibility requirements," *Energy*, vol. 69, pp. 236–246, 2014.

[5] U. Desideri, F. Zepparelli, V. Morettini, and E. Garroni, "Comparative analysis of concentrating solar power and photovoltaic technologies: technical and environmental evaluations," *Applied Energy*, vol. 102, pp. 765–784, 2013.

[6] Technology Roadmap, *Solar Photovoltaic Energy*, International Energy Agency, Paris, 2014.

[7] IEA, *Solar Energy Perspectives*, IEA, Paris, 2011.

[8] Technology Roadmap, *Solar Thermal Energy*, International Energy Agency, Paris, 2014.

[9] C. Parrado, A. Girard, F. Simon, and E. Fuentealba, "2050 LCOE (Levelized Cost of Energy) projection for a hybrid PV (photovoltaic)-CSP (concentrated solar power) plant in the Atacama Desert, Chile," *Energy*, vol. 94, pp. 422–430, 2016,.

[10] F. Dominio, *Techno-Economic Analysis of Hybrid PV-CSP Power Plants: Advantages and disadvantages of intermediate and peak load operation*, Universitat Politècnica de Catalunya, 2014.

[11] D. Cocco, L. Migliari, and M. Petrollese, "A hybrid CSP–CPV system for improving the dispatchability of solar power plants," *Energy Conversion and Management*, vol. 114, pp. 312–323, 2016.

[12] G. Cau, D. Cocco, and M. Petrollese, "Optimal energy management strategy for CSP-CPV integrated power plants with energy storage," in *Proceedings of the 28th International*

Conference on Efficiency, Cost, Optimization, Simulation and Environmental Impact of Energy Systems, Pau, France, 2015.

[13] A. Green, C. Diep, R. Dunn, and J. Dent, "High capacity factor CSP-PV hybrid systems," *Energy Procedia*, vol. 69, pp. 2049–2059, 2015.

[14] M. Hlusiak, M. Götz, and H. A. B. Díaz, "Hybrid photovoltaic (PV)-concentrated solar thermal power (CSP) power plants: modelling, simulation and economics," in *Proceedings of the 29th European Photovoltaic Solar Energy Conference and Exhibition*, Amsterdam, Netherlands, September 2014.

[15] J. P. N. Bootello, S. V. Rubia, L. S. Gallar, and L. F. CAlderon, "Manageable hybrid plant using photovoltaic and solar thermal technology and associated operating method," 2015, U.S. Patent 9140241B.

[16] K. Larchet, *Solar PV-CSP Hybridisation for Baseload Generation: A Techno-economic Analysis for the Chilean Market*, 2015.

[17] L. R. Castillo Ochoa, "Techno-economic analysis of combined hybrid concentrating solar and photovoltaic power plants: a case study for optimizing solar energy integration into the South African electricity grid," 2014.

[18] "Mathworks," June 2018, https://ww2.mathworks.cn.

[19] "System Advisor Model (SAM)," June 2018, https://sam.nrel.gov/.

[20] "Weather data of System Advisor Model," August 2017, https://sam.nrel.gov/weather/.

[21] M. Petrollese and D. Cocco, "Optimal design of a hybrid CSP-PV plant for achieving the full dispatchability of solar energy power plants," *Solar Energy*, vol. 137, pp. 477–489, 2016.

[22] A. M. Patnode, *Simulation and performance evaluation of parabolic trough solar power plants*, University of Wisconsin-Madison, 2006.

[23] S. Shijin, *Simulation and Calculation of Trough Solar Thermal Power Generation*, Inner Mongolia University of Technology, 2014.

[24] H. Suyi and H. Shuhong, *Principle and Technology of Solar Thermal Power Generation*, vol. 204, China Electric Power Press, Beijing, 2012.

[25] R. L. Bartlett, *Steam Turbine Performance and Economics*, McGraw-Hill, New York, 1958.

[26] R. Messenger, J. Ventre, and T. R. Mancini, "Solar electric systems," in *The Engineering Handbook*, CRC Press, Second edition, 2004.

[27] Q. Yin, W. J. Du, and L. Cheng, "Optimization design of heat recovery systems on rotary kilns using genetic algorithms," *Applied Energy*, vol. 202, pp. 153–168, 2017.

[28] T. Gentils, L. Wang, and A. Kolios, "Integrated structural optimisation of offshore wind turbine support structures based on finite element analysis and genetic algorithm," *Applied Energy*, vol. 199, pp. 187–204, 2017.

[29] F. F. Li and J. Qiu, "Multi-objective optimization for integrated hydro–photovoltaic power system," *Applied Energy*, vol. 167, pp. 377–384, 2016.

[30] G. B. Leyland, "Multi-objective optimisation applied to industrial energy problems," in *Section De Génie Mécanique Pour L'obtention Du Grade De Docteur Ès Sciences Techniques par Master of Engineering in Mechanical Engineering*, University of Auckland, 2002.

[31] A. R. Starke, J. M. Cardemil, R. A. Escobar, and S. Colle, "Assessing the performance of hybrid CSP+ PV plants in northern Chile," *Solar Energy*, vol. 138, pp. 88–97, 2016.

[32] S. Ong, C. Campbell, P. Denholm, R. Margolis, and G. Heath, *Land-Use Requirements for Solar Power Plants in the United States*, vol. 140, National Renewable Energy Laboratory, Golden, CO, USA, 2013.

8

A Hybrid Maximum Power Point Tracking Approach for Photovoltaic Systems under Partial Shading Conditions Using a Modified Genetic Algorithm and the Firefly Algorithm

Yu-Pei Huang⑩, Xiang Chen⑩, and Cheng-En Ye⑩

Department of Electronic Engineering, National Quemoy University, Kinmen County, Taiwan

Correspondence should be addressed to Yu-Pei Huang; tim@nqu.edu.tw

Academic Editor: Francesco Riganti-Fulginei

This paper proposes a modified maximum power point tracking (MPPT) algorithm for photovoltaic systems under rapidly changing partial shading conditions (PSCs). The proposed algorithm integrates a genetic algorithm (GA) and the firefly algorithm (FA) and further improves its calculation process via a differential evolution (DE) algorithm. The conventional GA is not advisable for MPPT because of its complicated calculations and low accuracy under PSCs. In this study, we simplified the GA calculations with the integration of the DE mutation process and FA attractive process. Results from both the simulation and evaluation verify that the proposed algorithm provides rapid response time and high accuracy due to the simplified processing. For instance, evaluation results demonstrate that when compared to the conventional GA, the execution time and tracking accuracy of the proposed algorithm can be, respectively, improved around 69.4% and 4.16%. In addition, in comparison to FA, the tracking speed and tracking accuracy of the proposed algorithm can be improved around 42.9% and 1.85%, respectively. Consequently, the major improvement of the proposed method when evaluated against the conventional GA and FA is tracking speed. Moreover, this research provides a framework to integrate multiple nature-inspired algorithms for MPPT. Furthermore, the proposed method is adaptable to different types of solar panels and different system formats with specifically designed equations, the advantages of which are rapid tracking speed with high accuracy under PSCs.

1. Introduction

The output power of photovoltaic (PV) systems is generally nonlinear, particularly when influenced by rapidly changing environmental conditions. Additionally, the power-voltage characteristic (*P-V* curve) of a PV system exhibits a unique maximum power point (MPP), which is affected by solar irradiance and ambient temperature. To ensure the maximum power is harvested, various MPP tracking (MPPT) algorithms have been designed to operate a PV system at the MPP. However, when partial shading conditions (PSCs) occur, PV modules receive different amounts of solar irradiance due to shadows. Under PSCs, the *P-V* curve exhibits multiple local maximum power points (LMPPs), thereby causing difficulties for conventional MPPT methods in identifying the global maximum power point (GMPP) from the LMPPs [1–3].

Numerous nature-inspired algorithms have been designed for MPPT under PSCs [4–6]. Evolutionary algorithms such as particle swarm optimization (PSO) methods have effectively addressed the problem of reaching GMPP [7, 8]. In addition, the firefly algorithm (FA) [9, 10], cuckoo search (CS) [11], flower pollination algorithm (FPA) [12], and intelligent monkey king evolution (IMKE) [13] have good potential to handle the MPPT optimization problem. Among the various evolutionary techniques, the genetic algorithm (GA) can solve nonlinear stochastic problems and accurately extract the GMPP [14, 15]. However, the performance of the GA is determined mainly by the random coefficient [3]. In addition, the conventional GA is not suitable

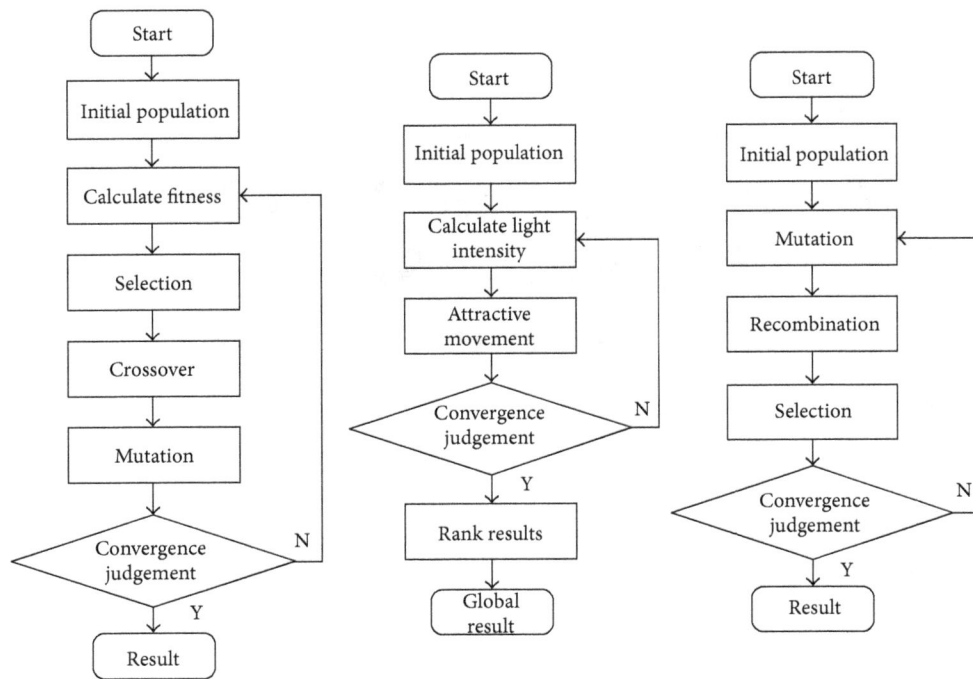

FIGURE 1: Process of GA, FA, and DE.

for MPPT due to its complicated calculation, which reduces the convergence speed and lowers the accuracy under PSCs [2, 16]. As a result, modified GA methods have been proposed to accelerate the processing speed based on

(1) Reduced population size [14]

(2) Simplified mutation processes [17]

(3) Simplified calculation of crossover and attractive processes [15]

However, the accuracy of these methods tends to decline with decreasing processing time. In order to improve the accuracy of the MPPT algorithms, in this paper, we propose integrating a modified GA with the FA and further improve the calculation process by the differential evolution (DE) algorithm. More specifically, the proposed algorithm employs the FA attractive process and DE mutation process and simplifies the calculation of GA and FA to decrease iterations. To our knowledge, this is the first study that evaluates the fusion of the GA and FA for MPPT [5, 6]. In addition, the proposed algorithm can overcome the high execution time and low convergence speed issues of the conventional GA. The advantages of the proposed algorithm are rapid response time with high accuracy under PSCs.

2. Methodology

The processes of the GA, FA, and DE algorithms are demonstrated in Figure 1. The GA is an optimization method for extracting solutions based on "survival of the fittest" evolutionary law, the primary processes of which are crossover,

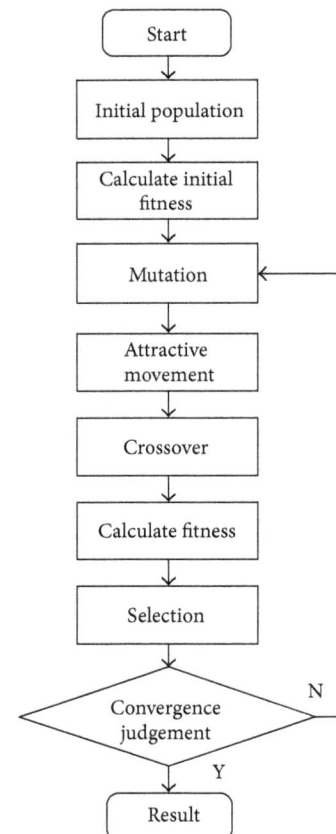

FIGURE 2: Process of the proposed algorithm.

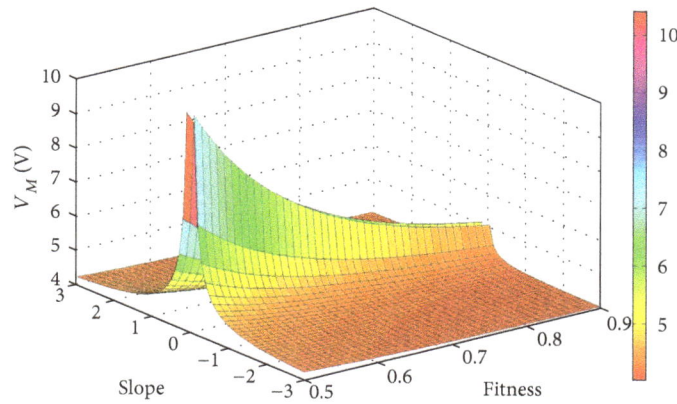

FIGURE 3: Relationship between the voltage adjustment V_M, fitness F, and slope S in the mutation process.

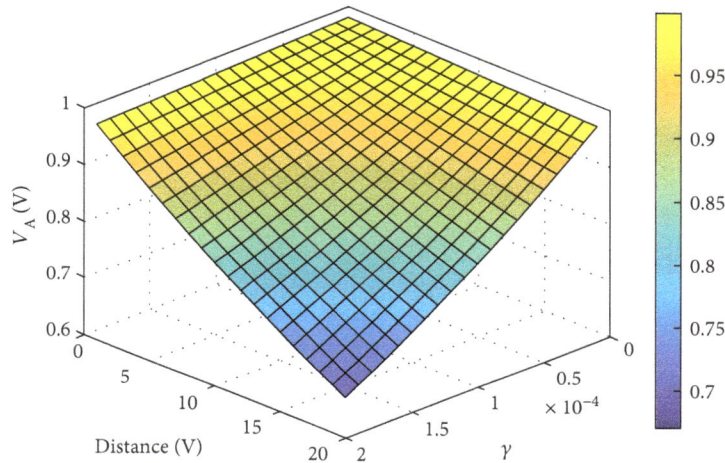

FIGURE 4: Relationship between the voltage movement V_A, distance between two individuals D, and parameter γ in the attractive process.

mutation, and selection. In addition, fitness of individuals is calculated for the selection mechanism. The FA is based on the behavior and flashing of fireflies, where the attractiveness is proportional to the brightness of a firefly. With the FA, fireflies converge into an optimal solution by the attractiveness [4, 9, 10]. With respect to the DE algorithm, although its architecture is similar to the GA, they differ in their mutation processes [16].

As demonstrated in Figure 2, the process of the proposed algorithm is based on the conventional GA. The differences between the proposed method and the GA are the mutation and attractive processes. Firstly, the individuals are initialized by open circuit voltage (V_{OC}), after which the initial fitness of each individual is calculated. In this process, the individuals are positioned in four voltage points of the P-V curve, as shown below:

$$V_{ini} = [0.2\,0.4\,0.6\,0.8] \times V_{OC}, \tag{1}$$

where V_{ini} is the voltage of the initial individuals. The corresponding output power of the positions is taken as the fitness. The fitness of the individual that has the maximum power

output is set to 1 for normalization, while fitness of the other individuals is calculated by

$$F_i = 1 \times \frac{P_i}{P_{i\_max}}, \tag{2}$$

where F_i and P_i are the fitness and output power of individual number i, respectively. In addition, P_{i_max} is the maximum power output of the individuals.

The main steps in an iteration include (1) mutation process, (2) attractive process. (3) crossover and reproduction of new individuals, (4) fitness calculation, (5) next-generation selection, and (6) convergence judgment. Iterations are repeated until convergence conditions are satisfied; that is, the GMPP is found.

In the mutation process in this study, the random procedure of the conventional GA algorithm is modified to avoid fitness decay or individuals with higher fitness from being discarded after the mutation process. The operating voltage of each individual is adjusted to move to a higher power point using the following equation:

$$V_M = \frac{M}{S \times F^2} - B, \tag{3}$$

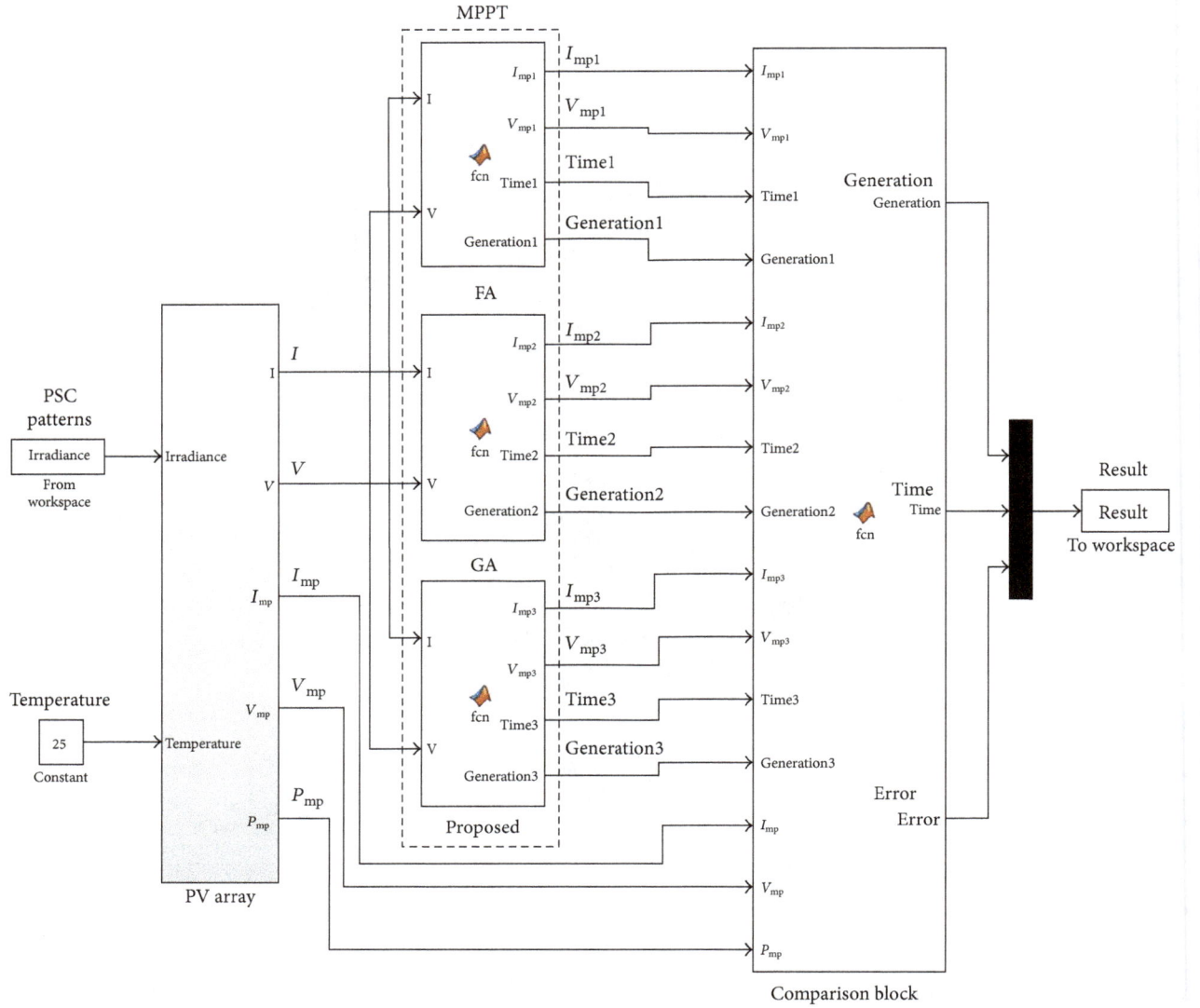

FIGURE 5: System blocks for evaluation.

where V_M is the voltage adjustment, S is the slope of the individual on the P-V curve, F is the fitness, and M and B are variables used for fine-tuning. The optimized values of M and B are set as $0.01\,V_{OC}$ and $0.07\,V_{OC}$, respectively, as determined from trial and error experiments. For the sake of simplicity, the slope S of individual i is derived from

$$S = \frac{\Delta V_i}{\Delta I_i} = \frac{V_i - V_i'}{I_i - I_i'}, \tag{4}$$

where V_i and I_i are the operating voltage and current of individual i, $V_i' = V_i - 0.1V$, and I_i is the current when operated at voltage V_i'. The relationship between slope S, fitness F, and voltage adjustment V_M is shown in Figure 3, where individuals with a low slope and low fitness lead to a greater voltage adjustment; in this manner, the convergence speed can be improved.

TABLE 1: Parameters of the experimental modules.

Characteristics	Spec.
V_{OC}	11.6 V
I_{SC}	1.25 A
V_{mp}	10.4 V
I_{mp}	1.19 A
P_{mp}	12.4 W

In the attractive process, each individual is attracted by others and move their position (operating voltage) according to attractiveness. The attractiveness between two individuals is calculated by

$$V_A = \alpha + \beta \times \exp^{-\gamma \times D}, \tag{5}$$

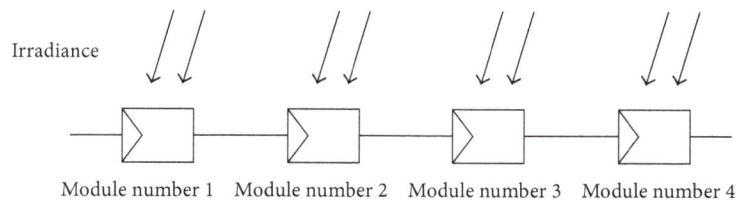

FIGURE 6: The block diagram of the experimental PV array.

TABLE 2: Irradiance setups of the 10 evaluation patterns (W/m^2).

Irradiance setup	LMPPs	Module number 1	Module number 2	Module number 3	Module number 4
Pattern number 1	1	1000	1000	1000	1000
Pattern number 2	2	1000	1000	1000	500
Pattern number 3	2	1000	1000	500	500
Pattern number 4	3	1000	200	200	100
Pattern number 5	3	1000	750	750	500
Pattern number 6	3	1000	1000	750	600
Pattern number 7	4	1000	300	200	100
Pattern number 8	4	1000	700	400	100
Pattern number 9	4	1000	750	500	250
Pattern number 10	4	1000	600	500	400

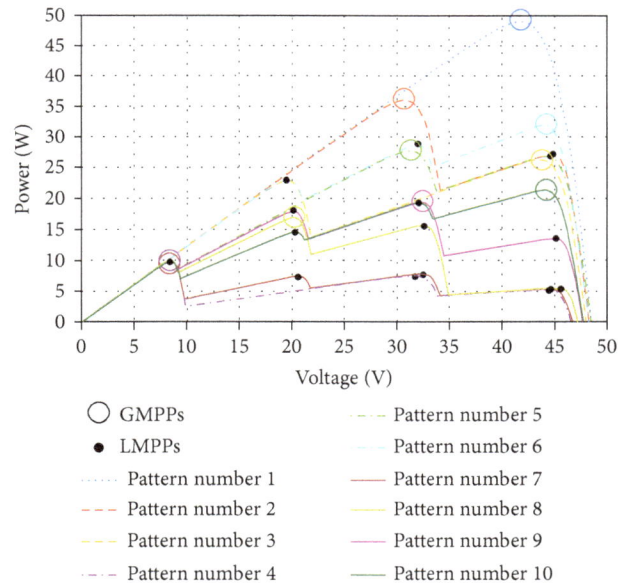

FIGURE 7: *P-V* curves of the 10 PSC patterns in the simulation.

where V_A is the position/voltage movement, α is a random coefficient, β is the attractiveness at zero distance ($D = 0$), γ is a variable for fine-tuning, and D is the distance between two individuals. To improve the convergence speed, the randomization parameter α is modified to vary within $0.1\,V_{OC}$, while the parameter β is set as $0.5\,V_{OC}$ in this study. To refine the other parameters, the voltage movement V_A, distance between two individuals D, and parameter γ in the attractive process are further analyzed. Figure 4 demonstrates that the position movement V_A is inversely proportional to γ and D. However, the distance between two individuals might be affected by other processes. As a result, the optimized value of γ is set as 1×10^{-4}, as determined by trial and error experiments for better convergence speed.

During the crossover process, new individuals are generated, the parents of which are selected randomly. The voltage of the offspring is calculated by

$$V_O = \frac{V_{RA} + V_{RB}}{2},\qquad(6)$$

where V_O represents the offspring individual, and V_{RA} and V_{RB} are the operating voltages of the parents.

The total population number should be kept the same before and after the crossover and selection processes. For example, if the population number is set as four, two individuals will be generated after the crossover process. As a result, in the selection process, four out of six individuals must be selected for the following procedure. The fitness of each individual is calculated for selection using (2).

In the convergence judgment process, successful convergence is judged by the following equations:

$$\sum F_i > N_P \times F_{i\_max} \times 98.75\%,\qquad(7)$$

$$V_{max} - V_{min} < V_{OC} \times 5\%,\qquad(8)$$

where F_i is the fitness of individual number i calculated by (2), N_P represents the population number, and F_{i_max} denotes the fitness of the individual which has the maximum power output. In addition, V_{max} and V_{min} represent the maximum and minimum voltages of the individuals, respectively. If the fitness summation of the individuals is larger than 98.75% of the maximum fitness multiplied by N_P, and the distribution range of the individuals' voltages is less than 5% V_{OC}, convergence is judged as satisfied. Then, the individual with the highest fitness is considered as the MPP. These two criteria were determined by experimental experience in this research. In this study, the major convergence judgement is based on (7), which uses fitness summation as the criterion. On the other hand, (8) is used as an auxiliary filter to ensure the distribution range of the individuals' voltage remains within one interval. The experimental PV array was composed of four series-connected modules; consequently, the maximum interval number of the array's P-V curve under PSC was 4, which corresponds to 25% V_{OC} for each interval. Based on the simulations, setting 1/5 of each interval (25%× 1/5 = 5%) as the filter's criterion could obtain acceptable results.

3. Simulation

The proposed algorithm was first simulated by MATLAB R2016a software, the system blocks for which are shown in Figure 5. The input irradiance patterns and the PV array for evaluation were carried out using the Simulink toolbox. The PV array was composed of four series-connected modules (module numbers 1–4). Parameters of the experimental modules are shown in Table 1. As illustrated in Figure 6, the irradiance for which could be individually programmed for each module to generate partial shading condition (PSC) patterns. The setup of the 10 different PSC patterns (numbers 1–10) is shown in Table 2. The P-V curves of the 10 PSC patterns and their corresponding GMPPs (numbers 1–10) are

```
Generate initial population
Initial fitness F_ini of initial individuals V_ini is determined
by Eq. (2).
while (limit the execution times (max generations))
    while (convergence judgement conditions unsatisfied)
        //mutation process
        for i=1:4  //individual number
            calculate slope S of individual i by Eq. (4).
            calculate voltage adjustment V_M of individual i by
            Eq. (3).
        end for i
        //attractive process
        for i=1:4  //individual number
          for j=1:4  //individual number
            if (i≠j)
                calculate distance D between i and j.
                calculate voltage movement V_A of individual i by
                Eq. (5).
            end if
          end for j
        end for i
        //crossover process
        select parents for new individuals randomly.
        calculate the voltage of offspring V_O from their
        parents'operating
        voltages V_RA & V_RB using Eq. (6).
        //selection process
        for i=1:6  //4 parents and 2 offspring
            calculate fitness of each individual using Eq. (2).
        end for i
        sort the fitness of the individuals
        define the individual with the highest fitness as MPP.
        discard 2 individuals with the lowest fitness.
        //convergence process
        for i=1:4  /the rest of the individuals
            execute convergence judgement using Eq. (7)
            and Eq. (8).
        end for i
    end while (max generations)
    iteration number = iteration number + 1  //record
    iteration number
end while (conditions are satisfied)
output the operating voltage of the MPP and the iteration
number.
```

PSEUDOCODE 1: MATLAB pseudocode of the proposed algorithm.

depicted in Figure 7. The purpose of the 10 irradiance setups is to simulate patterns that have 1–4 LMPPs with their respective GMPP located at different intervals.

The MPPT algorithms implementation was carried out using MATLAB M-file programming. The process flow of the GA, FA, and proposed algorithm, illustrated in Figures 1 and 2, could be implemented by MATLAB code. Pseudocode 1 demonstrates the MATLAB pseudocode of the proposed algorithm and denotes the explanations of the program after the double slashes. The PSC patterns are fed to the MPPT programs. Figure 8 demonstrates the iteration process of the proposed algorithm, for which pattern number

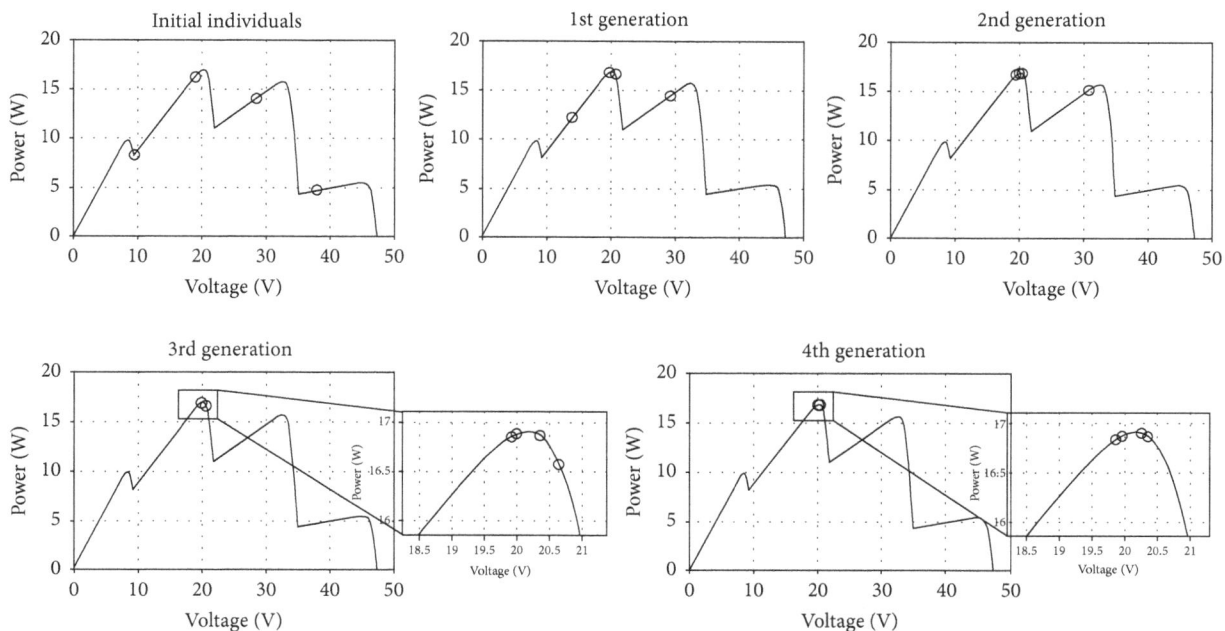

FIGURE 8: Position variation of each individual (pattern number 9).

TABLE 3: Simulation results of GA with different population numbers under uniform irradiance.

Population number	Iterations	Time (s)	Error (%)
3	10.64	0.044	9.20
4	7.91	0.031	3.80
5	10.14	0.042	1.01
6	13.26	0.055	0.12

TABLE 6: Simulation results of the proposed algorithm with different population numbers under PSCs.

Population number	Iterations	Time (s)	Error (%)
3	3.32	0.006	7.98
4	3.09	0.006	0.12
5	3.79	0.007	0.24
6	5.71	0.01	0.17

TABLE 4: Simulation results of FA with different population numbers under uniform irradiance.

Fireflies	Iterations	Time (s)	Error (%)
3	25.3	0.111	2.10
4	17.39	0.07	0.61
5	14.17	0.059	0.09
6	12.25	0.048	0.06

TABLE 7: Parameter settings of the GA, FA, and the proposed method.

GA	FA	Proposed method
$B = 0.07\,V_{OC}$	$\alpha_0 = 0.1\,V_{OC}$	$M = 0.01\,V_{OC}$
—	$\beta = 0.25\,V_{OC}$	$B = 0.07\,V_{OC}$
—	$\gamma = 0.0001$	$\beta = 0.5\,V_{OC}$
—	$\theta = 0.95$	$\gamma = 0.0001$

TABLE 5: Simulation results of the proposed algorithm with different population numbers under uniform irradiance.

Population number	Iterations	Time (s)	Error (%)
3	2.8	0.006	3.38
4	3.04	0.006	0.14
5	4.31	0.008	0.11
6	6.54	0.011	0.16

9 was used for evaluation. It was observed that the MPP was found by the proposed algorithm after 4 iterations.

It is important to determine the number of data points (population) for the searching processes because the data amount impacts the response time and accuracy of a searching algorithm. Although using less data points reduces the response time, the accuracy decreases. Conversely, using more data points improves the searching accuracy, but the measurement and calculation time increases. As a result, in order to decide the optimal population number for locating the MPPT using the proposed method, we first simulated the tracking performance of the conventional GA, FA, and

TABLE 8: Simulation results of the tracking speed and tracking accuracy.

Pattern	Algorithm	Iterations		Time (s)		Error (%)	
		Average	SD	Average	SD	Average	SD
	GA	12.17	3.4	0.048	0.014	3.75	0.66
Pattern number 1	FA	12.12	6.11	0.1	0.05	0.29	0.35
	Proposed method	3.31	0.76	0.02	0.004	0.17	0.14
	GA	8.75	2.79	0.085	0.029	0.73	0.99
Pattern number 2	FA	5.87	3.56	0.051	0.03	0.15	0.21
	Proposed method	3.45	0.83	0.022	0.009	0.1	0.23
	GA	15.83	6.58	0.151	0.063	9.75	0.54
Pattern number 3	FA	12.95	4.48	0.109	0.039	5.68	3.86
	Proposed method	4.95	0.9	0.025	0.008	1.02	1.51
	GA	26.88	18.05	0.274	0.183	18.75	12.6
Pattern number 4	FA	24.75	5.03	0.226	0.07	23.37	0.07
	Proposed method	19.77	20.03	0.095	0.072	0.76	1.05
	GA	9.16	2.87	0.091	0.03	0.81	1.23
Pattern number 5	FA	5.56	2.35	0.048	0.021	0.1	0.15
	Proposed method	3.46	0.81	0.024	0.009	0.08	0.12
	GA	14.41	8.87	0.147	0.093	10.11	0.44
Pattern number 6	FA	8.4	6.61	0.078	0.062	9.67	1.74
	Proposed method	5.61	2.06	0.027	0.008	1.19	1.83
	GA	20.47	10.77	0.065	0.034	17.47	11.33
Pattern number 7	FA	28.33	12.13	0.236	0.101	21.14	1.15
	Proposed method	19.55	13.78	0.061	0.035	0.73	1.34
	GA	10.53	4.05	0.103	0.04	0.4	0.78
Pattern number 8	FA	13.69	7.02	0.116	0.057	0.08	0.14
	Proposed method	5.12	1.85	0.025	0.006	0.37	0.49
	GA	12.78	5.78	0.128	0.069	5.21	3.17
Pattern number 9	FA	9.79	4.35	0.087	0.044	0.29	0.77
	Proposed method	5.31	1.43	0.034	0.075	0.28	0.45
	GA	13.03	5.64	0.051	0.023	10.14	0.78
Pattern number 10	FA	7.63	1.09	0.063	0.009	9.7	0.24
	Proposed method	4.95	1.18	0.026	0.014	2.68	2.49

the proposed algorithm under uniform irradiance (pattern number 1) with different population numbers. Evaluations were conducted under the same conditions for the three algorithms, and all experiments were performed with uniformly distributed initial positions of the individuals. In addition, the convergence judgement criteria, according to (7) and (8), were the same for all algorithms. The simulation was executed 100 times. The difference between the power of the estimated MPP and the power of the actual MPP is considered an error, which can be calculated by

$$\mathrm{ERR} = 1 - \frac{P_{\mathrm{MPP}}}{P_{\mathrm{MPP\_true}}} \times 100\%, \qquad (9)$$

where ERR is the error, while P_{MPP} and $P_{\mathrm{MPP\_true}}$ are the estimated MPP power and the actual MPP power, respectively. In addition, the number of iterations and execution time required for each algorithm to reach convergence were

TABLE 9: Tracking speed and tracking accuracy comparison of the GA, FA, and proposed method.

Algorithm	Averaged iterations	Averaged time (s)	Averaged error (%)
GA	14.4	0.114	7.72
FA	12.91	0.111	7.05
Proposed method	7.55	0.036	0.74

counted for tracking speed comparison. All simulations were conducted using a computer with a standard Intel i7-4790K CPU and Windows 7 operating system.

Tables 3–5 list the simulation results of the tracking performance under uniform irradiance using GA, FA, and the proposed algorithm, respectively. The population number was evaluated from 3 to 6 for these three algorithms. From

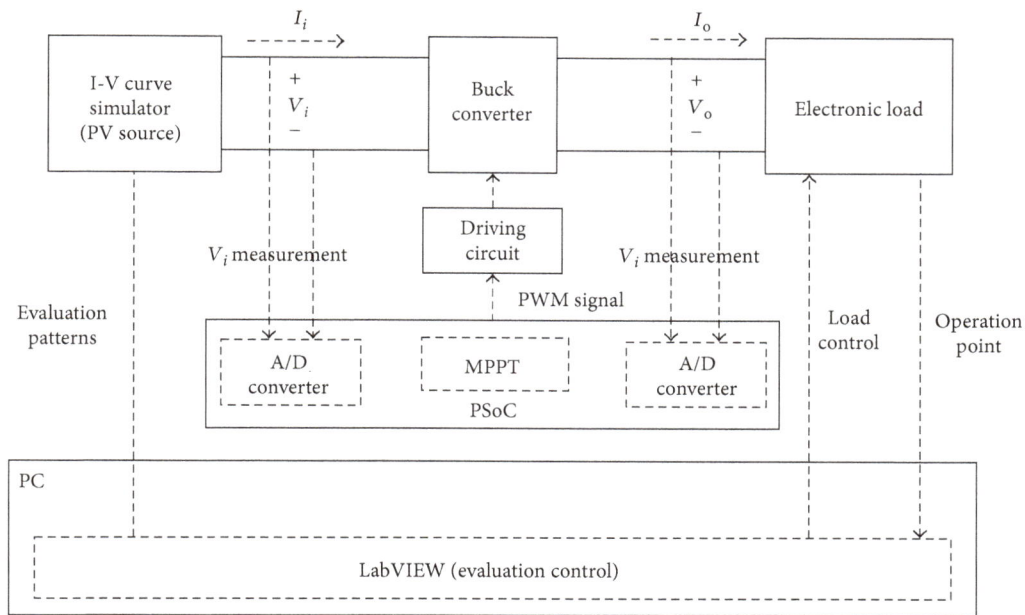

FIGURE 9: Evaluation system structure.

Table 3, it can be observed that the GA takes the shortest execution time using 4 data points. On the other hand, Table 4 shows that the most data points (6 fireflies) used by FA lead to the shortest execution time. However, as shown in Table 5, with the proposed method, more data points increase the iterations, and so more time is needed for searching. In addition, the accuracy generally improved by increasing the population number for all three algorithms. In order to further investigate the tracking performance of the proposed method with different population numbers under PSCs, irradiance pattern number 9 was used for evaluation. Table 6 lists the averaged iterations, execution time, and corresponding tracking error with different population numbers. It can be found that with 4 data points, the proposed algorithm has the best execution time and tracking accuracy.

Consequently, considering the shorter execution time and reasonable tracking accuracy, the population number of 4 was chosen for further evaluation experiments. All 10 irradiance patterns were used for further experiments under different PSCs with the proposed algorithm. In addition, the conventional GA and FA were employed for comparison. The parameter settings for each method are presented in Table 7. All experiments were conducted under the same conditions for the three algorithms, with uniformly distributed initial positions and the same convergence judgement criteria. The simulations were executed 100 times for iterations, execution time, and error calculation.

The simulation results of the tracking speed and accuracy using the GA, FA, and proposed algorithm are shown in Table 8. From the results, the maximum tracking errors of the GA and FA occurred with pattern numbers 4 and 7 and were caused by being trapped at the LMPPs. The trapping at LMPPs phenomenon of the GA might be the result of its random mutation process, while the trapping phenomenon

TABLE 10: Specifications of the buck converter.

Parameter	Value
Switching frequency	20 kHz
Inductor	35 μH
Capacitor	16 V, 1000 μF

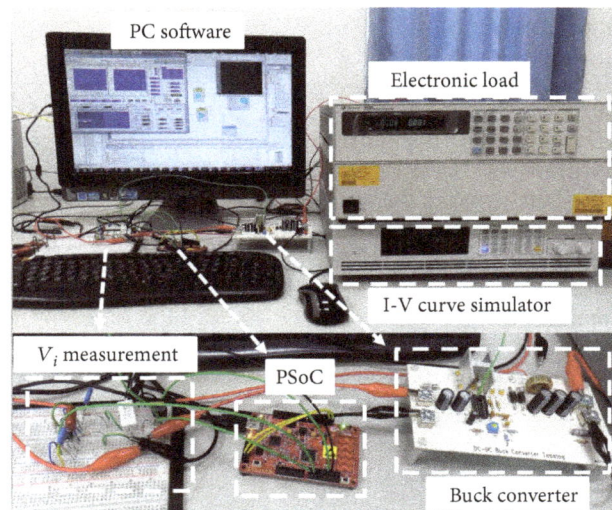

FIGURE 10: Photograph of the evaluation system.

of the FA might be due to its random attractive process. In the proposed algorithm, these two random processes are modified; therefore, the tracking accuracy is improved.

Table 9 summarizes and compares the averaged iterations, execution time, and errors of the 10 simulation patterns. The differences between the averaged iterations

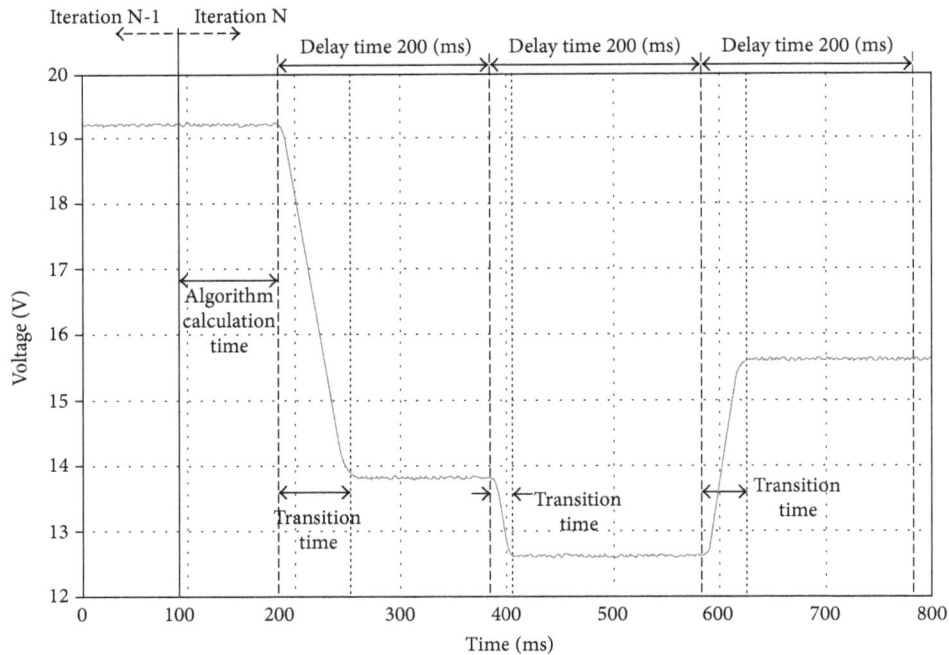

FIGURE 11: Execution procedure of one iteration and delay time setting for each voltage adjustment.

TABLE 11: Irradiance setup of the 6 evaluation patterns (W/m^2).

Irradiance setup	LMPPs	Module number 1	Module number 2	Module number 3	Module number 4
Pattern number 1	1	1000	1000	1000	1000
Pattern number 2	2	1000	1000	1000	500
Pattern number 3	2	1000	1000	500	500
Pattern number 4	3	1000	200	200	100
Pattern number 5	3	1000	750	750	500
Pattern number 9	4	1000	750	500	250

required for the 10 simulation patterns using the proposed algorithm and GA/FA were around 6.85 and 5.36, respectively. Furthermore, the averaged reduction of the execution time using the proposed algorithm was around 0.078 (s) compared with the GA and 0.075 (s) compared with the FA. In other words, the averaged time reduction using the proposed method was around 68.4% of the GA and 67.6% of the FA. Moreover, the averaged tracking error of the proposed algorithm was reduced around 6.98% compared with GA and 6.31% compared with FA.

4. Evaluation Results

To further verify the MPP tracking speed and tracking accuracy of the proposed method, a hardware evaluation system was installed, as illustrated in Figure 9. An I-V curve simulator (Chroma, 62020H) was utilized to program different PSC patterns as the PV source. A DC-DC buck converter controlled by a pulse width modulation (PWM) signal was used as the interface to feed the output power of the PV source to a programmable electronic load (Keysight, N3300A). The specifications of the buck converter are listed

in Table 10. The MPPT algorithms for evaluation were coded and executed using a programmable system-on-chip controller (Cypress, PSoC4), which has an ARM Cortex-M0 core running at 48 MHz. The input/output voltage and current (V_i, I_i, V_o, and I_o) of the buck converter were measured by the PSoC's onboard programmable analog-to-digital (A/D) converters. All evaluation experiments were conducted and controlled by a PC software programmed in LabVIEW (National Instruments). Figure 10 presents a photograph of the evaluation system.

The slew rate of the adopted converter was approximately 0.112 V/ms. In addition, the maximum voltage adjustment needed for the three experimental algorithms to move the operating points was around 11.6 V (across one interval). As a result, the maximum transition time needed to reach steady state was approximately 96 ms. Consequently, a 200 ms delay time was set to ensure a stable output for each voltage adjustment. Figure 11 illustrates the execution procedure of one iteration for the experimental algorithms. At the beginning of an iteration, the algorithms calculate the operating voltages of the individuals. Then, the duty cycle of the PWM signal is adjusted sequentially for each individual so

TABLE 12: Evaluation results of the tracking speed and tracking accuracy.

Pattern	Algorithm	Iterations		Time (s)		Error (%)	
		Average	SD	Average	SD	Average	SD
Pattern number 1	GA	14.61	0.81	0.032	0.033	5.18	1.42
	FA	13.5	0.98	0.028	0.058	3.51	0.69
	Proposed method	5.25	2.19	0.026	0.011	2.46	4.61
Pattern number 2	GA	11.54	1.29	0.374	0.064	3.44	1.9
	FA	12.76	2.12	0.116	0.041	2.83	1.66
	Proposed method	6.62	2.52	0.107	0.007	4.06	3.48
Pattern number 3	GA	18.73	2.2	0.301	0.22	10.98	2.11
	FA	9.97	1.54	0.257	0.032	6.03	1.84
	Proposed method	7.22	2.15	0.041	0.027	3.84	6.06
Pattern number 4	GA	24.52	10.09	0.497	0.385	9.35	2.02
	FA	13.82	4.06	0.314	0.08	1.53	0.08
	Proposed method	6.29	6.88	0.242	0.067	1.25	0.77
Pattern number 5	GA	12.03	0.75	0.367	0.071	4.34	2.53
	FA	10.9	1.25	0.131	0.013	4.54	2.71
	Proposed method	6.06	2.1	0.079	0.018	1.86	4.78
Pattern number 9	GA	17.17	1.73	0.174	0.268	8.12	4.65
	FA	18.6	2.93	0.091	0.218	9.08	1.73
	Proposed method	6.2	1.86	0.04	0.297	2.99	11.42

that the PV array operates at the designated voltage. Since different conversion rates and time constants of the converter will significantly influence the total MPPT time, we summarized the net calculation times of the algorithms as the execution times for comparison.

Settings of the PV array and the irradiance patterns for evaluation were the same as the simulation. The simulation results of pattern numbers 6, 7, 8, and 10 were similar to the other 6 patterns. For simplification, these 4 patterns were eliminated and the other 6 representative patterns, shown in Table 11, were conducted for evaluation experiments.

Table 12 shows and compares the three evaluation results (iterations required, execution time, and tracking error) of the three methods (GA, FA, and the proposed method). From the results, the maximum tracking errors of the GA and FA occurred with pattern number 4 and were found to become trapped at the LMPPs, which also occurred in the simulations. In the proposed algorithm, the random processes are modified; therefore, the tracking accuracy is improved. Table 13 summarizes and compares the averaged iterations, execution time, and tracking errors from the 6 evaluation patterns. The differences between the averaged iterations required for the 6 simulation patterns using the proposed algorithm and GA/FA were around 10.16 and 6.99, respectively. Furthermore, the averaged reduction of the execution time using the proposed algorithm was around 0.202 (s) compared with the GA and 0.067 (s) compared with the FA. In other words, the averaged time reduction using the proposed method was around 69.4% of the GA and 42.9% of the FA. In addition, the average tracking error of the proposed algorithm was reduced around 4.16% compared with the GA and 1.85% compared with the FA.

TABLE 13: Comparison of evaluation results for the GA, FA, and the proposed method.

Algorithm	Averaged iterations	Averaged time (s)	Averaged error (%)
GA	16.43	0.291	6.9
FA	13.26	0.156	4.59
Proposed method	6.27	0.089	2.74

5. Discussion and Conclusion

Although the conventional GA can accurately extract the GMPP under PSCs, implementation is challenging due to its complicated calculation. Though the calculation could be simplified, accuracy tends to decline with decreasing processing time. In addition, simplified calculation processes might cause search results to become trapped at LMPPs. In order to solve this issue, this work proposes a fusion algorithm that integrates three nature-inspired algorithms for MPPT. The proposed algorithm simplifies the calculation of the GA with the integration of the mutation process of DE and modifies the attractive process of FA. The simulation and evaluation results demonstrate that the proposed algorithm offers rapid tracking speed and high accuracy despite PSCs due to the simplified processing. In addition, the long execution time and low convergence speed issues of the conventional GA can be improved using the proposed processes. Table 14 presents a qualitative comparison of the proposed method with the GA, FA, and other algorithms. The

TABLE 14: Qualitative comparison of the proposed method with other algorithms.

Number	Parameters	P&O [18]	ABC [19]	Cuckoo [11]	FPA [12]	FA [4]	GA [17]	Proposed
1	Tracking speed	Slow	Fast	Fast	Fast	Moderate	Moderate	Fast
2	Tracking accuracy	Low	Moderate	High	High	Moderate	Moderate	High
3	Trapping at LMPPs	High	Low	Low	Low	Low	Moderate	Low
4	Steady state oscillation	Large	Zero	Zero	Zero	Zero	Moderate	Low
5	Complexity	Low	High	Moderate	Moderate	Low	High	Moderate
6	Execution time	Slow	Moderate	Moderate	Fast	Moderate	Moderate	Fast

performance evaluation of the MPPT techniques are qualitatively weighed using (1) tracking speed, (2) tracking accuracy, (3) trapping at LMPPs, (4) steady-state oscillation, (5) complexity, and (6) execution time. In addition, the parameters are gauged qualitatively as "high," "moderate," and "low." As can be seen, the tracking speed and tracking accuracy of the proposed method are found to be superior to the conventional GA and FA. Furthermore, the proposed method reduces the complexity and accelerates the execution time. Consequently, the proposed method could be considered one of the most promising substitutes for existing methods. Moreover, this research provides a framework to integrate multiple nature-inspired algorithms for MPPT.

However, the proposed approach has not been optimized and future studies could investigate the feasibility of process optimization and fusing other algorithms using this framework. In addition, in order to enable the proposed method to accommodate different types of solar panels, the calculation equations and parameter settings were specifically designed to be related to V_{OC}. Furthermore, in this work, the experimental PV array was composed of four series-connected modules; consequently, the maximum interval number of the array's P-V curve under PSC was 4, and so the optimal population number was also 4. Accordingly, this model could be used and expanded to different 4-group PV arrays. However, for different system formats, for example, if 8 or more PV groups are connected as an array, the population number, parameters V_{ini}, M, B, β, and γ, as well as the convergence criteria of (7) and (8), should be refined for better performance. Future studies could explore the optimal number of parameters for different system formats.

Conflicts of Interest

The authors declare that there is no conflict of interests regarding the publication of this paper.

Acknowledgments

The authors would like to thank the Ministry of Science and Technology, R.O.C., for financially supporting this research under Contract no. MOST 106-2221-E-507-005.

References

[1] M. A. M. Ramli, S. Twaha, K. Ishaque, and Y. A. Al-Turki, "A review on maximum power point tracking for photovoltaic systems with and without shading conditions," *Renewable and Sustainable Energy Reviews*, vol. 67, pp. 144–159, 2017.

[2] M. Seyedmahmoudian, B. Horan, T. K. Soon et al., "State of the art artificial intelligence-based MPPT techniques for mitigating partial shading effects on PV systems – a review," *Renewable and Sustainable Energy Reviews*, vol. 64, pp. 435–455, 2016.

[3] L. Liu, X. Meng, and C. Liu, "A review of maximum power point tracking methods of PV power system at uniform and partial shading," *Renewable and Sustainable Energy Reviews*, vol. 53, pp. 1500–1507, 2016.

[4] X. S. Yang, *Nature-Inspired Optimization Algorithms*, Elsevier, London, UK, 2014.

[5] G. Li, Y. Jin, M. W. Akram, X. Chen, and J. Ji, "Application of bio-inspired algorithms in maximum power point tracking for PV systems under partial shading conditions – a review," *Renewable and Sustainable Energy Reviews*, vol. 81, pp. 840–873, 2018.

[6] M. A. Danandeh and S. M. Mousavi G, "Comparative and comprehensive review of maximum power point tracking methods for PV cells," *Renewable and Sustainable Energy Reviews*, vol. 82, Part 3, pp. 2743–2767, 2018.

[7] R.-M. Chao, A. Nasirudin, I.-K. Wang, and P.-L. Chen, "Multi-core PSO operation for maximum power point tracking of a distributed photovoltaic system under partially shading condition," *International Journal of Photoenergy*, vol. 2016, Article ID 9754514, 19 pages, 2016.

[8] K.-H. Chao, "A high performance PSO-based global MPP tracker for a PV power generation system," *Energies*, vol. 8, no. 7, pp. 6841–6858, 2015.

[9] Y. Jin, W. Hou, G. Li, and X. Chen, "A glowworm swarm optimization-based maximum power point tracking for photovoltaic/thermal systems under non-uniform solar irradiation and temperature distribution," *Energies*, vol. 10, no. 4, p. 541, 2017.

[10] W. Hou, Y. Jin, C. Zhu, and G. Li, "A novel maximum power point tracking algorithm based on glowworm swarm optimization for photovoltaic systems," *International Journal of Photoenergy*, vol. 2016, Article ID 4910862, 9 pages, 2016.

[11] J. Ahmed and Z. Salam, "A maximum power point tracking (MPPT) for PV system using cuckoo search with partial shading capability," *Applied Energy*, vol. 119, pp. 118–130, 2014.

[12] J. Prasanth Ram and N. Rajasekar, "A novel flower pollination based global maximum power point method for solar maximum power point tracking," *IEEE Transactions on Power Electronics*, vol. 32, no. 11, pp. 8486–8499, 2017.

[13] N. Kumar, I. Hussain, B. Singh, and B. K. Panigrahi, "Maximum power peak detection of partially shaded PV panel by using intelligent monkey king evolution algorithm," *IEEE*

Transactions on Industry Applications, vol. 53, no. 6, pp. 5734–5743, 2017.

[14] S. Daraban, D. Petreus, and C. Morel, "A novel MPPT (maximum power point tracking) algorithm based on a modified genetic algorithm specialized on tracking the global maximum power point in photovoltaic systems affected by partial shading," *Energy*, vol. 74, pp. 374–388, 2014.

[15] Y. Shaiek, M. Ben Smida, A. Sakly, and M. F. Mimouni, "Comparison between conventional methods and GA approach for maximum power point tracking of shaded solar PV generators," *Solar Energy*, vol. 90, pp. 107–122, 2013.

[16] Y.-H. Liu, J.-H. Chen, and J.-W. Huang, "A review of maximum power point tracking techniques for use in partially shaded conditions," *Renewable and Sustainable Energy Reviews*, vol. 41, pp. 436–453, 2015.

[17] M. Zagrouba, A. Sellami, M. Bouaïcha, and M. Ksouri, "Identification of PV solar cells and modules parameters using the genetic algorithms: application to maximum power extraction," *Solar Energy*, vol. 84, no. 5, pp. 860–866, 2010.

[18] J. Ahmed and Z. Salam, "An improved perturb and observe (P&O) maximum power point tracking (MPPT) algorithm for higher efficiency," *Applied Energy*, vol. 150, pp. 97–108, 2015.

[19] A. s. Benyoucef, A. Chouder, K. Kara, S. Silvestre, and O. A. sahed, "Artificial bee colony based algorithm for maximum power point tracking (MPPT) for PV systems operating under partial shaded conditions," *Applied Soft Computing*, vol. 32, pp. 38–48, 2015.

9

Review on Substrate and Molybdenum Back Contact in CIGS Thin Film Solar Cell

Kam Hoe Ong[ID],[1] Ramasamy Agileswari,[1] Biancamaria Maniscalco,[2] Panagiota Arnou,[3] Chakrabarty Chandan Kumar,[1] Jake W. Bowers,[2] and Marayati Bte Marsadek[4]

[1]*Department of Electronics & Communication Engineering, Universiti Tenaga Nasional, Selangor, Malaysia*
[2]*Centre for Renewable Energy Systems Technology (CREST), Wolfson School of Mechanical, Electrical and Manufacturing Engineering, Loughborough University, Loughborough, Leicestershire LE11 3TU, UK*
[3]*Laboratory for Energy Materials, University of Luxembourg, L-4422 Belvaux, Luxembourg*
[4]*Institute of Power Engineering, Universiti Tenaga Nasional, Selangor, Malaysia*

Correspondence should be addressed to Kam Hoe Ong; wil_ng-okh@hotmail.com

Academic Editor: K. R. Justin Thomas

Copper Indium Gallium Selenide- (CIGS-) based solar cells have become one of the most promising candidates among the thin film technologies for solar power generation. The current record efficiency of CIGS has reached 22.6% which is comparable to the crystalline silicon- (c-Si-) based solar cells. However, material properties and efficiency on small area devices are crucial aspects to be considered before manufacturing into large scale. The process for each layer of the CIGS solar cells, including the type of substrate used and deposition condition for the molybdenum back contact, will give a direct impact to the efficiency of the fabricated device. In this paper, brief introduction on the production, efficiency, etc. of a-Si, CdTe, and CIGS thin film solar cells and c-Si solar cells are first reviewed, followed by the recent progress of substrates. Different deposition techniques' influence on the properties of molybdenum back contact for CIGS are discussed. Then, the formation and thickness influence factors of the interfacial $MoSe_2$ layer are reviewed; its role in forming ohmic contact, possible detrimental effects, and characterization of the barrier layers are specified. Scale-up challenges/issues of CIGS module production are also presented to give an insight into commercializing CIGS solar cells.

1. Introduction

Copper Indium Selenide ($CuInSe_2$ or CIS) is a ternary compound p-type absorber material belonging to the I-III-VI_2 family [1]. The very first CIS material being synthesized was in 1953, and then, an efficiency of 12% has been reported for single crystal $CuInSe_2$-based solar cells [2]. In 1976, the first CIS thin film solar cell with buffer layer CdS was fabricated with an efficiency of 4–5% by evaporating $CuInSe_2$ powder in the presence of excess Se vapor (coevaporation) [3]. CIG-based thin film solar cell started to receive even more attention in 1981 when Mickelsen and Chen achieved an efficiency of 9.4% by using coevaporation technique from elemental sources [4]. From that onwards, numbers of emerged technologies such as alloying CIS with gallium

(Ga) to become Copper Indium Gallium Selenide (CIGS), incorporating sodium (Na) into the CIGS absorber layer, and replacing thick cadmium sulfide (CdS) buffer layer with thin CdS layer have boosted the efficiency significantly. Copper Indium Gallium Selenide (CIGS) thin film solar cell currently holds a record efficiency of 22.6% since 2016 [5]. To accomplish the record efficiency, research institute ZSW has developed a new method to deposit a layer of potassium fluoride (KF) between CIGS and buffer layer (CdS) to improve the electrical properties of the solar cells, in particular the V_{oc}.

The cell structure of CIGS is known as substrate configuration where the light enters the cell through Transparent Conducting Oxide (TCO), passes through the buffer layer, is absorbed by the CIGS, and then reaches the

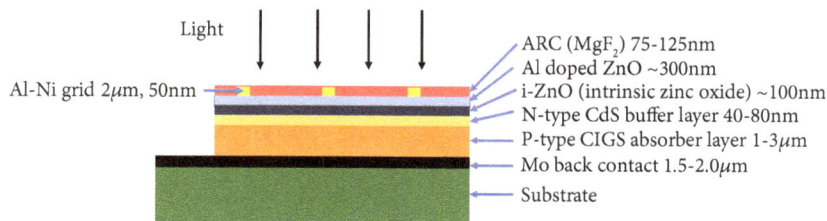

FIGURE 1: Structure of CIGS thin film solar cell.

back contact, usually molybdenum, which is deposited on the substrate. The typical structure of CIGS solar cell consisting substrate/Mo/p-type CIGS/n-type CdS/i-ZnO/ZnO:Al/ARC/metal-grid is shown in Figure 1 [1]. The reason CIGS has been one of the most promising absorber layer for thin film photovoltaic devices is due to its high absorption coefficient for solar radiation and compatibility of its bandgap (1.6 eV–1.0 eV) [6]. The advantages of CIGS-based solar cells over CIS-based solar cells are as follows: (i) the bandgap can be tuned by adjusting the Ga/In ratio to match the solar spectrum. If all indium (In) is replaced by gallium (Ga), the CIGS bandgap increases from about 1.04 eV to 1.68 eV [7]. It has been stated that CIGS absorber layer can absorb most parts of the solar spectrum with a thickness of $1\,\mu m$ [1]. Hence, a layer thickness of ~2.0–2.5 μm will be sufficient for the completed device, and a thinner layer device means reduction in raw material usage and lower production cost incurred. (ii) Ga incorporation can also improve the open-circuit voltage V_{oc} of CIGS since $V_{oc} \sim E_g/2$ (E_g is referring to bandgap) [1].

Moreover, CIGS thin film solar cell has very high potential to overcome the cost level of conventional PV crystalline silicon (c-Si) technology [8]. The c-Si modules with efficiencies of 19–23% will have a production cost of \$0.6–\$0.7/Wp [9]. Whereas for CIGS modules, manufacturing cost of \$0.75/Wp can be achieved at 50 MW/yr production capacity with an average efficiency of 12% [10].

The substrate in CIGS has a crucial role in the development of the whole device. Deposition of the molybdenum back contact on rigid or flexible substrate will define the selenization condition. It has to be considered that flexible substrate cannot withstand high process temperature over 500°C, but high process temperature is required in crystallizing the CIGS absorber layer. Lower temperature processes have to be developed when dealing with flexible materials. Recently, an efficiency as high as 20.4% was achieved on flexible polymer substrate [11]. This has revealed that the flexible solar cells with performance close to rigid solar cells can be developed.

The following layer in CIGS after substrate is the molybdenum (Mo) back contact which acts as an optical reflector to reflect the light back to the absorber layer in CIGS solar cell [12, 13]. Molybdenum (Mo) is a preferred back contact material for CIGS solar cells because it does not react strongly with CIGS; it forms low-resistivity ohmic contact to CIGS, and the conductivity of Mo does not degrade during deposition of CIGS at high substrate temperature [14–17]. Mo has high conductivity and is more chemically stable

and mechanically stable during CIGS growth (selenization) than other materials such as W, Ta, Nb, Cr, V, Ti, and Mn [18–23].

The layer after Mo back contact is Copper Indium Gallium (CIG) before going through the process of selenization. During the selenization process, selenium (Se) vapor will react with CIG to become CIGS and react with Mo to form the $MoSe_2$ layer. This interfacial layer between Mo and CIGS is beneficial in terms of having a wider bandgap (1.35–1.41 eV) than CIGS, hence it can absorb more near-infrared light to improve the cell performance [24]. The formation of $MoSe_2$ layer does not depend only on the selenization condition but also on the properties of the Mo film [25, 26]. Therefore, improving the properties of the Mo film can promote the growth of $MoSe_2$ layer. Recently, substrates used in CIGS, either rigid or flexible, together with the properties of the Mo back contact and the $MoSe_2$ interface layer were discussed in various papers [8, 25, 27, 28]. This paper aims to focus on the mentioned area by first providing an overview of comparison between conventional PV and thin film solar cells followed by reporting the recent progress of the substrates in CIGS, specifically regarding the available substrates. This paper will then converge towards the Mo layer in CIGS and further discuss about the deposition techniques and effect of deposition condition on the properties of Mo back contact. Then, the formation and thickness influence factors of the interfacial $MoSe_2$ layer will be reviewed in this paper. Scale-up issues of CIGS module production will also be presented to give an insight into commercializing CIGS solar cells.

2. Brief Introduction on the Production, Efficiency, etc. of a-Si, CdTe, and CIGS Thin Film Solar Cells and c-Si Solar Cells

Global production of photovoltaics (PV) has been expanding drastically in the past decades, moving from 202 MW in 1999, 17 GW in 2010, until the recent production over 78 GW in 2016 [29–31]. About 92% of the commercial modules are made from Si while thin film modules contributed 8% in the market share of which <1% for amorphous silicon (a-Si) modules, 5% for Cadmium Telluride (CdTe) modules, and 2% for CIGS modules [32, 33]. Considering the 8% market share in 78 GW global PV production, the total annual production for thin film solar cell will be ~10.26 GW. The estimated total energy world consumption in years 2050 and 2100 are 28 and 46 TW, respectively [34].

Looking at the current PV technologies with an average growth rate of 30–40% per year, significant fraction of the future world energy demand can be satisfied [35]. The time where thin film PV technologies started to grow rapidly was during the silicon feedstock shortage that happened back in mid 2000s [30]. This scenario has caused the price of the conventional PV module to be increased and thus opened up an opportunity for the researchers and investors to explore further in the thin film technologies. According to the latest research cell record efficiency chart reported by the National Renewable Energy Laboratory (NREL), crystalline silicon (c-Si) has a record efficiency (lab scale $2 \times 2\,cm^2$) of 25.7% and 24.4% efficiency for commercially available module [36, 37]. For thin film solar cell, amorphous silicon (a-Si), CdTe, and CIGS, each cell respectively has a record efficiency (lab scale) of 14%, 22.1%, and 22.6%; at the same time, the best module efficiency for each cell falls at 12.3%, 18.6%, and 15.7% [36, 38, 39].

Crystalline silicon (c-Si) currently plays a major part in thin film energy production with the highest module efficiency. Thick and rigid c-Si wafer (180 μm) is required for a good absorber (high absorption) in module production [1]. This is because silicon (Si) is a poor absorber due to its indirect bandgap nature and its low absorption coefficient ($10^4\,cm^{-1}$) [40]. Whereas for CIGS, its absorption coefficient is beyond $10^5\,cm^{-1}$ making the thickness of CIGS to be 100 times less than the thickness of c-Si wafer. In addition, the module production of CIGS solar cell requires a lower thermal budget (~550°C) than the c-Si solar cell (~1100°C) [1, 41]. CIGS solar cells provide an alternative to Si solar cells, and it is highly competitive as less raw material, time, and cost are involved in module production. On the basis of material, less material usage will lead to lower manufacturing cost for CIGS solar cell and hence inducing a shorter energy payback time (~1 year) as compared to the c-Si solar cell (~2 years) [10, 42]. Aside from the conventional c-Si cells, there are also other commercially available thin film modules in the market, including amorphous silicon (a-Si), Cadmium Telluride (CdTe), and, as mentioned, Copper Indium Gallium Selenide (CIGS).

Amorphous silicon (a-Si) has a direct bandgap and only uses 1% of the material (Si) needed for crystalline silicon cells production [43]. a-Si module can be made flexible and lightweight which later enable various possibilities when it comes to application such as mounting on uneven surface, incorporating into small devices, and being portable. One of the a-Si solar cell advantages is the high actual power output in hot climate by having a low temperature coefficient [44, 45]. However, a-Si solar cells have difficulty in the solar market because the price of conventional PV (c-Si) has been decreasing dramatically [46]. The main issue in a-Si solar cell technology is having a low conversion efficiency. The 14% stabilized research efficiency (by National Institute of Advanced Industrial Science and Technology, AIST) and 12.3% module efficiency have clearly shown its limitation to compete in the PV sector. Until now, a-Si technology is matured and commonly being used in the application of consumer products (e.g., calculators, watches, and other noncritical outdoor applications) [43].

Cadmium Telluride (CdTe) currently holds the highest module efficiency at 18.6% due to one of its advantages of nearly ideal bandgap (1.45 eV) for solar terrestrial photoconversion [36, 47]. The module production status of CdTe is currently ahead of a-Si and CIGS thin film solar cell as it can be produced at a cost of \$0.75/Wp (watt peak), and the production cost at year 2020 is expected to be around €0.5/Wp [48, 49]. Amorphous silicon (a-Si) module can be produced at €0.75/Wp but in a lower efficiency (~9%) while CIGS is still trying to lower its production cost to below €1/Wp [49]. Cadmium Telluride (CdTe) is a favourable technology in terms of prospects in PV sector except the public has raised concern about the core material used in CdTe, cadmium, which is extremely toxic. It has been stated that cadmium forms a very stable compound with tellurium and hence is not banned as a hazardous substance, but before forming a compound, cadmium itself can lead to a variety of adverse health effects including cancer [50]. More advance technology is needed to handle the cadmium and hence creating another challenge for CdTe in lowering the production cost.

Copper Indium Gallium Selenide (CIGS) solar cell is one of the best thin film candidate to look into because its lab scale efficiency (22.6%) has just surpassed CdTe's lab scale efficiency (22.1%) based on the latest research cell record efficiency chart reported by NREL [37]. For CIGS solar cells, a large efficiency gap occurred between the lab scale cells (22.6%) and commercially available modules (15.7%). Although the CIGS module production presently lags behind of CdTe module due to process complexity, nonetheless, CIGS technologies still have a higher efficiency potential versus CdTe, and this can potentially make them more cost-effective than CdTe solar cells [1, 48]. The theoretical efficiency limit for CIGS bandgap of 1.14 eV is 33.5% [51]. According to the CIGS current record efficiency (22.6%), cell efficiency as high as 25% can be reached in the near future. In terms of module efficiency, CIGS (15.7%) is definitely not far behind from CdTe (18.6%) (Table 1).

3. Recent Progress of Substrates

Soda-lime glass (SLG) is a type of rigid substrate being used widely in the CIGS thin film industry due to its material properties which can supply sufficient amount of Na to the absorber during coevaporation or selenization process [53–56]. The supply of sodium (Na) at 0.1 at% (atomic percentage) is reported to be beneficial for CIGS solar cells in terms of increasing the open-circuit voltage and fill factor that lead to an enhancement in solar cell efficiency [53, 57]. The improvement in device efficiency is mainly due to sodium (Na) that passivates the defects at the cadmium sulfide (CdS) and CIGS junction [58–60]. In addition, SLG meets most of the requirements needed such as good adhesion, low weight, and able to work on suitable temperature. SLG substrate also has an optimal coefficient of thermal expansion (CTE) for CIGS which is desired between 5×10^{-6} and $12 \times 10^{-6}\,K^{-1}$ to avoid adhesion problem or crack formation during deposition of CIGS at high temperature [27]. Solar cell company like Solibro has achieved a

TABLE 1: Brief introduction on the production, efficiency, etc. of conventional PV (c-Si) and thin film solar cells (a-Si, CdTe, and CIGS) [10, 32, 33, 36, 37, 40, 42, 51, 52].

| | Conventional PV | Thin film solar cells | | |
	c-Si	a-Si	CdTe	CIGS
Best research cell efficiency	25.7% [37]	14% [37]	22.1% [37]	22.6% [37]
Best module efficiency	24.4% [36]	12.3% [36]	18.6% [36]	15.7% [36]
Theoretical efficiency limit	29.43% [52]	20% [51]	32.8% [51]	33.5% [51]
Absorption coefficient	10^4 cm^{-1} [40]	(5×10^4) cm^{-1} [40]	10^5 cm^{-1} [40]	$>10^5$ cm^{-1} [40]
Current PV market share	92% [32]	<1% [32]	5% [32]	2% [32]
Annual production	~71.76 GW	~0.78 GW	~3.9 GW	~1.56 GW
Energy payback time	~2 years [42]	~1.5 years [42]	~7 months [42]	~1 year [42]
Major manufacturer	Jinko Solar [33]	Sharp [10]	First Solar [10]	Solar Frontier [10]

current record efficiency of 21% (single-junction terrestrial cell) and 18.7% on minimodule using SLG as a substrate [61, 62]. The first time ever, in 2010, the substrate involved in pushing the efficiency of CIGS beyond 20% was SLG [63]. Other than the standard glass SLG, specialty glasses have been explored by several research groups which mainly focus on high-temperature glasses to avoid softening of the substrate (SLG) during CIGS absorber deposition [64, 65]. High temperature condition will not only enhance the growth of absorber layer but also favour alkali-diffusion from substrate to absorbers. Thus, specialty glass serves as a medium to provide fine control of alkali-diffusion without softening at high temperature [66, 67]. The CIGS world record efficiency back in 2014, 21.7%, and current world record efficiency since 2016, 22.6%, were both achieved by the same research group ZSW on a specialty alkali-aluminosilicate glass that incorporate Na during the CIGS growth process [5, 61, 68].

In CIGS solar cells, the most commonly used flexible substrates are metals and polymers [27, 60]. The essential aspects to be considered for a suitable flexible substrate for CIGS films are dependent on different physical and chemical properties, such as thermal stability, vacuum compatibility, suitable coefficient of thermal expansion (CTE), humidity barrier function, chemical inertness, and surface smoothness [57]. Metals are able to withstand very high deposition temperature but they possess a rather high roughness, density, and CTE especially Al and Cu [27]. Furthermore, most of the metals like steel contains metallic impurities (Fe and Cr) that affect the device performance [27, 69, 70]. Therefore, a metal oxide barrier is used in order to provide electrical insulation between the substrate and the Mo back contact at the same time serves as a diffusion barrier against impurities from the metal substrate [27, 69, 71, 72]. Typical examples of barrier layer materials are Al_2O_3 and SiO_x [73–75]. It has been announced that the highest cell efficiencies reached so far using stainless steel (SS) and titanium (Ti) foil as substrate are 17.7% and 17.9% [71, 76, 77]. Whereas the maximum cell efficiency (area of 0.5cm^2 with antireflective coating) achieved so far on an enamelled steel is 18.6% by ZSW research group [78]. These results prove the potential of metal as an alternative substrate to rigid glass.

Polymers as substrates have a much lower density, roughness than metals [27]. It allows direct monolithic integration of solar cells and roll-to-roll deposition process that can reduce manufacturing cost [27, 60]. Other than that, polymer has a high power-to-weight ratio and excellent radiation-hardness which makes polymer an ideal candidate for space application, but polymers cannot sustain high temperature of 550–660°C due to their limited thermal stability [79, 80]. Thus, low process temperature is required, and this generally leads to deterioration of absorber quality [81, 82]. Polyimide films are one of the few polymer films that can sustain temperature close to or above 450°C but not more than 500°C for a short period of time, but this type of polymer has high CTE and it varies depending on suppliers [27, 60]. Despite the challenges faced in using polymer foils as a substrate, the Swiss Federal Laboratories for Material Science and Technology (Empa) has successfully developed thin film solar cells, CIGS, on flexible polyimide foil with an efficiency of 20.4% [83]. The Empa research group was able to modify the properties of the CIGS layer so that it can be grown at low temperature without compensating the light absorption of the CIGS layer which contributes to the photo-current in solar cells [11]. Since 2010 until 2016, ceramics have also been used as flexible substrate [75, 84]. A submodule efficiency of 15.9% has been achieved by AIST in Japan using flexible zirconia-based ceramic sheet as a substrate [75]. Ceramic substrate is able to withstand higher temperature than the soda-lime glass and polymers due to its higher chemical, mechanical, and thermal stability and low porosity [84, 85]. However, its brittle behaviour might be an issue for industrial production on large scale (Table 2).

4. Different Deposition Techniques' Influence on the Properties of Molybdenum Back Contact for CIGS

Current deposition techniques available in the thin film industry for Mo are Ion-beam sputtering, Direct Current (DC) sputtering, Radio Frequency (RF) sputtering, and also High-target-utilization sputtering (HiTUS) [25, 86–88]. The common techniques being used are DC and RF sputtering as they have been discussed profoundly in several papers [12, 86]. Since the characteristics of Mo thin film depend strongly on deposition method and deposition parameters, therefore, a comparison has been made to investigate the

TABLE 2: Summary of current record efficiency on different substrates (CIGS solar cell) [5, 11, 61, 62, 71, 75, 76, 78].

Substrate	Current record efficiency	Description
	Rigid	
Soda-lime glass	21% [61]	Solibro (single-junction terrestrial cell) [61]
	18.7% [62]	Solibro (minimodule) [62]
Alkali-aluminosilicate glass	22.6% [5]	ZSW [5]
	Flexible	
Titanium	17.9% [76]	Aoyama Gakuin University, Japan [76]
Stainless steel	17.7% [71]	Empa [71]
Enamelled steel	18.7% [78]	ZSW [78]
Polymer	20.4% [11]	Empa [11]
Zirconia-based ceramic sheet	15.9% [75]	AIST (17-cell-integrated submodules) [75]

properties of Mo layers based on DC and RF sputtering techniques. DC sputtering is a well-established industrial process with high throughput capability and requires a cheaper set up cost, whereas RF sputtering deposits a more reflective Mo thin films but with added expense on RF power supplies and impedance matching networks [12, 14, 18, 88, 89]. DC sputtering technique has higher deposition rate than the RF sputtering technique, and DC sputtered films possess good uniformity and adhesion properties over large surface areas [24]. Low deposition rate that arises in RF sputtering technique is due to the number of cycles involved during deposition where RF sputtering only deposits in the second cycle of the AC supply. However, RF sputtered films are found to be more conductive with improvements in open-circuit voltage (V_{oc}) and short circuit current (I_{sc}) [90].

A desirable Mo back contact for CIGS solar cells is addressed to conductive, stress-free, well-adherent, uniform, and crystalline molybdenum (Mo) thin films with preferred orientation (110) on large area glass substrates [91]. By altering the deposition condition on discharge power, working pressure, substrate temperature, and target-to-substrate distance, the desired properties of Mo back contact in terms of physical, optical, and electrical can be obtained [20, 91, 92]. In this paper, the properties of the Mo back contact on both DC and RF sputtering will be further reviewed. The focus deposition parameters will be on discharge power and working pressure, as these parameters have major effect on the properties of the DC and RF sputtered films.

In DC sputtering at low working pressure, surface with dense microstructures is observed while surface with loosened microstructures is observed at high working pressure [18, 55]. At low working pressure, atoms that obtained high kinetic energy are able to travel and bombard onto the substrate surface to form a compact Mo layer, whereas at high working pressure, insufficient kinetic energy has caused the number of atom bombardment towards the substrate surface to reduce, thus resulting in porous microstructure. The high-pressure Mo with loosened microstructure will eventually lead to greater inclusion of impurities such as oxygen that can increase the resistivity of the Mo film. The direct effect of working pressure of RF sputtering on Mo microstructure has not been reported yet, but low RF power was stated necessary to minimize stress and obtain a compact

Mo microstructure despite high RF power deposits good quality of Mo [93]. The Mo deposited with high RF power caused delamination of the absorber layer after the selenization process, and this can be due to the presence of microstresses that existed on the Mo layer.

Another physical property to evaluate is the grain size of Mo. The trend of grain size is found to increase by increasing sputtering power and decreasing sputtering pressure in both DC and RF sputtering Mo films [88, 94–96]. In high power and low working pressure, the increase of Mo grain size causes space between grains to reduce and thus correlates well to the formation of densely packed Mo microstructure. Additionally, along all applied pressures, the grain size of DC sputtered Mo film is always larger than the RF sputtered Mo film [97]. One possible way to explain the formation of large grain size is probably higher power will induce higher flux, and high deposition rate of DC sputtering tends to increase the probability of the Mo particles to nucleate with each other.

Crystal structure of sputtered Mo back contact also plays a fundamental role in Mo quality determination. The lower the working gas pressure, the better the crystallinity of the DC sputtered film [88]. At lower working gas pressure (higher gas power), the atoms gain higher energy due to lesser scattering, and the atoms will then impact the substrate surface with sufficient energy which enhance the atoms mobility in order to facilitate atom diffusion and microvoid fill up, thus creating a conducive requirement for large grain growth and better crystallinity. The same crystallite behaviour was observed in RF sputtered films at low working pressure, but the degree of crystallization appeared to be lower as compared to the DC sputtered films [95, 97]. This can be related to the deposition rate of both RF and DC sputtering technique. Lower deposition rate of RF sputtering generally requires longer deposition time to achieve the same thickness as DC sputtered films, and within the deposition period, impurities such as oxygen can be introduced which can potentially restrain the process of atom diffusion and deteriorates the crystallite property of Mo film.

Surface morphology is one of the properties to be investigated, as it will affect the adhesion, optical, and electrical properties of the solar cell [20]. The surface roughness of DC and RF sputtered Mo films were found to be increased

TABLE 3: Effect of DC and RF sputtering techniques on molybdenum properties for CIGS [12, 86, 95, 97].

DC sputtering	Mo properties	RF sputtering
Larger grain size [97]	Grain size	
	Optical reflectance	Higher optical reflectance [86]
Better crystallization [95]	Crystallization	
Better surface morphology [97]	Surface morphology	
	Resistivity	Lower resistivity [12]

as the working pressure increased [24, 95]. At lower working pressure, the average roughness was due to large grain size formed under high kinetic energy gained by the atoms whereas the roughness at higher working pressure was attributed to loosened microstructure with voids formed under low energy excitation. Although both DC and RF sputtering portray similar surface roughness trend, RF sputtered films possess smoother surface morphology than the DC sputtered films at same working pressure [86]. Smoother morphology also indicates less void formation and porosity effect which contribute to a higher optical reflection in the Mo film; hence, RF sputtered films are more reflective than DC sputtered films as proven in several literatures [93, 98]. In other words, Mo films deposited under low working pressure will have higher optical reflection and conductivity due to densely packed microstructure and large grain size formed (lesser void formation yield to reduced incorporation of foreign atoms).

Furthermore, sputtering pressure has a major influence on the sheet resistance of DC and RF sputtered Mo films over sputtering power [12, 88]. As the working pressure increases, the sheet resistance increases and vice versa, whereas, the sheet resistivity of the Mo film is inversely proportional to the sputtering power [24]. The DC sputtered films also have higher resistivity as compared to the RF sputtered one in all applied working pressures [86]. This phenomenon is interrelated to the microstructure and surface morphology properties. As DC sputtered Mo films are rougher, it verifies the formation of larger voids (more porous microstructure) along the grain boundary, and this will allow impurities such as oxygen to occupy the voids resulting in an increase in resistivity. However, low-resistivity Mo films deposited under low working pressure often gets delaminated, and Mo deposited at high working pressure has better adhesion but less conductive [24]. Mo back contact that exhibits low resistivity and adhesive properties are challenging to simultaneously achieve in single Mo layer [20]. Thus, Mo multilayers like bilayer structure are usually applied to deposit a well-adhered and conductive Mo layer [99–101]. This was done by depositing the first bottom layer with high working pressure (to achieve better adhesion) and the second top layer at lower working pressure (to achieve lower resistivity).

Besides that, single Mo layer can also cause excessive tensile or compressive stress which lead to surface cracking and delamination between the Mo film and the substrate [18]. Bilayer structure is used as a solution to reduce excessive residual stress that exists on the Mo layer [24, 102]. The bottom Mo layer of tensile stress (formed under high pressure) and top Mo layer of compressive stress (formed under low

pressure) contribute an overall residual stress compensation to avoid delamination of the Mo film from the substrate. It is an undeniable fact that depositing two layers of Mo can resolve adhesion and resistivity problem until certain extent, but taking into consideration of other properties such as microstructure, grain size, crystal structure, and surface morphology will further improve the quality (electrical and optical) of the Mo films. Referring to Table 3, RF sputtered Mo films are more reflective and conductive while DC sputtered Mo films have larger grain size, better crystallization, and better surface morphology. A combination mode of both sputtering techniques to form multilayer molybdenum (DC/RF) can be realised to optimise the Mo back contact accordingly. The potential of simultaneous DC and RF sputtering on Mo deposition can also be examined as this technique has been used by employing an induction coil in the DC path (to avoid short circuit to the RF voltage) for depositing ZnO/ZnO:Al window and contact layer with an improvement in V_{oc} and I_{sc} [90] (Table 3).

5. Formation and Thickness Influence Factors of MoSe₂ Interface in CIGS Solar Cell

$MoSe_2$ is an interface formed between molybdenum back contact and CIGS absorber layer during the process of selenization at high temperature (above 723 K) [103, 104]. The selenium diffuses into the Mo back contact and reacts to form $MoSe_2$ along the process of selenization [105]. The $MoSe_2$ layers consist of polycrystalline grains with columnar structure and lattice spacing which improve the adhesion between Mo and CIGS [106–108]. Instead of that, $MoSe_2$ has a wider bandgap of 1.41 eV than the CIGS absorber that forms a back surface field which can hinder the recombination of electrons and holes [109–111]. The CIGS/Mo heterocontact including the $MoSe_2$ layer leads to a favourable ohmic-type contact by the evaluation of dark I-V measurement at lower temperature [25, 109]. Without the interface layer, a Schottky contact will be formed at the Mo/CIGS contact, causing significant problem in resistive losses [112]. However, excessive formation of $MoSe_2$ can lead to the delamination of the film and adverse effect on V_{oc} and FF of the completed CIGS solar cells due to high resistance of the $MoSe_2$ [113, 114]. Therefore, a range of specific $MoSe_2$ thickness between 100 nm and 200 nm is required to ensure good adhesion and electrical contact between Mo/CIGS [115–117].

The thickness of the $MoSe_2$ layer can be influenced by several factors such as sputtering conditions [117], residual stress in Mo layer [118], characterization of the barrier layer (TiN, MoN_x, and MoO_x) [115, 116, 119] and selenization conditions [120–122]. The quality of the Mo crystals improves with sputtering power, thereby reducing the resistance of the Mo back contact. As the sputtering power increases, the thickness of $MoSe_2$ also increases [117]. This is due to the fact that increase in sputtering power enhances the diffraction intensity and crystallinity of Mo (110), (211) which facilitate the transformation of the cubic crystal structure Mo back contact into a hexagonal crystal structure $MoSe_2$ layer [123, 124]. However, varying the sputtering power to achieve the desired $MoSe_2$ thickness is not practical since high and low sputtering power are required to deposit a well-adhered and conductive Mo layer (bilayer structure) [99]. The $MoSe_2$ layer thickness is also dependent on the in-grain density of Mo which is interrelated to the residual stress. In a typical sputtered Mo layer, the tensile residual stress of Mo increases with the sputtering pressure up to 10 mTorr and decreases with further increase of pressure [118]. The increase in pressure reduces the mean free path and decreases the ion energy of the Mo particles, causing an increase in tensile stress (Region I). As the pressure further increases beyond 10 mTorr (Region II), the atomic attraction across gain boundary increases, creating porosity which eliminates the grain boundary attraction, thus reducing tensile stress. The reduction of the tensile stress in Region II will increase the in-grain density of the Mo back contact and decrease the $MoSe_2$ reactivity (thickness) during the selenization process.

In recent years, barrier layers are used to control the excessive formation of $MoSe_2$. A thin Titanium Nitride (TiN) barrier layer with a thickness of ~20 nm was grown on a Mo-coated soda-lime glass under vacuum condition prior to CZTSe deposition (CZTS has the same solar cell structure and works similarly to CIGS) [115]. Transition metal nitrides are used as a barrier layer because several researches showed that transition metal nitrides worked better in preventing diffusion at higher temperature than transition metals (Ta, Ti, W, and Mo) [125–127]. The TiN barrier layer has successfully suppressed the growth of $MoSe_2$ from ~1300 nm to ~200 nm significantly improving the V_{oc}, J_{sc}, and FF of the device. Other than TiN, the same concept has been applied by forming a molybdenum nitride (MoN_x) barrier layer to passivate the Mo back contact against selenization [116]. With MoN_x thickness of ~120 nm, the thickness of $MoSe_2$ formed was ~150 nm without consuming the entire upper thin Mo layer (20–30 nm left). One previous study has also demonstrated that the MoN_x barrier layer is able to control the formation of $MoSe_2$ and improve FF and current-voltage characteristic of the device, but thicker barrier layer will increase the series resistance of the device [128]. Other than the transition metal nitride barrier layer, molybdenum oxide (MoO_x) was proposed to control the growth of $MoSe_2$ [119]. It has been reported that a thin layer of MoO_2 as low as 10 nm is able to prevent the Mo back contact from overselenization and improve the V_{oc}, FF, and shunt resistance (R_{sh}) of the device [129]. In addition, oxygen allows diffusion of alkali metals during selenization which will enhance the performance of the device [130–134]. The behaviour of the MoO_2 barrier layer to control the formation of $MoSe_2$ can be explained with their corresponding Gibbs free energy. At 900 K, the Gibbs free energy of the reaction between Mo and oxygen (–423 kJ/mol) is lower than the one of Mo with selenium (–129 kJ/mol) causing the reaction of MoO_2 with selenium (+294 kJ/mol) to be thermodynamically unfavoured, validating the passivation effect of MoO_2 on Mo back contact against selenization [119].

A well-crystallised CIGS absorber layer requires a process temperature of at least 500°C and above for the selenization process [118]. Nonetheless, at high process temperature, the thickness of the $MoSe_2$ layer will increase significantly [117]. If the process temperature is lowered to reduce the thickness of $MoSe_2$, deterioration of the absorber layer (electrical properties) will occur [135]. Besides the mentioned thickness-influencing factors, sodium (Na) might be the dominant factor to control formation of $MoSe_2$ layer [25, 123]. An experiment has been conducted by varying the amount of Na (using SiO_x barrier layer) diffuses from the substrate across the Mo back contact to the CIGS absorber layer [123]. The experiment concluded that Na aids in the formation of $MoSe_2$, but if the Na content is too high, it passivates the grain boundaries at CIGS layer to form Na_2Se_x; thus, lesser Se atom is available to react with Mo, and this will retard the formation of $MoSe_2$. Comparing these thickness-influencing factors, barrier layer appeared to be the most effective method to control the thickness of $MoSe_2$, allowing sodium (Na) diffusion while minimizing the adverse effect on cell performance.

6. Scale-up Challenges/Issues of CIGS Thin Film Modules

Since 2006, thin film solar cell production in the U.S. has outperformed the production of c-Si solar cell, becoming the least expensive technology to be manufactured [35, 48]. The existing thin film PV technologies especially CIGS have reached over 1.6 GW of cumulative module production in 2015 and is reported to have a high cost reduction potential at high production volumes [136, 137]. For a production capacity of 1000 MW/yr with 15% module efficiency, the CIGS module production cost as low as $0.34/Wp can be achieved [10]. Due to photovoltaics cost declining with maturity of the conventional PV technologies and new entry of China into the market in 2010, the future challenges for CIGS production will be combining high production volumes with high throughput, sufficient yield, and superior quality. Meaning that the production cost has to be brought down to make it competitive with conventional sources.

Currently, the leading CIGS module manufacturer Solar Frontier claims that their CIS module is able to output more electricity than the conventional crystalline silicon (c-Si) in real-world conditions such as better performance at high temperatures, low-light condition, light-soaking effect, and also shadow tolerance [138]. With mass production of modules having efficiencies ranging from 11.8% to 13.8%, Solar Frontier was able to achieve an annual production of ~1 GW and over 3 GW of shipments worldwide in 2015 [139, 140]. The statement made on surpassing the performance of c-Si solar cells is according to the cadmium- and lead-free CIS module (SF 150-170S Series) which provides an efficiency of 13.8% for a total area of 12,280 cm^2 [140, 141]. The current technology offered by Solar Frontier with high energy yield leads to a shorter energy payback time which fulfil the requirement to be competitive towards c-Si solar cells by producing high-quality module at lower cost. On the other hand, Solar Frontier has also attained a higher efficiency of 19.2% on cadmium-free CIGS minimodule [36, 142]. This research direction affirms the potential of cadmium-free CIGS module with consideration on environmental issues (reduce usage of cadmium) as the energy production increases.

The production of CdTe PV modules is ahead of CIGS PV modules at present stage due to simplicity of the process. However, different approaches have been used to boost up the CIGS cell efficiency such as applying Post Deposition Treatment (PDT) on the CIGS surface with alkali elements, incorporating more gallium into the absorber layer (CIGS) and also combining with other materials such as perovskites for multijunctions [1]. PDT is the process developed by ZSW to reach 22.6% efficiency in CIGS solar cells, but the compatibility and practicality between new technologies and scaling it up into mass production should be taken into account earnestly. Therefore, the challenges/issues in CIGS solar cell production are discussed as below:

(i) Uniformity of CIGS absorber film over large areas: Uniformity is essential for electrical and optical properties of high-efficiency solar cells, and it directly influences the yield in production. The yield determines actual production volume in MW/yr and production cost in ¢/Wp [35]. Coevaporation can be an appropriate technique for large area substrate because the coevaporated CuGaIn precursor has a higher selenization rate than the cosputtered one, and it has been actively developed by many manufacturing companies [122]

(ii) Standardizing the cell fabrication process: In the industry, Mo and ZnO films were sputtered (vacuum system); buffer layer CdS was deposited using Chemical Bath Deposition (nonvacuum system), and absorber layer CIGS was deposited with the coevaporation method (vacuum system) [143]. Fabricating the cell in and out from vacuum and nonvacuum process will cause difficulty in troubleshooting once a problem occurs. Substrate handling in open air environment and between different tools can contaminate the substrate. This can be avoided by using a vacuum process along the fabrication of the whole cell

(iii) Presence of moisture in CIGS modules: Water vapor will oxidise the back contact molybdenum (Mo) causing Mo to degrade [8]. To solve this issue, a robust encapsulation technology with the properties of durability, adhesion, thermal stability, etc. is required

(iv) Long processing time for CIGS and TCO layers: The processing time for CIGS (including deposition and selenization) and Transparent Conducting Oxide (TCO) layers should be reduced to meet the required industrial production time (10 min [1]) when it involves large volume production. The processing time can be improved by using a thinner CIGS absorber layer and high-speed deposition technique for TCO

(v) Cadmium used in the CIGS buffer layer (CdS) is toxic, and disposal of the cadmium-containing product causes detrimental effect to human health [144]. Moreover, the use of CdS buffer layer leads to optical absorption loss [145]. Hence, the CdS buffer layer can be replaced by other appropriate wider bandgap buffer materials to improve the short-circuit current (J_{sc}) of the device

(vi) Indium scarcity on CIGS module: Production of indium currently relies in by-products of mining and refining of other material (in particular, zinc) [146]. There is a concern raised upon scarcity of indium might escalate the price and can be a threat to CIGS ambitions for production cost reduction and cost competitiveness in the wider PV market. Addressing this issue, CIGS layer of $\leq 1\ \mu m$ should be used without compromising J_{sc} of the device [1].

To sustain a profitable business in the PV market, PV module manufacturers will have to adopt or innovate technologies that offer a production cost <$0.5/watt with the capability to match c-Si module performance. The manufacturing cost of CIGS solar cell can be further reduced by increasing process yield, production capacity, and efficiency while the raw material scarcity can be eased by improving material utilisation (thinner absorber layer) during the deposition process. Nonetheless, due to the maturation of solar PV sector, the cost effectiveness of solar cell can no longer be determined solely on manufacturing cost ($/Wp) and efficiency. Aspects such as annual electricity yield, cost of PV modules, Balance of System (BOS) ($/W), and cost incurred for the PV system are critical to compute the minimum price at which energy must be sold to break-even the solar PV project [147–149]. This concept is also known as Levelized Cost of Energy (LCOE). Between 2010 and 2017, the LCOE for utility scale crystalline silicon (c-Si) PV plants have reduced from 0.36 $/kWh to 0.10 $/kWh, mainly driven by the reduction of module prices [150]. From the year 2010 to 2012, LCOE at the price of 0.12–0.20 $/kWh to 0.11 $/kWh was achieved by the thin film PV plants, and the LCOE price is estimated to be decreased to about 0.06–0.10 $/kWh by 2020 [151–153].

The major manufacturer of CdTe thin films, First Solar, was able to obtain a LCOE at 0.15 $/kWh, and the LCOE was targeted to be brought down to 0.08 $/kWh with the system cost of 2 $/W [154]. Whereas CIGS (14% efficiency) has attained a decent LCOE at 0.084 $/kWh for a 100 MW scale PV system [155]. Besides that, the CIGS LCOE reported at 0.084 $/kWh was being compared with c-Si solar cell (16% efficiency) in an identical location and system, and the LCOE of the c-Si solar cell was at 0.80 $/kWh, lower than CIGS but comparable to each other [155]. This can be explained by the fact that c-Si solar cell has better module efficiency causing the Balance of System (BOS) to be reduced meanwhile CIGS offers better performance under high temperature and low-light condition thus diminishing the LCOE differences between c-Si and CIGS solar cell. The LCOE of CIGS is also believed to be higher than the CdTe thin films at this moment; it is because CIGS thin film possesses lower module efficiency and requires more complex system that tends to increase both cost on BOS and PV system. Nevertheless, CIGS technology is still feasible and promising as its record efficiency (22.6%) has surpassed CdTe (21.6%), and the theoretical efficiency limit of CIGS (33.5%) is higher than the c-Si (29.43%) solar cells [37, 51, 52]. This implies that CIGS solar cell with greater efficiency potential will result in lower material usage, lower system cost, and higher energy yield which eventually contribute to an exceptional Levelized Cost of Energy (LCOE).

7. Conclusion

CIGS solar cells are believed to have a very high potential against c-Si and CdTe solar cells in achieving low production cost with high module efficiency as the CIGS possesses better absorption coefficient (lower material usage), requires lower thermal budget than c-Si solar cells, and its record efficiency (22.6%) has just surpassed CdTe (22.1%). The rigid substrates such as soda-lime glass and alkali-aluminosilicate glass give rise to higher efficiency with direct Na incorporation. However, flexible substrates have proven its capability to be an alternative to rigid substrates by achieving a decent efficiency comparable to the rigid substrate solar cells by applying Post Deposition Treatment. DC sputtered Mo films are favouring physical properties while RF sputtered Mo films are favouring electrical properties. Thus, multilayer molybdenum (DC/RF) or simultaneous DC and RF sputtering mode can be explored to optimise the molybdenum back contact. On the other hand, the $MoSe_2$ layer is important in forming ohmic-type contact and improving the adhesion between CIGS and Mo layers. Excessive formation of $MoSe_2$ can cause delamination problem and increase in resistivity of the solar cell. Addressing this issue, the most effective method to control the thickness of $MoSe_2$ is forming a barrier layer in between the bilayer structure Mo back contact. The scale-up challenges/issues of CIGS thin film modules discussed are possible to overcome as current CIGS technology has started to offer better performance at high temperature and low-light condition than the c-Si solar cells in real-world conditions. Lastly, with greater efficiency potential and higher total lifetime power produced, CIGS technology will eventually attain an outstanding Levelized Cost of Energy (LCOE) against c-Si and CdTe solar cells.

Conflicts of Interest

The authors declare that there is no conflict of interest regarding the publication of this paper.

References

[1] J. Ramanujam and U. P. Singh, "Copper indium gallium selenide based solar cells – a review," *Energy & Environmental Science*, vol. 10, no. 6, pp. 1306–1319, 2017.

[2] S. Wagner, J. L. Shay, P. Migliorato, and H. M. Kasper, "CuInSe2/CdS heterojunction photovoltaic detectors," *Applied Physics Letters*, vol. 25, no. 8, pp. 434-435, 1974.

[3] L. L. Kazmerski, F. R. White, and G. K. Morgan, "Thin-film CuInSe2/CdS heterojunction solar cells," *Applied Physics Letters*, vol. 29, no. 4, pp. 268–270, 1976.

[4] R. A. Mickelsen and W. S. Chen, "Development of a 9.4% efficient thin-film CuInSe2/CdS solar cell," in *15th Photovoltaic Specialists Conference*, pp. 800–880, New York, FL, USA, 1981.

[5] P. Jackson, R. Wuerz, D. Hariskos, E. Lotter, W. Witte, and M. Powalla, "Effects of heavy alkali elements in Cu(In,Ga)Se$_2$ solar cells with efficiencies up to 22.6%," *Physica Status Solidi (RRL) - Rapid Research Letters*, vol. 10, no. 8, pp. 583–586, 2016.

[6] T. D. Lee and A. Ebong, "Thin film solar technologies: a review," in *2015 12th International Conference on High-capacity Optical Networks and Enabling/Emerging Technologies (HONET)*, pp. 1–10, Islamabad, Pakistan, December 2015.

[7] Y.-C. Wang and H.-P. D. Shieh, "Double-graded bandgap in Cu(In,Ga)Se$_2$ thin film solar cells by low toxicity selenization process," *Applied Physics Letters*, vol. 105, no. 7, article 073901, 2014.

[8] M. Powalla and B. Dimmler, "Scaling up issues of CIGS solar cells," *Thin Solid Films*, vol. 361-362, pp. 540–546, 2000.

[9] A. Goodrich, P. Hacke, Q. Wang et al., "A wafer-based monocrystalline silicon photovoltaics road map: utilizing known technology improvement opportunities for further reductions in manufacturing costs," *Solar Energy Materials & Solar Cells*, vol. 114, pp. 110–135, 2013.

[10] V. K. Kapur, V. K. Kapur, A. Bansal, and S. Roth, "Roadmap for manufacturing cost competitive CIGS modules," in *2012 38th IEEE Photovoltaic Specialists Conference*, pp. 3343–3348, Austin, TX, USA, October 2012.

[11] Empa takes thin film solar cells to a new level, "A new world record for solar cell efficiency," January 2017, https://www.empa.ch/web/s604/weltrekord.

[12] X. Ma, D. Liu, L. Yang, S. Zuo, and M. Zhou, "Molybdenum (Mo) back contacts for CIGS solar cells," in *Eighth International Conference on Thin Film Physics and Applications*, vol. 9068, pp. 906811–906814, Osceola, FL, USA, 2013.

[13] P. J. Rostan, J. Mattheis, G. Bilger, U. Rau, and J. H. Werner, "Formation of transparent and ohmic ZnO:Al/MoSe$_2$ contacts for bifacial Cu(In,Ga)Se$_2$ solar cells and tandem structures," *Thin Solid Films*, vol. 480-481, pp. 67–70, 2005.

[14] P. Pradhan, P. Aryal, A. Ibdah et al., "Effect of molybdenum deposition temperature on the performance of CuIn 1- x Ga x Se 2 solar cells," in *2015 IEEE 42nd Photovoltaic Specialist Conference (PVSC)*, pp. 1–4, New Orleans, LA, USA, June 2015.

[15] S. Ashour, S. Alkuhaimi, H. Moutinho, R. Matson, and F. Abou-Elfotouh, "Junction formation and characteristics of CdS/CuInSe$_2$/metal interfaces," *Thin Solid Films*, vol. 226, no. 1, pp. 129–134, 1993.

[16] S. Raud and M. A. Nicolet, "Study of the CuInSe$_2$/Mo thin film contact stability," *Thin Solid Films*, vol. 201, no. 2, pp. 361–371, 1991.

[17] E. Moons, T. Engelhard, and D. Cahen, "Ohmic contacts to p-CuInSe$_2$ crystals," *Journal of Electronic Materials*, vol. 22, no. 3, pp. 275–280, 1993.

[18] Y. Huang, S. Gao, Y. Tang, J. Ao, W. Yuan, and L. Lu, "The multi-functional stack design of a molybdenum back contact prepared by pulsed DC magnetron sputtering," *Thin Solid Films*, vol. 616, pp. 820–827, 2016.

[19] K. Orgassa, H. W. Schock, and J. H. Werner, "Alternative back contact materials for thin film Cu(In,Ga)Se$_2$ solar cells," *Thin Solid Films*, vol. 431-432, pp. 387–391, 2003.

[20] W. Li, X. Yan, A. G. Aberle, and S. Venkataraj, "Analysis of microstructure and surface morphology of sputter deposited molybdenum back contacts for CIGS solar cells," *Procedia Engineering*, vol. 139, pp. 1–6, 2016.

[21] A. Romeo, M. Terheggen, D. Abou-Ras et al., "Development of thin-film Cu(In,Ga)Se$_2$ and CdTe solar cells," *Progress in Photovoltaics: Research and Applications*, vol. 12, no. 23, pp. 93–111, 2004.

[22] A. N. Tiwari, M. Krejci, F.-J. Haug, and H. Zogg, "12.8% efficiency Cu(In,Ga)Se$_2$ solar cell on a flexible polymer sheet," *Progress in Photovoltaics: Research and Applications*, vol. 7, no. 5, pp. 393–397, 1999.

[23] P. Jackson, R. Würz, U. Rau et al., "High quality baseline for high efficiency, Cu(In$_{1-x}$,Ga$_x$)Se$_2$ solar cells," *Progress in Photovoltaics: Research and Applications*, vol. 15, no. 6, pp. 507–519, 2007.

[24] Z. H. Li, E. S. Cho, and S. J. Kwon, "Molybdenum thin film deposited by in-line DC magnetron sputtering as a back contact for Cu(In,Ga)Se$_2$ solar cells," *Applied Surface Science*, vol. 257, no. 22, pp. 9682–9688, 2011.

[25] T. Wada, N. Kohara, S. Nishiwaki, and T. Negami, "Characterization of the Cu(In,Ga)Se$_2$/Mo interface in CIGS solar cells," *Thin Solid Films*, vol. 387, no. 1-2, pp. 118–122, 2001.

[26] S. Chaisitsak, A. Yamada, and M. Konagai, "Preferred orientation control of Cu(In$_{1-x}$Ga$_x$)Se$_2$ (x ≈0.28) thin films and its influence on solar cell characteristics," *Japanese Journal of Applied Physics*, vol. 41, pp. 507–513, 2002.

[27] P. Reinhard, A. Chirilă, P. Blösch et al., "Review of progress toward 20% efficiency flexible CIGS solar cells and manufacturing issues of solar modules," *IEEE Journal of Photovoltaics*, vol. 3, no. 1, pp. 572–580, 2013.

[28] V. Mohanakrishnaswamy, "The effect of Mo deposition conditions on defect formation and device performance for CIGS solar cells," in *Photovoltaic Specialists Conference, 2005. Conference Record of the Thirty-first IEEE*, pp. 3–6, Lake Buena Vista, FL, USA, January 2005.

[29] J. Waldau, *PV Status Report 2010*, Joint Research Centre, Renewable Energy Unit. EUR 24344 EN -2010, European Commission Joint Research Centre, 2010.

[30] B. P. Rand, J. Genoe, P. Heremans, and J. Poortmans, "Solar cells utilizing small molecular weight organic semiconductors," *Progress in Photovoltaics: Research and Applications*, vol. 15, no. 8, pp. 659–676, 2007.

[31] J. S. Hill, "GTM forecasting more than 85 gigawatts of solar PV to be installed in 2017," July 2017, https://cleantechnica.com/2017/04/05/gtm-forecasting-85-gw-solar-pv-installed-2017/.

[32] J. Ramanujam, A. Verma, B. González-Díaz et al., "Inorganic photovoltaics – planar and nanostructured devices," *Progress in Materials Science*, vol. 82, pp. 294–404, 2016.

[33] A. Jager-waldau, *PV Status Report 2016*, European Commission Joint Research Centre, 2016.

[34] N. G. Dhere, "Toward GW/year of CIGS production within the next decade," *Solar Energy Materials & Solar Cells*, vol. 91, no. 15-16, pp. 1376–1382, 2007.

[35] N. G. Dhere, "Scale-up issues of CIGS thin film PV modules," *Solar Energy Materials & Solar Cells*, vol. 95, no. 1, pp. 277–280, 2011.

[36] M. A. Green, Y. Hishikawa, W. Warta et al., "Solar cell efficiency tables (version 50)," *Progress in Photovoltaics: Research and Applications*, vol. 25, no. 7, pp. 668–676, 2017.

[37] "Best research-cell efficiencies," July 2017, https://www.nrel.gov/pv/assets/images/efficiency-chart.png.

[38] "First solar achieves world record 18.6% thin film module conversion efficiency," July 2017, http://investor.firstsolar.com/releasedetail.cfm?ReleaseID=917926.

[39] "MiaSolé," July 2017, http://miasole.com/.

[40] B. J. Stanbery, "Copper indium selenides and related materials for photovoltaic devices," *Critical Reviews in Solid State and Materials Sciences*, vol. 27, no. 2, pp. 73–117, 2002.

[41] F. Staiss, J. Springer, and H.-W. Schock, *New Developments in the Field of Photovoltaics Cells*, European Parliament, 1999.

[42] S. Energy, *Photovoltaics Report*, Fraunhofer Institute for Solar Energy Systems, 2016.

[43] World Energy Council, *World Energy Resources: Solar 2016*, World Energy Council, 2016.

[44] C. N. Jardine, G. J. Conibeer, and K. Lane, "PV-COMPARE: direct comparison of eleven PV technologies at two locations in northern and southern Europe," in *17th European Photovoltaic Solar Energy and Conference*, pp. 724–727, Munich, Germany, 2015.

[45] M. van Cleef, P. Lippens, and J. Call, "Superior energy yields of UNI-SOLAR triple junction thin film silicon solar cells compared to crystalline silicon solar cells under real outdoor conditions," in *17th European Photovoltaic Solar Energy Conference and Exhibition*, pp. 22–26, Munich, Germany, 2001.

[46] D. J. You, S. H. Kim, H. Lee et al., "Recent progress of high efficiency Si thin-film solar cells in large area," *Progress in Photovoltaics: Research and Applications*, vol. 23, no. 8, pp. 973–988, 2015.

[47] J. Britt and C. Ferekides, "Thin-film CdS/CdTe solar cell with 15.8% efficiency," *Applied Physics Letters*, vol. 62, no. 22, pp. 2851-2852, 1993.

[48] C. Candelise, M. Winskel, and R. Gross, "Implications for CdTe and CIGS technologies production costs of indium and tellurium scarcity," *Progress in Photovoltaics: Research and Applications*, vol. 20, no. 6, pp. 816–831, 2012.

[49] G. Arrowsmith, *A Strategic Research Agenda for Photovoltaic Solar Energy Technology*, European Photovoltaic Technology Platform, 2007.

[50] "Safety and health topics | cadmium - health effects | Occupational Safety and Health Administration," August 2017, https://www.osha.gov/SLTC/cadmium/healtheffects.html.

[51] A. V. Shah, M. Vaněček, J. Meier et al., "Basic efficiency limits, recent experimental results and novel light-trapping schemes in a-Si:H, μc-Si:H and 'micromorph tandem' solar

cells," *Journal of Non-Crystalline Solids*, vol. 338-340, pp. 639–645, 2004.

[52] A. Richter, M. Hermle, and S. W. Glunz, "Reassessment of the limiting efficiency for crystalline silicon solar cells," *IEEE Journal of Photovoltaics*, vol. 3, no. 4, pp. 1184–1191, 2013.

[53] P.-P. Choi, O. Cojocaru-Mirédin, R. Wuerz, and D. Raabe, "Comparative atom probe study of Cu(In,Ga)Se$_2$ thin-film solar cells deposited on soda-lime glass and mild steel substrates," *Journal of Applied Physics*, vol. 110, no. 12, article 124513, 2011.

[54] D. Rudmann, *Effects of Sodium on Growth and Properties of Cu(In,Ga)Se$_2$ Thin Films and Solar Cells*, Thesis, 2004.

[55] J. H. Scofield, "Sodium diffusion, selenization, and micro-structural effects associated with various molybdenum back contact layers for CIS-based solar cells," in *Proceedings of 1994 IEEE 1st World Conference on Photovoltaic Energy Conversion - WCPEC (A Joint Conference of PVSC, PVSEC and PSEC)*, pp. 164–167, Waikoloa, HI, USA, December 1994.

[56] J. Li, S. Glynn, L. Mansfield et al., "Density profiles in sputtered molybdenum thin films and their effects on sodium diffusion in Cu(InxGa1-x)Se2 photovoltaics," in *Photovoltaic Specialists Conference (PVSC), 2011 37th IEEE*, pp. 2749–2752, Seattle, WA, USA, June 2011.

[57] F. Kessler and D. Rudmann, "Technological aspects of flexible CIGS solar cells and modules," *Solar Energy*, vol. 77, no. 6, pp. 685–695, 2004.

[58] D. Rudmann, A. F. Da Cunha, M. Kaelin et al., "Efficiency enhancement of Cu(In,Ga)Se$_2$ solar cells due to post-deposition Na incorporation," *Applied Physics Letters*, vol. 84, no. 7, pp. 1129–1131, 2004.

[59] S.-H. Wei, S. B. Zhang, and A. Zunger, "Effects of Na on the electrical and structural properties of CuInSe2," *Journal of Applied Physics*, vol. 85, no. 10, pp. 7214–7218, 1999.

[60] R. Caballero, C. A. Kaufmann, T. Eisenbarth et al., "The influence of Na on low temperature growth of CIGS thin film solar cells on polyimide substrates," *Thin Solid Films*, vol. 517, no. 7, pp. 2187–2190, 2009.

[61] M. A. Green, K. Emery, Y. Hishikawa et al., "Solar cell efficiency tables (version 49)," *Progress in Photovoltaics: Research and Applications*, vol. 25, no. 1, pp. 3–13, 2017.

[62] E. Wallin, U. Malm, T. Jarmar, O. L. M. Edoff, and L. Stolt, "World-record Cu(In,Ga)Se$_2$-based thin-film sub-module with 17.4% efficiency," *Progress in Photovoltaics: Research and Applications*, vol. 20, no. 7, pp. 851–854, 2012.

[63] B. P. Rand, J. Genoe, P. Heremans, and J. Poortmans, "New world record efficiency for Cu(In,Ga)Se$_2$ thin-film solar cells beyond 20%," *Progress in Photovoltaics: Research and Applications*, vol. 19, no. 7, pp. 894–897, 2011.

[64] J. Haarstrich, H. Metzner, M. Oertel et al., "Increased homogeneity and open-circuit voltage of Cu(In,Ga)Se$_2$ solar cells due to higher deposition temperature," *Solar Energy Materials & Solar Cells*, vol. 95, no. 3, pp. 1028–1030, 2011.

[65] P. M. P. Salomé, A. Hultqvist, V. Fjällström et al., "The effect of high growth temperature on Cu(In,Ga)Se$_2$ thin film solar cells," *Solar Energy Materials & Solar Cells*, vol. 123, pp. 166–170, 2014.

[66] S. Ishizuka, A. Yamada, K. Matsubara, P. Fons, K. Sakurai, and S. Niki, "Development of high-efficiency flexible Cu(In,Ga)Se$_2$ solar cells: a study of alkali doping effects on CIS, CIGS, and CGS using alkali-silicate glass thin layers," *Current Applied Physics*, vol. 10, no. 2, pp. S154–S156, 2010.

[67] S. Ishizuka, A. Yamada, and S. Niki, "Efficiency enhancement of flexible CIGS solar cells using alkali-silicate glass thin layers as an alkali source material," in *2009 34th IEEE Photovoltaic Specialists Conference (PVSC)*, pp. 2349–2353, Philadelphia, PA, USA, June 2009.

[68] P. Jackson, D. Hariskos, R. Wuerz et al., "Properties of Cu(In,Ga)Se$_2$ solar cells with new record efficiencies up to 21.7%," *Physica Status Solidi (RRL) - Rapid Research Letters*, vol. 9, no. 1, pp. 28–31, 2015.

[69] W. K. Batchelor, M. E. Beck, R. Huntington et al., "Substrate and back contact effects in CIGS devices on steel foil," in *Conference Record of the Twenty-Ninth IEEE Photovoltaic Specialists Conference, 2002*, pp. 716–719, 2002, New Orleans, LA, USA, May 2002.

[70] M. Hartmann, M. Schmidt, A. Jasenek et al., "Flexible and light weight substrates for Cu(In,Ga)Se$_2$ solar cells and modules," in *Photovoltaic Specialists Conference, 2000. Conference Record of the Twenty-Eighth IEEE*, pp. 638–641, Anchorage, AK, USA, September 2000.

[71] B. P. Rand, J. Genoe, P. Heremans, and J. Poortmans, "CIGS stainless steel foil as substrate," *Progress in Photovoltaics: Research and Applications*, vol. 17, pp. 659–676, 2013.

[72] R. Wuerz, A. Eicke, M. Frankenfeld et al., "CIGS thin-film solar cells on steel substrates," *Thin Solid Films*, vol. 517, no. 7, pp. 2415–2418, 2009.

[73] S. Gledhill, A. Zykov, N. Allsop et al., "Spray pyrolysis of barrier layers for flexible thin film solar cells on steel," *Solar Energy Materials & Solar Cells*, vol. 95, no. 2, pp. 504–509, 2011.

[74] T. Satoh, Y. Hashimoto, S. Shimakawa, S. Hayashi, and T. Negami, "Cu(In,Ga)Se solar cells on stainless steel substrates covered with insulating layers," *Solar Energy Materials & Solar Cells*, vol. 75, no. 1-2, pp. 65–71, 2003.

[75] N. Murakami, K. Moriwaki, M. Nangu et al., "Monolithically integrated CIGS sub-modules fabricated on new-structured flexible substrates," in *Photovoltaic Specialists Conference (PVSC), 2011 37th IEEE*, pp. 1310–1313, Seattle, WA, USA, June 2011.

[76] "CIGS thin film solar cells on flexible foils," in *24th European Photovoltaic Solar Energy Conference*, pp. 2425–2428, Hamburg, Germany, September 2009.

[77] R. Wuerz, A. Eicke, F. Kessler, S. Paetel, S. Efimenko, and C. Schlegel, "CIGS thin-film solar cells and modules on enamelled steel substrates," *Solar Energy Materials & Solar Cells*, vol. 100, pp. 132–137, 2012.

[78] M. Powalla, W. Witte, P. Jackson et al., "CIGS cells and modules with high efficiency on glass and flexible substrates," *IEEE Journal of Photovoltaics*, vol. 4, no. 1, pp. 440–446, 2014.

[79] T. Nakada, T. Kuraishi, T. Inoue, and T. Mise, "CIGS thin film solar cells on polyimide foils," in *Photovoltaic Specialists Conference (PVSC), 2010 35th IEEE*, pp. 330–334, Honolulu, HI, USA, June 2010.

[80] A. Chirilă, P. Bloesch, S. Seyrling et al., "Cu(In,Ga)Se$_2$ solar cell grown on flexible polymer substrate with efficiency exceeding 17%," *Progress in Photovoltaics: Research and Applications*, vol. 19, no. 5, pp. 560–564, 2011.

[81] D. Rudmann, D. Brémaud, H. Zogg, and A. N. Tiwari, "Na incorporation into Cu (In,Ga) Se2 for high-efficiency flexible solar cells on polymer foils," *Journal of Applied Physics*, vol. 97, no. 8, article 084903, 2005.

[82] S. Niki, M. Contreras, I. Repins et al., "CIGS absorbers and processes," *Progress in Photovoltaics: Research and Applications*, vol. 18, no. 6, pp. 453–466, 2010.

[83] "Empa -207- thin films and photovoltaics-research topics," September 2017, https://www.empa.ch/web/s207/tfsc.

[84] D. Fraga, T. Stoyanova Lyubenova, R. Martí, I. Calvet, E. Barrachina, and J. B. Carda, "Ecologic ceramic substrates for CIGS solar cells," *Ceramics International*, vol. 42, no. 6, pp. 7148–7154, 2016.

[85] M. P. Seabra, L. Grave, C. Oliveira, A. Alves, A. Correia, and J. A. Labrincha, "Porcelain stoneware tiles with antimicrobial action," *Ceramics International*, vol. 40, no. 4, pp. 6063–6070, 2014.

[86] M. Jubault, L. Ribeaucourt, E. Chassaing, G. Renou, D. Lincot, and F. Donsanti, "Optimization of molybdenum thin films for electrodeposited CIGS solar cells," *Solar Energy Materials & Solar Cells*, vol. 95, Supplement 1, pp. S26–S31, 2011.

[87] M. Kaelin, D. Rudmann, F. Kurdesau, H. Zogg, T. Meyer, and A. N. Tiwari, "Low-cost CIGS solar cells by paste coating and selenization," *Thin Solid Films*, vol. 480-481, pp. 486–490, 2005.

[88] S. A. Pethe, E. Takahashi, A. Kaul, and N. G. Dhere, "Effect of sputtering process parameters on film properties of molybdenum back contact," *Solar Energy Materials & Solar Cells*, vol. 100, pp. 1–5, 2012.

[89] T. J. Vink, M. A. J. Somers, J. L. C. Daams, and A. G. Dirks, "Stress, strain, and microstructure of sputter-deposited Mo thin films," *Journal of Applied Physics*, vol. 70, no. 8, pp. 4301–4308, 1991.

[90] K. Ellmer, R. Wendt, and R. Cebulla, "ZnO/ZnO:Al window and contact layer for thin film solar cells: high\nrate deposition by simultaneous RF and DC magnetron sputtering," in *Conference Record of the Twenty Fifth IEEE Photovoltaic Specialists Conference – 1996*, pp. 881–884, Washington, DC, USA, May 1996.

[91] A. C. Badgujar, S. R. Dhage, and S. V. Joshi, "Process parameter impact on properties of sputtered large-area Mo bilayers for CIGS thin film solar cell applications," *Thin Solid Films*, vol. 589, pp. 79–84, 2015.

[92] G. Gordillo, M. Grizález, and L. C. Hernandez, "Structural and electrical properties of DC sputtered molybdenum films," *Solar Energy Materials & Solar Cells*, vol. 51, no. 3-4, pp. 327–337, 1998.

[93] M. A. Martinez and C. Guillen, "Effect of r.f.-sputtered Mo substrate on the microstructure of electrodeposited CuInSe2 thin films," *Surface and Coatings Technology*, vol. 110, no. 1-2, pp. 62–67, 1998.

[94] J. H. Jiang and S. Y. Kuo, "Optimization of DC-sputtered molybdenum back contact layers," in *2014 International Symposium on Next-Generation Electronics (ISNE)*, pp. 5-6, Kwei-Shan, Taiwan, May 2014.

[95] E. Takahashi, S. A. Pethe, and N. G. Dhere, "Correlation between preparation parameters and properties of molybdenum back contact layer for CIGS thin film solar cell," in *2010 35th IEEE Photovoltaic Specialists Conference*, pp. 2478–2482, Honolulu, HI, USA, June 2010.

[96] H. Khatri and S. Marsillac, "The effect of deposition parameters on radiofrequency sputtered molybdenum thin films," *Journal of Physics: Condensed Matter*, vol. 20, no. 5, article 55206, 2008.

[97] K. Aryal, H. Khatri, R. W. Collins, and S. Marsillac, "In situ and ex situ studies of molybdenum thin films deposited by rf and dc magnetron sputtering as a back contact for CIGS solar cells," *International Journal of Photoenergy*, vol. 2012, Article ID 723714, 7 pages, 2012.

[98] J.-H. Yoon, S. Cho, W. M. Kim et al., "Optical analysis of the microstructure of a Mo back contact for Cu(In,Ga)Se$_2$ solar cells and its effects on Mo film properties and Na diffusivity," *Solar Energy Materials & Solar Cells*, vol. 95, no. 11, pp. 2959–2964, 2011.

[99] L. Assmann, J. C. Bernède, A. Drici, C. Amory, E. Halgand, and M. Morsli, "Study of the Mo thin films and Mo/CIGS interface properties," *Applied Surface Science*, vol. 246, no. 1–3, pp. 159–166, 2005.

[100] T. Wada, "Microstructural characterization of high-efficiency Cu (In, Ga)Se$_2$ solar cells," *Solar Energy Materials and Solar Cells*, vol. 49, no. 1-4, pp. 249–260, 1997.

[101] J. H. Scofield, A. Duda, D. Albin, B. L. Ballard, and P. K. Predecki, "Sputtered molybdenum bilayer back contact for copper indium diselenide-based polycrystalline thin-film solar cells," *Thin Solid Films*, vol. 260, no. 1, pp. 26–31, 1995.

[102] K. H. Yoon, S. K. Kim, R. B. V. Chalapathy et al., "Characterization of a molybdenum electrode deposited by sputtering and its effect on Cu(In,Ga)Se solar cells," *Journal of the Korean Physical Society*, vol. 45, no. 4, pp. 1114–1118, 2004.

[103] Y. C. Lin, Y. T. Hsieh, C. M. Lai, and H. R. Hsu, "Impact of Mo barrier layer on the formation of MoSe$_2$ in Cu(In,Ga)Se$_2$ solar cells," *Journal of Alloys and Compounds*, vol. 661, pp. 168–175, 2016.

[104] Q. Cao, O. Gunawan, M. Copel et al., "Defects in Cu(In,Ga)Se$_2$ chalcopyrite semiconductors: a comparative study of material properties, defect states, and photovoltaic performance," *Advanced Energy Materials*, vol. 1, no. 5, pp. 845–853, 2011.

[105] C. Amory, J. C. Bernède, E. Halgand, and S. Marsillac, "Cu(In,Ga)Se$_2$ films obtained from γ-In$_2$Se$_3$ thin film," *Thin Solid Films*, vol. 431-432, pp. 22–25, 2003.

[106] A. Urbaniak and K. Macielak, "Chemical and structural characterization of Cu(In ,Ga)Se2/Mo interface in Cu(In, Ga)Se2 solar cells," *Japanese Journal of Applied Physics*, vol. 35, 1996.

[107] R. Würz, D. Fuertes Marrón, A. Meeder et al., "Formation of an interfacial MoSe$_2$ layer in CVD grown CuGaSe$_2$ based thin film solar cells," *Thin Solid Films*, vol. 431-432, pp. 398–402, 2003.

[108] P. M. P. Salomé, V. Fjallstrom, A. Hultqvist, P. Szaniawski, U. Zimmermann, and M. Edoff, "The effect of Mo back contact ageing on Cu(In,Ga)Se$_2$ thin-film solar cells," *Progress in Photovoltaics: Research and Applications*, vol. 22, no. 1, pp. 83–89, 2014.

[109] N. Kohara, S. Nishiwaki, Y. Hashimoto, T. Negami, and T. Wada, "Electrical properties of the Cu(In,Ga)Se$_2$/MoSe$_2$/Mo structure," *Solar Energy Materials & Solar Cells*, vol. 67, no. 1–4, pp. 209–215, 2001.

[110] K.-J. Hsiao, J.-D. Liu, H.-H. Hsieh, and T.-S. Jiang, "Electrical impact of MoSe$_2$ on CIGS thin-film solar cells," *Physical Chemistry Chemical Physics*, vol. 15, no. 41, pp. 18174–18178, 2013.

[111] J. B. Pang, Y. A. Cai, Q. He et al., "Preparation and characteristics of MoSe$_2$ interlayer in bifacial Cu(In,Ga)Se$_2$ solar cells," *Physics Procedia*, vol. 32, pp. 372–378, 2012.

[112] F. S. Hasoon and H. A. Al-Thani, "The formation of the MoSe$_2$ layer at Mo/CIGS interface and its effect on the CIGS device performance," in *2016 IEEE 43rd Photovoltaic Specialists Conference (PVSC)*, pp. 1469–1473, Portland, OR, USA, June 2016.

[113] A. Polizzotti, I. L. Repins, R. Noufi, S.-H. Wei, and D. B. Mitzi, "The state and future prospects of kesterite photovoltaics," *Energy & Environmental Science*, vol. 6, no. 11, p. 3171, 2013.

[114] Y. C. Lin, D. H. Hong, Y. T. Hsieh, L. C. Wang, and H. R. Hsu, "Role of Mo:Na layer on the formation of MoSe$_2$ phase in Cu(In,Ga)Se$_2$ thin film solar cells," *Solar Energy Materials & Solar Cells*, vol. 155, pp. 226–233, 2016.

[115] B. Shin, Y. Zhu, N. A. Bojarczuk, S. Jay Chey, and S. Guha, "Control of an interfacial MoSe2layer in Cu2ZnSnSe4thin film solar cells: 8.9% power conversion efficiency with a TiN diffusion barrier," *Applied Physics Letters*, vol. 101, no. 5, article 53903, 2012.

[116] C. W. Jeon, T. Cheon, H. Kim, M. S. Kwon, and S. H. Kim, "Controlled formation of MoSe$_2$ by MoN$_x$ thin film as a diffusion barrier against Se during selenization annealing for CIGS solar cell," *Journal of Alloys and Compounds*, vol. 644, pp. 317–323, 2015.

[117] Y.-C. Lin, M.-T. Shen, Y.-L. Chen, H.-R. Hsu, and C.-H. Wu, "A study on MoSe$_2$ layer of Mo contact in Cu(In,Ga)Se$_2$ thin film solar cells," *Thin Solid Films*, vol. 570, pp. 166–171, 2014.

[118] J. H. Yoon, K. H. Yoon, J. K. Kim et al., "Effect of the Mo back contact microstructure on the preferred orientation of CIGS thin films," in *2010 35th IEEE Photovoltaic Specialists Conference*, pp. 2443–2447, Honolulu, HI, USA, January 2010.

[119] A. Duchatelet, G. Savidand, R. N. Vannier, and D. Lincot, "Optimization of MoSe$_2$ formation for Cu(In,Ga)Se$_2$-based solar cells by using thin superficial molybdenum oxide barrier layers," *Thin Solid Films*, vol. 545, pp. 94–99, 2013.

[120] T. T. Wu, F. Hu, J. H. Huang et al., "Improved efficiency of a large-area Cu(In,Ga)Se$_2$ solar cell by a nontoxic hydrogen-assisted solid Se vapor selenization process," *ACS Applied Materials & Interfaces*, vol. 6, no. 7, pp. 4842–4849, 2014.

[121] S. Ahn, K. Kim, and K. Yoon, "MoSe2 formation from selenization of Mo and nanoparticle derived Cu(In,Ga)Se2/Mo films," in *2006 IEEE 4th World Conference on Photovoltaic Energy Conference*, vol. 1, pp. 506–508, Waikoloa, HI, USA, May 2006.

[122] J. Han, J. Koo, H. Jung, and W. K. Kim, "Comparison of thin film properties and selenization behavior of CuGaIn precursors prepared by co-evaporation and co-sputtering," *Journal of Alloys and Compounds*, vol. 552, pp. 131–136, 2013.

[123] X. Zhu, Z. Zhou, Y. Wang, L. Zhang, A. Li, and F. Huang, "Determining factor of MoSe$_2$ formation in Cu(In,Ga)Se$_2$ solar cells," *Solar Energy Materials & Solar Cells*, vol. 101, pp. 57–61, 2012.

[124] D. Abou-Ras, G. Kostorz, D. Bremaud et al., "Formation and characterisation of MoSe$_2$ for Cu(In,Ga)Se$_2$ based solar cells," *Thin Solid Films*, vol. 480-481, pp. 433–438, 2005.

[125] S.-. Q. Wang, I. Raaijmakers, B. J. Burrow, S. Suthar, S. Redkar, and K.-. B. Kim, "Reactively sputtered TiN as a diffusion barrier between Cu and Si," *Journal of Applied Physics*, vol. 68, no. 10, pp. 5176–5187, 1990.

[126] T. Oku, E. Kawakami, M. Uekubo, K. Takahiro, S. Yamaguchi, and M. Murakami, "Diffusion barrier property of TaN between Si and Cu," *Applied Surface Science*, vol. 99, no. 4, pp. 265–272, 1996.

[127] B.-S. Suh, Y.-J. Lee, J.-S. Hwang, and C.-O. Park, "Properties of reactively sputtered WN$_x$ as Cu diffusion barrier," *Thin Solid Films*, vol. 348, no. 1-2, pp. 299–303, 1999.

[128] B. J. Mueller, A. Fotler, V. Haug, F. Hergert, S. Zweigart, and U. Herr, "Influence of Mo-N as diffusion barrier in Mo back contacts for Cu(In,Ga)Se2 solar cells," *Thin Solid Films*, vol. 612, pp. 186–193, 2016.

[129] S. Lopez-Marino, M. Espíndola-Rodríguez, Y. Sánchez et al., "The importance of back contact modification in Cu$_2$ZnSnSe$_4$ solar cells: the role of a thin MoO$_2$ layer," *Nano Energy*, vol. 26, pp. 708–721, 2016.

[130] R. V. Forest, E. Eser, B. E. McCandless, R. W. Birkmire, and J. G. Chen, "Understanding the role of oxygen in the segregation of sodium at the surface of molybdenum coated soda-lime glass," *AICHE Journal*, vol. 60, no. 6, pp. 2365–2372, 2014.

[131] E. Cadel, N. Barreau, J. Kessler, and P. Pareige, "Atom probe study of sodium distribution in polycrystalline Cu(In,Ga)Se$_2$ thin film," *Acta Materialia*, vol. 58, no. 7, pp. 2634–2637, 2010.

[132] J.-H. Yoon, T.-Y. Seong, and J. Jeong, "Effect of a Mo back contact on Na diffusion in CIGS thin film solar cells," *Progress in Photovoltaics: Research and Applications*, vol. 21, no. 1, pp. 58–63, 2013.

[133] M. B. Zellner, R. W. Birkmire, E. Eser, W. N. Shafarman, and J. G. Chen, "Determination of activation barriers for the diffusion of sodium through CIGS thin-film solar cells," *Progress in Photovoltaics: Research and Applications*, vol. 11, no. 8, pp. 543–548, 2003.

[134] M. Bodegård, K. Granath, L. Stolt, and A. Rockett, "The behaviour of Na implanted into Mo thin films during annealing," *Solar Energy Materials & Solar Cells*, vol. 58, no. 2, pp. 199–208, 1999.

[135] M. Theelen and F. Daume, "Stability of Cu(In,Ga)Se$_2$ solar cells: a literature review," *Solar Energy*, vol. 133, pp. 586–627, 2016.

[136] M. Powalla and B. Dimmler, "CIGS solar cells on the way to mass production: process statistics of a 30cm×30cm module line," *Solar Energy Materials & Solar Cells*, vol. 67, no. 1–4, pp. 337–344, 2001.

[137] "PV manufacturing capacity expansion announcements Top 55GW in 2015," July 2017, https://www.pv-tech.org/editors-blog/pv-manufacturing-capacity-expansion-announcements-top-55gw-in-2015.

[138] "Performance of CIGS in real world conditions," April 2018, http://www.solar-frontier.com/eng/technology/Performance/index.html.

[139] G. Schulz, *Corporate Profile*, vol. 9, no. 1, pp. 7–10, 2017.

[140] Solar Frontier, "Solar Frontier's CIS technology," *SportRadar*, pp. 1–18, 2014.

[141] F. Rated, P. Sticks, and P. Pads, *Product Data Sheet*, vol. 1, pp. 1-2, 2009.

[142] T. Kato, A. Handa, T. Yagioka et al., "Enhanced efficiency of Cd-free Cu(In,Ga)(Se,S)2 minimodule via (Zn,Mg)O second buffer layer and alkali metal post-treatment," *IEEE Journal of Photovoltaics*, vol. 7, no. 6, pp. 1773–1780, 2017.

[143] M. Powalla and B. Dimmler, "Process development of high performance CIGS modules for mass production," *Thin Solid Films*, vol. 387, no. 1-2, pp. 251–256, 2001.

[144] "All the information on cadmium - cadmium exposure and human health," July 2017, http://www.cadmium.org/environment/cadmium-exposure-and-human-health.

[145] T. Nakada, K. Furumi, and A. Kunioka, "High-efficiency cadmium-free Cu(In,Ga)Se/sub 2/ thin-film solar cells with chemically deposited ZnS buffer layers," *IEEE Transactions on Electron Devices*, vol. 46, no. 10, pp. 2093–2097, 1999.

[146] M. A. Green, "Estimates of te and in prices from direct mining of known ores," *Progress in Photovoltaics: Research and Applications*, vol. 17, no. 5, pp. 347–359, 2009.

[147] C. P. Cameron and A. C. Goodrich, "The levelized cost of energy for distributed PV: a parametric study," in *2010 35th IEEE Photovoltaic Specialists Conference*, pp. 529–534, Honolulu, HI, USA, June 2010.

[148] "Simple levelized cost of energy (LCOE) calculator documentation," April 2018, https://www.nrel.gov/analysis/tech-lcoe-documentation.html.

[149] J. Hernández-Moro and J. M. Martínez-Duart, "Analytical model for solar PV and CSP electricity costs: present LCOE values and their future evolution," *Renewable and Sustainable Energy Reviews*, vol. 20, pp. 119–132, 2013.

[150] International Renewable and Energy Agency, *Power Generation Costs*, International Renewable Energy Agency, 2017.

[151] Photovoltaic Sector, "World energy resources," *Solar Photovoltaics*, vol. 1, no. 4, 2012.

[152] K. Branker, M. J. M. Pathak, and J. M. Pearce, "A review of solar photovoltaic levelized cost of electricity," *Renewable and Sustainable Energy Reviews*, vol. 15, no. 9, pp. 4470–4482, 2011.

[153] S. D. Sadatian and H. Abolghasemi, *The Solar Energy Industry (PV) and It's Future*, vol. 52, pp. 195–206, 2016.

[154] W. A. Wohlmuth, *Thin Film CdTe Module Manufacturing*, First Solar, 2009.

[155] K. A. W. Horowitz and M. Woodhouse, *Cost and Potential of Monolithic CIGS Photovoltaic Modules*, vol. 2, National Renewable Energy Laboratory (NREL), 2015.

10

Computational Prediction of Electronic and Photovoltaic Properties of Anthracene-Based Organic Dyes for Dye-Sensitized Solar Cells

Hongbo Wang,[1] Qian Liu,[2] Dejiang Liu,[3] Runzhou Su ®,[1] Jinglin Liu ®,[4] and Yuanzuo Li ®[1]

[1]College of Science, Northeast Forestry University, Harbin, Heilongjiang 150040, China
[2]Department of Applied Physics, Xi'an University of Technology, Xi'an 710054, China
[3]Life Science College, Jiamusi University, Jiamusi, Heilongjiang 154007, China
[4]College of Science, Jiamusi University, Jiamusi, Heilongjiang 154007, China

Correspondence should be addressed to Runzhou Su; 13503631076@163.com, Jinglin Liu; jinglinliujms@yeah.net, and Yuanzuo Li; yzli@nefu.edu.cn

Academic Editor: K. R. Justin Thomas

Three kinds of anthracene-based organic dyes for dye-sensitized solar cells (DSSCs) were studied, and their structures are based on a push–pull framework with anthracenyl diphenylamine as the donor connected to a carboxyphenyl or carboxyphenyl-bromothiazole (BTZ) as the acceptor via an acetylene bridge. The photoelectric properties of the three dyes were investigated using density functional theory (DFT). The simulations indicate that the improvement of anthracene-based dyes (the addition of BTZ and the change of alkyl groups to alkoxy chains) can reduce the energy gap and produce a red shift. This structural modification also improves the light capturing and the electron injection capability, making it excellent in photoelectric conversion efficiency (PCE). In addition, twelve molecules have been designed to regulate photovoltaic performance.

1. Introduction

With the depletion of traditional fossil fuels and environmental pollution, green energy has aroused widespread concern in academia [1]. Therefore, nonpolluting solar energy has become the most promising alternative energy source [2]. Compared with traditional inorganic solar cells based on silicon crystal, dye-sensitized solar cells (DSSCs) have the advantages of easy synthesis, low cost, and high conversion efficiency [3, 4]. Since the first report in 1991, DSSCs have a high PCE [5]. In general, a typical DSSC device consists of a titania semiconductor film, a dye sensitizer, a redox electrolyte, a counterelectrode, and a transparent conductive substrate [6–8]. The dye is mainly divided into metal-containing ruthenium dyes [9], porphyrin dyes [10], and metal-free organic dyes [11]. As an important part of DSSCs, sensitizers play an important role in capturing sunlight and the electron transfer. Among them, ruthenium (II) polypyridyl complexes are considered to be efficient and stable

sensitizers with a power conversion efficiency (PCE) above 11% at AM1.5G [12, 13]. However, the scarcity, high cost, and toxicity of ruthenium metal limit the widespread use of such sensitizers in DSSCs. In addition, the zinc porphyrin dye is more than 12% efficient in Co^{II}/Co^{III} electrolytes under standard conditions, which is considered to be a very promising sensitizer [14, 15]. In recent years, perovskite solar cells have become another potential photovoltaic approach with efficiency of over 20% under AM1.5G light sources and dim light irradiation [16, 17]. However, solution of the instability of devices and pollution for environment caused from raw materials are still a challenge [18–20]. Metal-free organic dyes are characterized by low cost, ease of purification, and flexible molecular design [21], and metal-free sensitizers are designed with donor-π-acceptor (D-π-A), D-π-A-A, or D-A-π-A, which can lead to light-induced charge separation, improvement in stability, and optimization of the energy level of the dye from structure modifications [22–25]. To date, PCE of such kind of organic dyes has reportedly reached

14% [26]. Due to a large number of functional groups available for molecular design, there still is much work for DSSC performance improvement.

Molecular materials with the anthracene structure have good stability and special luminescent properties, showing a bright blue electroluminescence [27–29]. However, there are relatively few studies on the photoelectric properties of anthracene-based molecules. There are few metal-free sensitizers featuring a 9,10-disubstituted anthracene entity as a conjugated spacer between electron donor and acceptor moieties, and the best PCE is 7.03% [30–33]. Recently, a dye containing 2,6-conjugated anthracene showed a photoconversion efficiency of 9.11% [34] at 1 sun condition, which is the highest PCE reported by anthracene dyes. Mai and coworkers [35] have synthesized a simple D-π-A sensitizer (MS3), which is based on a 9,10-disubstituted anthracene entity with an optical efficiency of 5.84% at AM1.5G. Based on the MS3 sensitizer, TY3 (D-π-A) and TY6 (D-A-π-A) were also synthesized. TY6 has the best PCE (up to 8.80%) [36]. To study the relationship between structure and properties, we used the theory of density functional theory (DFT) and time-dependent density functional theory (TD-DFT) to calculate the three-molecular-geometry, electron injection, dye regeneration, and optical properties, and results confirmed that its excellent performance was due to its excellent J_{SC} and V_{OC} characteristics. In addition, a series of design molecules based on TY6 were investigated to improve optical response and electron injection.

2. Computational Details

The ground-state geometries of three molecules were optimized by DFT//B3LYP/6-31G(d) level [37–40]. In order to simulate the more realistic performance of dye-sensitized solar cells, the related calculations were performed in the solvent condition (THF) by using the C-PCM [41] method. Frequency calculations showed the minima on the potential energy surface for optimization. The bond lengths, dihedral angles, energy gaps, frontier molecular orbitals, electron injection, and recombination of the optimized molecules were calculated. The absorption spectra, transition energies, and oscillator strengths of molecules were obtained with TD-DFT [42] by using the CAM-B3LYP [43] functional at the same basis set as the ground state. Three excited states are calculated, including the first excited state (S1), the second excited state (S2), and the third excited state (S3). The natural bond orbital (NBO) analysis [44] for the charge difference between the ground state and the excited state was carried out at the B3LYP/6-31G(d) level using the NBO 3.1 program. By introduction of an electron-withdrawing group in the acceptor, it hoped that the molecular modification can attract electrons and promote an intramolecular charge transfer from donor to acceptor, further leading to better electron injection into the conduction band of TiO_2. Therefore, twelve new dye molecules were designed by introducing CN, F, and CF_3 into the acceptor of the TY6 molecule, and the correlation calculations were made using the same method as the original dye molecule. All calculations are made through the Gaussian 09 package [45].

3. Analysis

3.1. Geometric Structures. Figure 1 shows the optimized ground-state molecular structure. As shown in Figure 1(a), MS3 is an original molecule, and TY3 is obtained by converting the C-6 alkyl chain of MS3 to the carbon alloy group. Based on the TY3 molecule, benzotriazole (BTZ) was introduced between the acetylene bridge and benzoic acid. In order to reduce the aggregation of sensitizer and improve the performance, a long alkyl chain was introduced into the N position of BTZ to obtain TY6. Figure 1(b) shows the ground-state structures of the three optimized molecules in the THF solvent. In order to facilitate the calculation, the long carbon chains on the donor were pruned appropriately. Both MS3 and TY3 are typical D-π-A structures; the donor and the π-bridge are the amino donor and the acetylene bridge, and the acceptor is benzoic acid. TY6 is the D-A-π-A structure, in which an additional acceptor is added between the acetylene bridge and benzoic acid. Table 1 shows the bond length and dihedral angle in gas and solvent (THF), respectively. For example, in gas, the dihedral angles of MS3 $\angle$C1-C2-N3-C4 and $\angle$C2-N3-C4-C5 were 33.16° and 70.94°, and the average value of the two warped dihedral angles is 52.05°. In the same way, the calculated mean values of the donor dihedral angles for TY3 and TY6 are 52.60° and 52.40°. It shows that the donor has a distorted structure. Dyes TY3 and TY6 are more largely distorted than the original molecule MS3, which can reduce the aggregation of dye molecules. At the same time, the three molecules have similar results in solvent: TY3 (52.66°) and TY6 (52.33°) are greater than the donor dihedral angle of MS3 (51.88°). In gas, the bond lengths of the three molecules N3-C4 and C10-C11 are smaller than those of the single bond (i.e., C-C: 1.530 Å [46], N-C: 1.471 Å [47]). In general, the stability of a molecule can be judged by the length of the bond [48]. The shorter the bond length is, the more stable the molecule becomes. By comparison, the bond lengths of TY3 and TY6 are less than that of MS3, and the same results are also observed in solvent condition; therefore, TY3 and TY6 molecules are more stable.

3.2. Energy Levels. Table 2 shows the HOMO, HOMO-1, LUMO, LUMO + 1, and energy gaps (Δ_{H-L}) for MS3, TY3, and TY6. In the gas phase, the HOMO order of the molecule is as follows: TY6 (-4.59 eV) > TY3 (-4.67 eV) > MS3 (-4.91 eV). LUMO energy is arranged in the following order: TY3 (-2.26 eV) > TY6 (-2.30 eV) > MS3 (-2.33 eV). The energy gap is arranged in the following order: MS3 (2.58 eV) > TY3 (2.41 eV) > TY6 (2.29 eV). As shown in Figure 2, we can find that LUMO of the three molecules has little change, and thus, the gradual increase in the HOMO energy level leads to the decrease in energy gap. The reduction in the energy gap also favors the red-shifted absorption peak in the UV absorption spectrum (data in Table 3 and spectra nature in Figure 3). Therefore, the absorption peak of TY6 will have a significantly red shift.

In solvent condition, for MS3 the energy of HOMO (-5.01 eV) is reduced by 0.10 eV compared with that in the gas condition, and the energy of LUMO (-2.44 eV) is

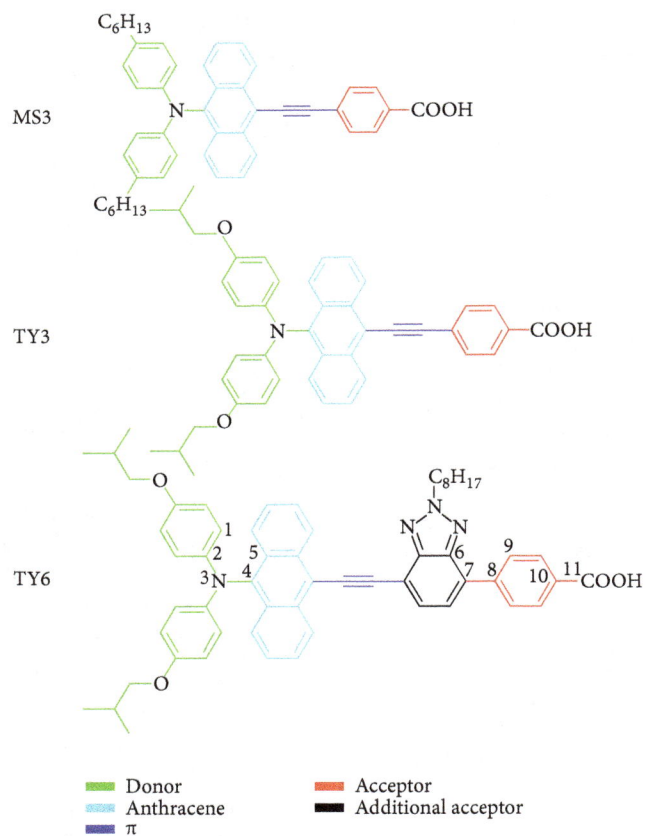

(a)

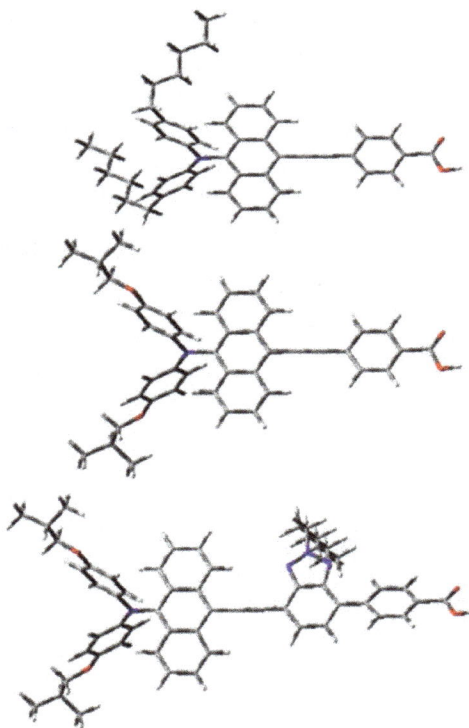

(b)

FIGURE 1: (a) Chemical structures of MS3, TY3, and TY6. (b) Side view for dyes optimized at the B3LYP/6-31G(d) level.

TABLE 1: Selected bond lengths (Å) and dihedral angles (°) of MS3, TY3, and TY6.

		MS3		TY3		TY6	
		Gas	Solvent	Gas	Solvent	Gas	Solvent
Dihedral angle	C1-C2-N3-C4	33.16	32.87	34.88	35.05	35.20	35.45
	C2-N3-C4-C5	70.94	70.89	70.32	70.27	69.60	69.20
Bond length	N3-C4	1.431	1.431	1.429	1.428	1.429	1.428
	C10-C11	1.484	1.484	1.483	1.484	1.483	1.484

TABLE 2: Frontier molecular orbital energies and energy gaps of MS3, TY3 and TY6.

	MS3		TY3		TY6	
	Gas (eV)	Solvent (eV)	Gas (eV)	Solvent (eV)	Gas (eV)	Solvent (eV)
H-1	−5.46	−5.56	−5.31	−5.46	−5.16	−5.37
H	−4.91	−5.01	−4.67	−4.81	−4.59	−4.77
L	−2.33	−2.44	−2.26	−2.42	−2.30	−2.49
L + 1	−1.41	−1.54	−1.36	−1.52	−1.63	−1.83
Gap	2.58	2.57	2.41	2.39	2.29	2.28

reduced by 0.11 eV compared with that in the gas condition. Therefore, the energy gap (2.58 eV) in solvent is lower than the energy gap in gas (2.57 eV). Similarly, dyes TY3 and TY6 also show the same change, and the results in solvent are better than those in the gas phase, and the energy gap is arranged in the following order: MS3 (2.57 eV) > TY3 (2.39 eV) > TY6 (2.28 eV). TY6 still has a smaller energy gap. The frontier molecular orbitals of three molecules are shown in Figure 4, which indicates the HOMO and LUMO of the three molecules and the distribution area of the electron density. As a whole, the electron density of HOMO is mainly located near the donor and the π-bridge, and the electron density of LUMO is mainly located near the π-bridge and the acceptor. HOMO-1 is lower than HOMO, and the electron density is dispersed in Figure 2. LUMO+1 is higher than LUMO, and its electron density is mainly concentrated on the π-bridge and the acceptor. Therefore, it can be concluded that the intramolecular charge transfer (ICT) exists in the dye molecule when the electron is transferred from the donor to the acceptor.

Figure 2 shows the orbital energy levels of three molecules. In the gas phase, the HOMO energy (−4.91 eV) of MS3 is lower than the energy of I^-/I_3^- (−4.85 eV), while the HOMO energy of TY3 (−4.67 eV) and TY6 (−4.59 eV) is higher than that of I^-/I_3^-. This shows that TY3 and TY6 can more easily recover electrons from electrolytes. The energies of LUMO of three dye molecules are lower than the conduction band energy of TiO_2 (−4.00 eV), which indicates that electrons can be successfully injected into TiO_2 from the excited state of dye molecules. One of the most important characteristics in the excellent ICT is charge separation. The charge distribution of HOMO and LUMO can promote the transfer of electrons (see Figure 4). In order to investigate the characteristics of ICT, we used the charge density difference (CDD) to show the change of the electron density

between the ground state and the excited state. It can characterize CT in organic molecular systems [49, 50], which clearly reflects the direction of the electrons. Taking the first excited state of MS3 as an example, the donor is covered with green holes, and the acceptor is covered with red electron (see Figure 3). Hole and electron are alternately distributed in the π-bridge moiety, and TY3 also shows similar results. For TY6, the S1 of CDD showed electron transfer is from donor to π-bridge, and for S2 week electron move to auxiliary acceptor BTZ and acceptor benzoic acid; wherefore, the S3 of CDD showed that the green holes make a shift from the donor to the π-bridge, and the red electrons transfer from the π-bridge to auxiliary acceptor BTZ and acceptor benzoic acid. The results show that there is a significant charge separation between the donor and the acceptor, which is considered as an ICT feature.

3.3. Absorption Spectrum. Table 3 lists the calculated absorption peaks, transition energies, and oscillator strengths (only discuss the state of $f > 0.1$). In the UV–Vis spectral region, the main absorption band was found to be the first excited state (S1 state). The first excited state (S1) of MS3 corresponds to the electron transition from HOMO to LUMO. It can be seen from Figure 4 that electrons transfer from the amino donor to the benzoic acid acceptor. In gas, the maximum absorption peak is 452 nm (463 nm in solvent), and the oscillator strength is 0.5880 (0.7021 in solvent). For the higher excited state (S2), the absorption intensity is lower than in S1, and the maximum absorption peak is 375 nm ($f = 0.1800$), which shows the electron transition process from HOMO-1 to LUMO. For TY3, the maximum absorption peak in gas S1 is 470 nm (481 nm in solvent), and the oscillator strength is 0.5112 (0.6182 in solvent), which shows the electron transition process from HOMO to LUMO. For the second excited state (S2), the maximum absorption peak is 384 nm ($f = 0.2814$), which shows the electron transition process from HOMO-1 to LUMO. For TY6, the maximum absorption peak of S1 in gas is 481 nm (494 nm in solvent). The oscillator strength is 0.8886 (0.9799 in solvent), which indicates the electron transition process from HOMO to LUMO. For the higher excited state (S2 and S3), the maximum absorption peak of S2 is 395 nm ($f = 0.3714$), showing an electron transition from HOMO-1 to LUMO. The maximum absorption peak of S3 is 344 nm ($f = 0.1763$), which shows the electron transition process from HOMO to LUMO+1. According to the data in Table 3, the maximum absorption peak value of the three molecules in gas is as follows: TY6 (481 nm) > TY3 (470 nm) > MS3 (452 nm). The UV–Vis spectra are given in Figure 5. Compared with

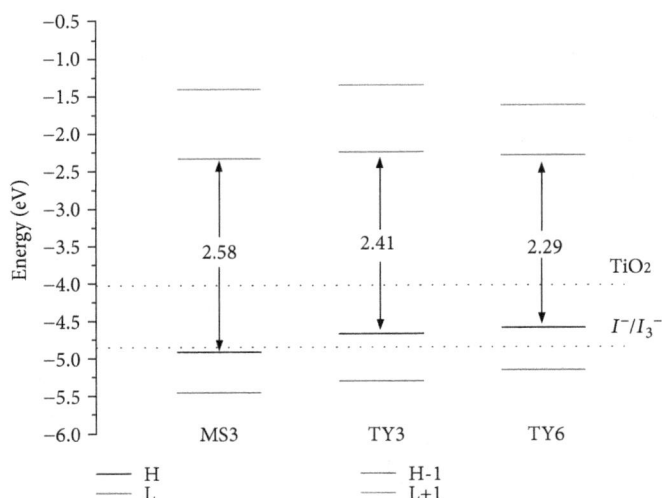

FIGURE 2: Frontier molecular orbital energies and energy gaps of MS3, TY3, and TY6.

TABLE 3: Calculated transition energies and oscillator strengths of MS3, TY3, and TY6.

Dye		State	Contribution Mo	E (eV)	Absorption peak λ (nm)	Strength f
	MS3	1	$0.68162/H \rightarrow L$	2.75	452	0.5880
		2	$0.67414/H\text{-}1 \rightarrow L$	3.31	375	0.1800
		3	$0.40992/H \rightarrow L+2$	3.88	320	0.0038
	TY3	1	$0.67550/H \rightarrow L$	2.64	470	0.5112
Gas		2	$0.67701/H\text{-}1 \rightarrow L$	3.23	384	0.2814
		3	$0.45755/H\text{-}3 \rightarrow L$	3.23	321	0.0019
	TY6	1	$0.63853/H \rightarrow L$	2.58	481	0.8886
		2	$0.62381/H\text{-}1 \rightarrow L$	3.14	395	0.3714
		3	$0.50506/H \rightarrow L+1$	3.61	344	0.1763
	MS3	1	$0.68199/H \rightarrow L$	2.68	463	0.7021
		2	$0.67305/H\text{-}1 \rightarrow L$	3.27	380	0.1855
		3	$0.40157/H \rightarrow L+2$	3.87	320	0.008
	TY3	1	$0.67467/H \rightarrow L$	2.58	481	0.6182
Solvent		2	$0.67454/H\text{-}1 \rightarrow L$	3.18	390	0.2912
		3	$0.46359/H\text{-}3 \rightarrow L$	3.86	321	0.0036
	TY6	1	$0.64236/H \rightarrow L$	2.51	494	0.9799
		2	$0.62799/H\text{-}1 \rightarrow L$	3.10	401	0.4138
		3	$0.46269/H \rightarrow L+1$	3.58	346.	0.2365

MS3 (452 nm) in gas, TY3 has a red shift (18 nm), and TY6 has a larger red shift (29 nm). This result indicates that the reduction in the energy gap favors the red-shifted absorption peak. The peak ranges are between 500 and 700 nm, which is helpful for effectively absorbing sunlight. It can be seen from Figure 5 that TY6 has the most pronounced red-shift absorption with the highest molar absorption coefficient. From the gas to solvent condition, absorption peaks of the red shift is approximately 11 nm, and the molar extinction coefficient is also increased (Figure 5). Performance in solvent is better than in gas. In summary, TY6 absorbs sunlight more efficiently than other molecules do, which may lead to higher PCE.

3.4. Chemical Reactivity Parameters. Ionization potential (IP) and electron affinity (EA) are important data for measuring the injection ability of holes and electrons [51–53]. The calculated IPs and EAs are listed in Table 4. In gas, the IPs of the three molecules are arranged in the order TY6 (5.49 eV) < TY3 (5.65 eV) < MS3 (5.89 eV). In addition, the reduction in IP means the increase in contribution ability. In gas, the EAs of the three molecules are arranged in the order TY6 (1.40 eV) > MS3 (1.34 eV) > TY3 (1.22 eV). The greater the value of EA becomes, the stronger the ability to receive the electronic have [54]. Therefore, TY6 has a higher ability to accept electrons. In solvent, the IP ordering of the three molecules is consistent with that in gas, and each

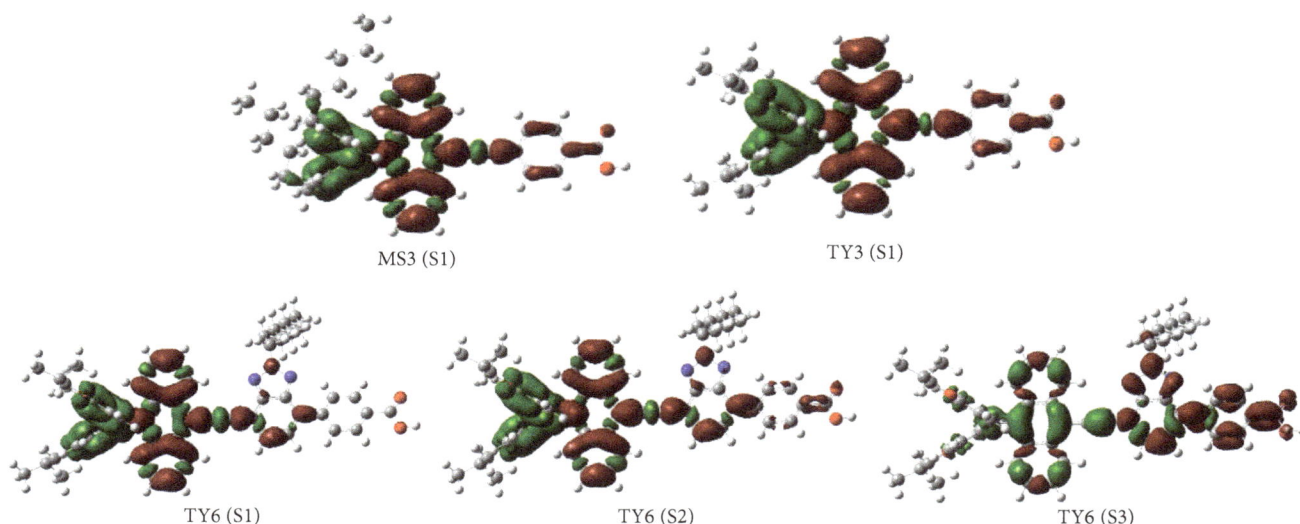

FIGURE 3: The charge difference density (CDD) for MS3, TY3, and TY6 in solvent (green and red stand for the hole and electron, respectively).

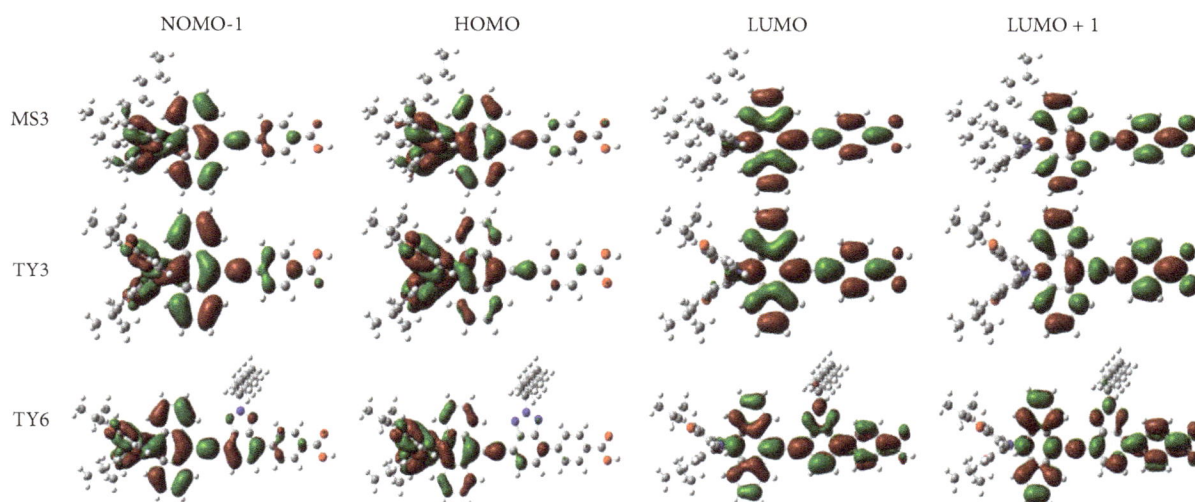

FIGURE 4: Frontier molecular orbitals of MS3, TY3, and TY6.

molecule decreases by about 0.9 eV. At the same time, EA increased by 1.2 eV compared with the gas phase. In general, the effect in solvent is better than that in gas. It is obvious that the electron transport performance of TY6 is better than that of TY3 and MS3.

Table 5 lists the chemical hardness (h), electrophilicity (ω), and electroaccepting power (ω^+). h represents the strength of resistance to ICT [55, 56]; a small h is helpful in reducing resistance to ICT. ω^+ represents the higher ability to accept electronics [57]. Many studies have shown that lower h and higher ω^+ can lead to higher short-circuit currents. In gas, h is arranged in the order MS3 (3.62 eV) > TY6 (3.44 eV) > TY3 (3.43 eV); ω^+ is arranged in the following order: TY6 (1.43 eV) > MS3 (1.35 eV) > TY3 (1.22 eV). There was no significant difference between TY3 (3.43 eV) and TY6 (3.44 eV) in h. However, TY6 (1.43 eV) was significantly higher than MS3 (1.35 eV) in terms of ω^+. In solvent, the orders of h and ω^+ are the same as those in the gas phase. However, ω^+ has a significant improvement, and MS3 increases by 2.75 eV, TY3 increases by 2.69 eV, and TY6 increases by 3.30 eV. The promotion of TY6 is the most obvious, and it represents a higher receiving capacity. ω represents the stability of the dye molecule system [58]. In gas, ω is arranged in the following order: TY6 (2.89 eV) > MS3 (2.87 eV) > TY3 (2.66 eV). In solvent conditions, MS3 increased by 2.94 eV, TY3 increased by 2.90 eV, and TY6 increased by 3.53 eV. It is clear that TY6 has not only the highest ω but also the highest increase in solvent. Therefore, TY6 has the highest energetic stability.

3.5. Performance of DSSCs Based on Dyes.
Normally, the photoelectric energy conversion efficiency of DSSCs is mainly affected by open-circuit photovoltage (V_{OC}), short-circuit current density (J_{SC}), fill factor (FF), and total

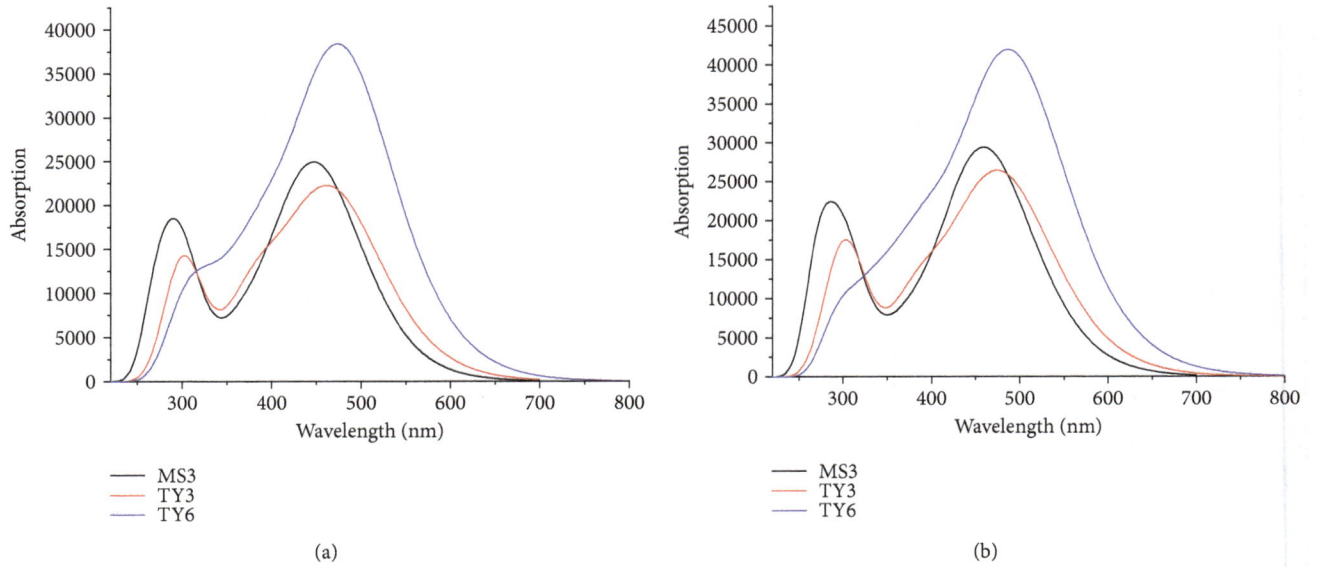

FIGURE 5: (a) The UV–Vis absorption spectra in gas and (b) the UV–Vis absorption spectra in solvent.

TABLE 4: Ionization potentials (IP) and electron affinities (EA) of three original molecules in gas and solvent (in eV).

		MS3	TY3	TY6
Gas	IP	5.89	5.65	5.49
	EA	1.34	1.22	1.40
Solvent	IP	4.94	4.74	4.70
	EA	2.54	2.43	2.62

TABLE 5: Chemical reactivity parameters (in eV) of MS3, TY3, and TY6 in gas and solvent, respectively.

	MS3		TY3		TY6	
	Gas	Solvent	Gas	Solvent	Gas	Solvent
h	3.62	3.74	3.43	3.58	3.44	3.66
ω^+	1.35	4.10	1.22	3.91	1.43	4.73
ω	2.87	5.81	2.66	5.56	2.89	6.42

incident solar energy (P_{in}). Calculated efficiency can be written as follows [59]:

$$\eta = \text{FF} \frac{V_{\text{OC}} J_{\text{SC}}}{P_{\text{in}}} \times 100\%. \tag{1}$$

From the formula, it can be seen that high V_{OC} and J_{SC} are the basis for producing photoelectric conversion efficiency. The J_{SC} in DSSCs can be calculated by the following equation:

$$J_{\text{SC}} = \int_{\lambda} \text{LHE}(\lambda) \Phi_{\text{inject}} \eta_{\text{collect}} (d\lambda), \tag{2}$$

where LHE (λ) is a light harvesting efficiency at maximum wavelength, Φ_{inject} is defined as the dye molecule exciting

TABLE 6: The electron injection and regeneration free energy, the light harvesting efficiency (LHE) and lifetime (t).

		ΔG^{inject}	$\Delta G_{\text{dye}}^{\text{regen}}$	$E_{\text{OX}}^{\text{dye}}$	$E_{\text{OX}}^{\text{dye*}}$	LHE	t (ns)
Gas	MS3	−1.84	−0.06	4.91	2.16	0.742	5.20
	TY3	−1.97	0.18	4.67	2.03	0.692	6.48
	TY6	−1.99	0.26	4.59	2.01	0.871	3.91
Solvent	MS3	−1.67	−0.16	5.01	2.33	0.801	4.57
	TY3	−1.77	0.04	4.80	2.23	0.759	5.60
	TY6	−1.74	0.08	4.77	2.26	0.895	3.73

electron injection efficiency, and η_{collect} is the charge collection efficiency. LHE can be expressed as [60]

$$\text{LHE} = 1 - 10^{-f}, \tag{3}$$

where f is the oscillator strength of the dye molecules; a large oscillator strength can contribute to the improvement of LHE. The electron injection-free energy (ΔG^{inject}) can be represented as [61]

$$\Delta G_{\text{inject}} = E_{\text{OX}}^{\text{dye*}} - E_{\text{CB}}. \tag{4}$$

E_{CB} is the reduction potential of the TiO$_2$ semiconductor and is equal to 4.0 eV [62] (versus vacuum) in this work. $E_{\text{OX}}^{\text{dye*}}$ is the oxidation reduction potential of the dye in the excited state. $E_{\text{OX}}^{\text{dye*}}$ can be expressed as [63]

$$E_{\text{OX}}^{\text{dye*}} = E_{\text{OX}}^{\text{dye}} - E_{00}, \tag{5}$$

where $E_{\text{OX}}^{\text{dye}}$ is the oxidation potential energy of the dye in the ground state, while E_{00} is the electronic vertical transition energy corresponding to λ_{max}.

The calculated values of LHE and ΔG^{inject} are listed in Table 6. ΔG^{inject} is an important factor affecting the electron

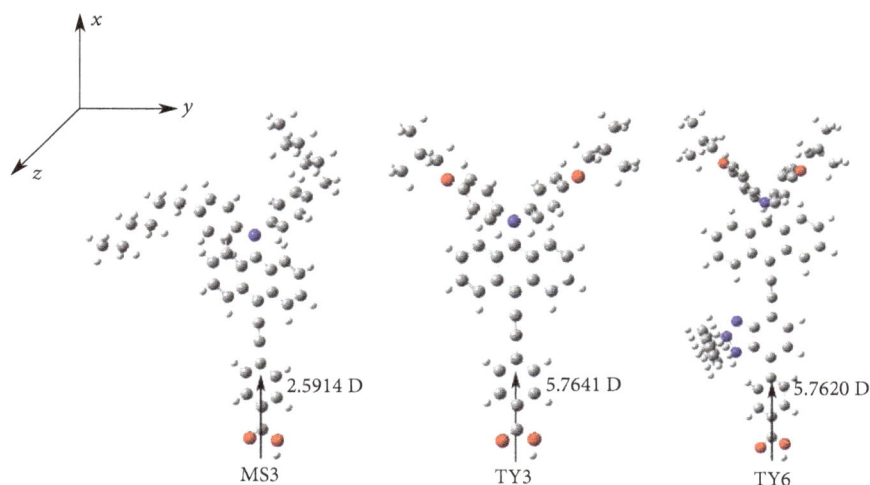

FIGURE 6: Calculated vertical dipole moment μ_{normal} (10^{-30} C·m) of MS3, TY3, and TY6 in solvent.

injection rate. In gas conditions, the ΔG^{inject} values of MS3, TY3, and TY6 are -1.83, -1.97, and -1.99 eV, respectively. Because ΔG^{inject} is negative, upon excitation of the molecule by light, electrons can be injected into TiO$_2$ more quickly. Higher LUMO can lead to a higher ΔG^{inject} (absolute value). As shown in Table 2, LUMO of TY3 (-2.26 eV) and TY6 (-2.30 eV) is higher than that of MS3 (-2.33 eV). The corresponding ΔG^{inject} is higher than that of MS3. Therefore, the improvement in DSSC's PCE is due to the high ΔG^{inject}, further resulting in high electron injection efficiency. The higher ΔG^{inject} is good for increasing the J_{SC} of the DSSC. In solvent conditions, ΔG^{inject} has a small change. However, TY6 still has the highest ΔG^{inject} (absolute value). To sum up, TY6 will have higher J_{SC}. Three molecules of light harvesting efficiency can be obtained from Table 6. In gas, the data is sorted in the order TY6 (0.871) > MS3 (0.742) > TY3 (0.692). The introduction of the side chains on BTZ will increase LHE, and a larger oscillator strength will lead to a higher LHE. From Table 3, it can be found that the oscillator strength of the three molecules in solvent was higher than that in gas. So the LHE in solvent is greatly improved, and TY6 (0.895) still has the highest LHE. Therefore, under the excitation of light, TY6 has higher solar light utilization, producing more photocurrent, and this result indicates that TY6 will have higher J_{SC}. Dye regeneration free energy ($\Delta G_{\text{dye}}^{\text{regen}}$) is a significant factor affecting photoelectric conversion efficiency which can be written as follows [64]:

$$\Delta G_{\text{dye}}^{\text{regen}} = E_{\text{redox}}^{\text{electrolyte}} - E_{\text{OX}}^{\text{dye}}. \tag{6}$$

where $E_{\text{redox}}^{\text{electrolyte}}$ is the redox potential of I^-/I_3^- (-4.85 eV) [65]. Table 6 shows $\Delta G_{\text{dye}}^{\text{regen}}$ of three molecules. In gas conditions, the data is sorted in the order TY6 (0.26 eV) > TY3 (0.18 eV) > MS3 (-0.06 eV). In solvent conditions, the decrease in HOMO results in the increase in $\Delta G_{\text{dye}}^{\text{regen}}$ (see Table 2). So, it has a great influence on TY6, which has the largest $\Delta G_{\text{dye}}^{\text{regen}}$ (0.08 eV). The larger $\Delta G_{\text{dye}}^{\text{regen}}$ can promote

dye regeneration and increase J_{SC}. This means that TY6 will have a better performance.

Another important factor affecting the efficiency of electron transfer is the lifetime (t) of the first excited state. If the molecule has a longer life span, it will contribute to the charge transfer of the molecule [66]. The lifetime (t) can be obtained by the latter formula: $t = 1.499/(fE^2)$, where f is the oscillator strength and E (cm^{-1}) is the excitation energy of the different electronic states [67]. Table 6 shows the corresponding data, and the result in the gas phase is in the order TY3 (6.48 ns) > MS3 (5.20 ns) > TY6 (3.91 ns). In solvent, the order does not change, but the gap of them decreases. TY6 has a lower t, which is owing to the fact that TY6 has a very high f (Table 3). As known, there are many complicated factors that affect J_{SC}. According to (5), although TY6 has a lower t, it has the highest LHE and ΔG^{inject}, and it does not affect the highest J_{SC}. As for V_{OC} in DSSCs, it is described by [68]

$$V_{\text{OC}} = \frac{E_{\text{CB}}}{q} + \frac{\text{KT}}{q} \ln \frac{n_c}{N_{\text{CB}}} - \frac{E_{\text{redox}}}{q}. \tag{7}$$

Here, E_{CB} is the conduction band edge of TiO$_2$, q is the unit charge, KT is the thermal energy, n_c is the number of electrons in the CB, N_{CB} is the accessible density of CB states, and E_{redox} is the redox potential of the electrolyte. Equation (7) shows that the energy E_{CB} and the number of electrons in the CB are important factors affecting V_{OC}. ΔE_{CB} is the displacement of CB when the dye absorbs on the surface of TiO$_2$, which can be expressed as [69]

$$\Delta E_{\text{CB}} = \frac{q\mu_{\text{normal}}\gamma}{\varepsilon\varepsilon_0}. \tag{8}$$

γ is the concentration of dyes in the surface, μ_{normal} is the dipole moment component perpendicular to the direction of the TiO$_2$ surface (where μ_{normal} is the x-axis direction), ε is the dielectric constant of the organic monolayer, and ε_0 is the dielectric constant of the vacuum. Figure 6 shows the values of the vertical dipole moment (μ_{normal}). TY3 (5.7641 D)

TABLE 7: Natural bond orbital analysis for the ground state (S0) and excited state (S1) of the dyes.

Dye		Donor	Anthracene	π	Acceptor
Ms3	S0	−0.1002	0.1376	0.0099	−0.0473
	S1	0.1806	−0.0839	−0.0005	−0.0963
	Δ_q	−0.2808	0.2215	0.0104	0.0490
TY3	S0	0.3139	0.1368	0.0099	−0.0433
	S1	0.6290	−0.1882	−0.0091	−0.1050
	Δ_q	−0.3151	0.3250	0.0190	0.0617
TY6	S0	−0.0947	0.1486	0.0258	−0.0796
	S1	0.1802	−0.0528	0.0155	−0.1429
	Δ_q	−0.2749	0.2014	0.0103	0.0633

and TY6 (5.7620 D) has a similar value. Compared with MS3 (2.5914 D), TY3 and TY6 have more μ_{normal}. We optimized the donor moiety through changing the C-6 alkyl chain of MS3 to the branched carbon alkoxy group, which proves that the change in the donor portion can improve the vertical dipole moment of dye molecules effectively. Meanwhile, it cause the nanocrystalline semiconductor conduction band E_{CB} mobile to the positive direction of x-axis, and then improve V_{OC}.

3.6. Natural Bond Orbital Analysis. In order to understand the mechanism of photoexcitation, we simulated the natural bond orbital (NBO) of the optimized structure of the ground state (S0) and the first excited state (S1) of MS3, TY3, and TY6. The acceptor of TY6 includes benzoic acid and the auxiliary acceptor (BTZ group), and the detailed data is listed in Table 7. The amount of charge difference (Δ_q) from S0 to S1 in the donor group shows that TY3 (−0.3151) can contribute more electrons compared to MS3 (−0.2808) and TY3 (−0.2749). But Δ_q in the anthracene group and acceptor groups shows that more of the contributing electrons in TY3 are concentrated in the anthracene group. So, only a small amount of electrons reach the acceptor part in TY3. Eventually, the acceptor group of TY6 has more electrons. It showed that the auxiliary acceptor (BTZ) in TY6 increased the total amount of electron in the acceptor. At the same time, Δ_q of MS3 (0.0104), TY3 (0.0190), and TY6 (0.0103) in the π region indicates that the π-linker is only a channel for the charge transfer. To sum up, TY6 can provide more efficient excitation electrons in the photoexcitation mechanism.

3.7. Hyperpolarizabilities and Reorganization Energies. The total static first hyperpolarisability can be expressed as [70]

$$\beta_{tot} = \sqrt{\beta_x^2 + \beta_y^2 + \beta_z^2}. \quad (9)$$

The static component is calculated by the following equation:

$$\beta_i = \beta_{iii} + \frac{1}{3}\sum_{i \neq j}\left(\beta_{iji} + \beta_{ijj} + \beta_{jji}\right), \quad (10)$$

where $\beta_{ijk}(i, j, k = x, y, z)$ are tenser components of hyperpolarizability. Finally, the equation is written as

$$\beta_{tot} = \left[\left(\beta_{xxx} + \beta_{xyy} + \beta_{xzz}\right)^2 + \left(\beta_{yyy} + \beta_{yzz} + \beta_{yxx}\right)^2 \right.$$
$$\left. + \left(\beta_{zzz} + \beta_{zxx} + \beta_{zyy}\right)^2\right]^{1/2}. \quad (11)$$

For DSSCs, the ICT process facilitates the aggregation of electrons in the acceptor moiety, and the enhanced electron density in the acceptor moiety can enhance the electronic coupling effect between acceptor and semiconductor. The first β is directly proportional to the transition dipole moment ((oscillator strength) (μ_{eg})) and the difference in the dipole moment between the ground and excited orbitals ($\Delta\mu_{eg}$), and it is inversely proportional to the transition energy (E_{eg}). β can be expressed as follows [71]:

$$\beta \propto \frac{\Delta\mu_{eg}\left(\mu_{eg}\right)^2}{E_{eg}^2}, \quad (12)$$

where $\Delta\mu_{eg}$ and μ_{eg} are difference in the dipole moment for ground state and excited state, and the transition dipole, E_{eg}, is the transition energy. The first hyperpolarizabilities are listed in Table 8. β_{xxx} (along the coordinate axis of the molecule) is a negative value, and the negative charge is far away from the nuclear charge of the molecule. β_{tot} of TY6 is much larger than MS3 and TY3 values. Table 3 and Figure 6 support the results of hyperpolarizabilities. Figure 6 shows that TY3 and TY6 have higher dipole moments, and TY6 has higher oscillator strengths (in Table 3); so, TY6 has a higher μ_{eg}. At the same time, TY6 has less excitation energy in Table 2. So, TY6 has larger hyperpolarizabilities with an obvious charge transfer.

The reorganization energy can affect CT, which is more beneficial to the improvement of CT [72]. On the basis of the Marcus theory, CT can be calculated by [73]

$$K_{ET} = A \exp\left[\frac{-\lambda}{4K_B T}\right], \quad (13)$$

where λ is the reorganization energy, A is the electronic coupling, K_B is the Boltzmann constant, and T is the temperature. Hole or electron reorganization energy is determined by the following equation [74]:

$$\lambda_h = (E_0^- - E_-) + (E_-^0 - E_0),$$
$$\lambda_e = (E_0^+ - E_+) + (E_+^0 - E_0), \quad (14)$$

where $E_-^0(E_0^+)$ represents the energy of the neutral molecule calculated at the anionic (cationic) state. $E_0^-(E_0^+)$ represents the anion (cation) energy calculated with the optimized structure of the neutral molecule. $E_-(E_+)$ represents the anion (cation) energy calculated with the optimized anion (cation) structure. E_0 represents the energy at the neutral molecule at the ground state. λ_h, λ_e, and λ_{total} values are listed in Table 9. In the gas phase, the λ_h/λ_e values of MS3, TY3,

TABLE 8: Hyperpolarizabilities of MS3, TY3, and TY6 in the gas phase.

Gas	β_{xxx}	β_{xxy}	β_{xyy}	β_{yyy}	β_{xxz}	β_{xyz}	β_{yyz}	β_{xzz}	β_{yzz}	β_{zzz}	β_{tot}	β_{xyy}
MS3	221,672	−0.5	−11,899	881	−39,351	35	1874	5715	−14	−934	218,887	−11,899
TY3	−236,112	8336	7172	2538	8946	−50	72	933	18	661	228,473	7172
TY6	754,632	7313	−8155	3421	32,781	322	−420	−97	24	−1037	747,115	−8155

TABLE 9: Reorganization energy of MS3, TY3, and TY6 in gas and solvent phases.

		MS3	TY3	TY6
	λ_h	0.18	0.21	0.20
Gas	λ_e	0.27	0.33	0.33
	λ_{total}	0.45	0.54	0.53
	λ_h	0.17	0.19	0.17
Solvent	λ_e	0.23	0.25	0.25
	λ_{total}	0.40	0.44	0.42

and TY6 are 0.18/0.27, 0.21/0.33, and 0.20/0.33. The λ_{total} values of the three molecules are in the order TY3 (0.54 eV) > TY6 (0.53 eV) > MS3 (0.45 eV). The results show that the λ_h values of three molecules are lower than those of λ_e, which indicates that the electron transfer rate is lower than the hole transfer rate. Obviously, MS3 has a higher hole and electron transfer rate, followed by TY6. From the gas phase to solvent, λ_h and λ_e have decreased. For λ_{total}, MS3, TY3, and TY6 are reduced to be 0.05 eV, 0.10 eV, and 0.11 eV (see Table 9). The above data shows that values of λ_h and λ_e have been reduced in the THF solvent. TY6 has the greatest reduction in solvent and is very close to MS3 (0.40 eV). As a result, MS3 and TY6 will lead to better hole/electron transport and efficient luminescent materials.

3.8. Analysis of Electrostatic Potential Distribution on the Molecular Surface. In order to determine the position of electrolyte ions, we can determine the reactive sites of dye molecules by the molecular surface electrostatic potential (ESP) [75]. The ESP of three molecules is listed in Figure 7(a) (the detailed values are marked). The red point represents the maximum point of the electrostatic potential of the molecular surface, and the blue point represents the minimum point of the electrostatic potential of the molecular surface. For the three molecules, the maximum values of the electrostatic potential on the molecular surface are distributed near the H atom of the acceptor, and the specific values are as follows: MS3 (53.56 kcal/mol), TY3 (52.61 kcal/mol), and TY6 (51.55 kcal/mol). It shows that the acceptor H atom is positively charged, which means that it has the most powerful ability to attract nucleophiles and is the most likely place to gather negative charges together. At the same time, the minimum values of electrostatic potential on the surface of molecules are distributed near the acceptor O atom, and the specific data are as follows: MS3 (−37.62 kcal/mol), TY3 (−38.43 kcal/mol), and TY6 (−39.81 kcal/mol). This indicates that the lone pair electrons of the acceptor O atom

negatively contribute to the electrostatic potential, and its position implies the ability to attract the electric reagents strongly, which is most likely to gather the position of positive charge together. By electrostatic interactions, it is possible to infer that the positions of O and H atoms of the three molecular acceptors are the most active regions of the molecular reactions.

In order to show the molecular surface area of different electrostatic potential intervals, the quantitative distribution chart of the electrostatic potential on the surface of molecules is plotted, which is listed in Figure 7(b). As shown, the distribution of the surface electrostatic potential at the maximum and minimum points of the three molecules is very small in this region. The electrostatic potential area of the larger region is as follows: MS3 (−17.83, 13.83 kcal/mol), TY3 (−18.07, 13.27 kcal/mol), and TY6 (−24.67, 18.27 kcal/mol). It seems that the electrostatic potential distribution of TY6 is wider and more homogeneous than that of MS3 and TY3. Therefore, the surface reaction region of TY6 is the largest, and the overall reaction activity is stronger.

3.9. Molecular Design. After the above discussion, the TY6 molecule has the best photoelectric conversion efficiency. On the basis of this molecule, a series of molecules were designed by inserting different electron-withdrawing groups (EWGs ($-CF_3$ (A), $-CN$ (B), and $-F$ (C))) into number 1, 2, 3, and 4 positions of the original molecule acceptor, respectively. The designed molecular structures are shown in Figure 8, and those dyes are named as TY6-X (X = 1A, 2A, 3A, 4A, 1B, …,4C). The following calculation results are obtained in the gas phase. The calculated bond lengths and dihedral angle are listed in Table 10. The dihedral angle (∠C1-C2-N3-C4 and ∠C2-N3-C4-C5) on the donor is not significantly changed compared with the original molecule, and bond length has almost no change. However, the dihedral angle (∠C6-C7-C8-C9) of the acceptor is changed obviously, and the change of the molecule with the insertion of the −CN group has the largest change. It is worth noting that the dihedral angles of TY6-1X and TY6-4X (∠C6-C7-C8-C9) have a significant increase, while TY6-2X and TY6-3X have no significant changes. It is due to the fact that the close distance between the introduction of EWGs and the long alkyl chain introduced on BTZ will result in mutual exclusion, and TY6-1X (X = A, B, C, D) is larger than TY6-4X (X = A, B, C, D). For example, the dihedral angles of TY6-1A and TY6-4A are 65.82° and 49.99°, respectively. The twisted structure of the donor blocks the movement of electrons on the donor; therefore, TY6-2X and TY6-3X may exhibit a better performance.

The MOs of the designed molecules are shown in Table 11. Compared with the original molecule TY6

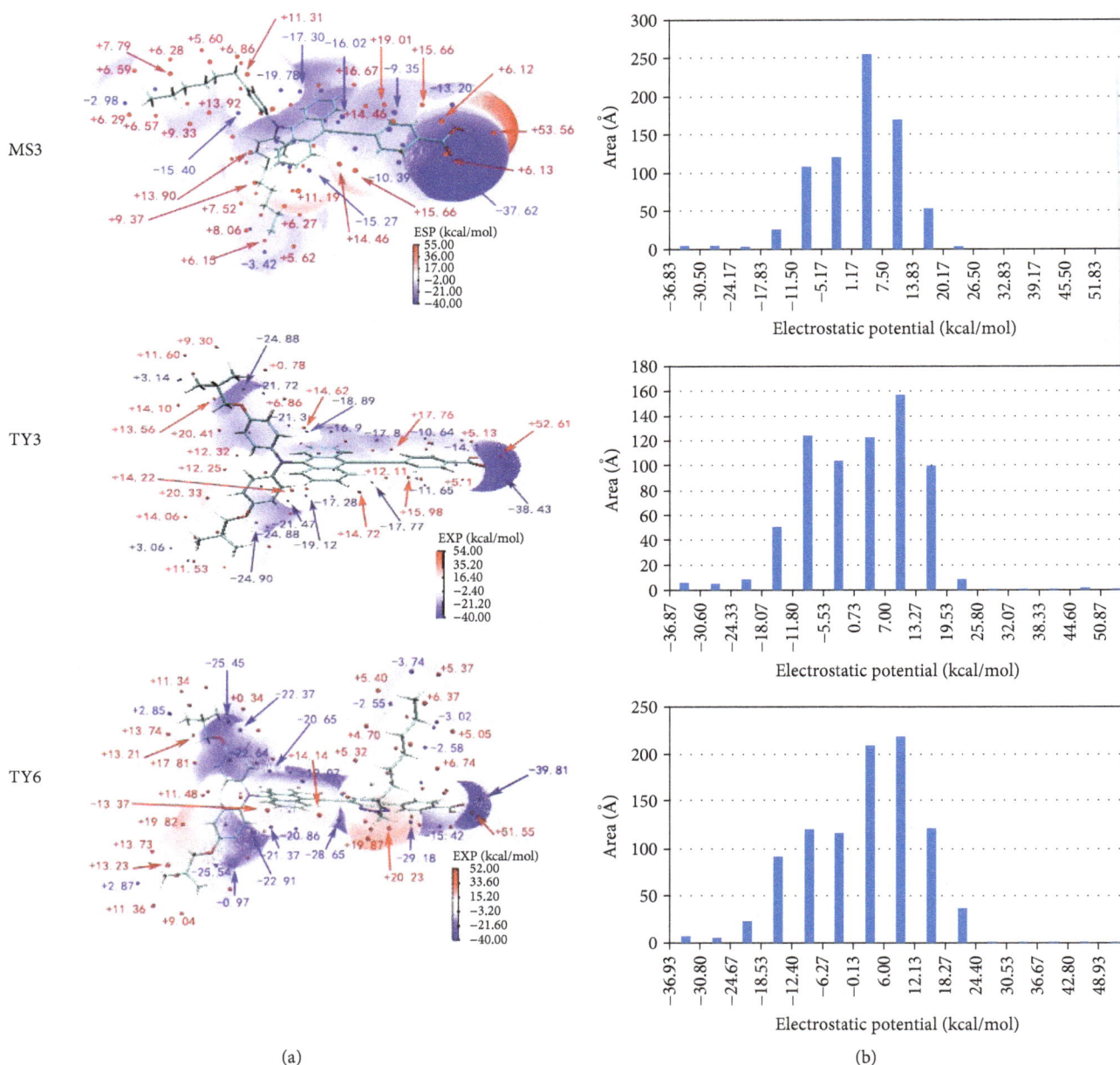

(a)   (b)

FIGURE 7: (a) The ESP on the VDW surface of MS3, TY3, and TY6; red and blue points represent maximum and minimum values, respectively. The extrema ESP (in kcal/mol) points on the molecular surface are marked. (b) The molecular surface area of different electrostatic potential interval; the electrostatic potential interval is divided into 15 equal parts.

(−4.59 eV), the HOMO values (−4.76 to −4.58 eV) of other design molecules are lower. For instance, the maximum value of HOMO is −4.58 eV, which adds an −F/−CF₃ group on position 1 of the acceptor (TY6-1C/TY6-1A). The minimum value of HOMO is −4.76 eV, which adds a −CF₃ group on position 3 of the acceptor (TY6-3A). However, for LUMO, most of the designed molecules have a better performance. Compared with the original molecule TY6 (−2.30 eV), the LUMO values (−2.54 to −2.23 eV) of the designed molecules are relatively low. The maximum value of LUMO is −2.23 eV, which adds a −CF₃ group on position 1 of the acceptor (TY6-1A). The minimum value of LUMO is −2.54 eV,

which adds a −CN group on position 2/3 of the acceptor (TY6-2B/TY6-3B). Compared with the energy gap of the original molecule TY6 (2.29 eV), the design molecules inserted into the −CN group are all reduced. The minimum energy gap is 2.13 eV, which adds a −CN group on position 2/3 of the acceptor (TY6-2B/TY6-3B). Interestingly, when EWG is inserted into the right side of the acceptor (positions 2 and 3), the energy gap is lower than in other positions. The −CN group can reduce the molecular LUMO and lead to the decrease in the energy gap. Through the analysis, it can be concluded that the −CN group is the most conducive to the reduction of the molecular energy gap, followed by

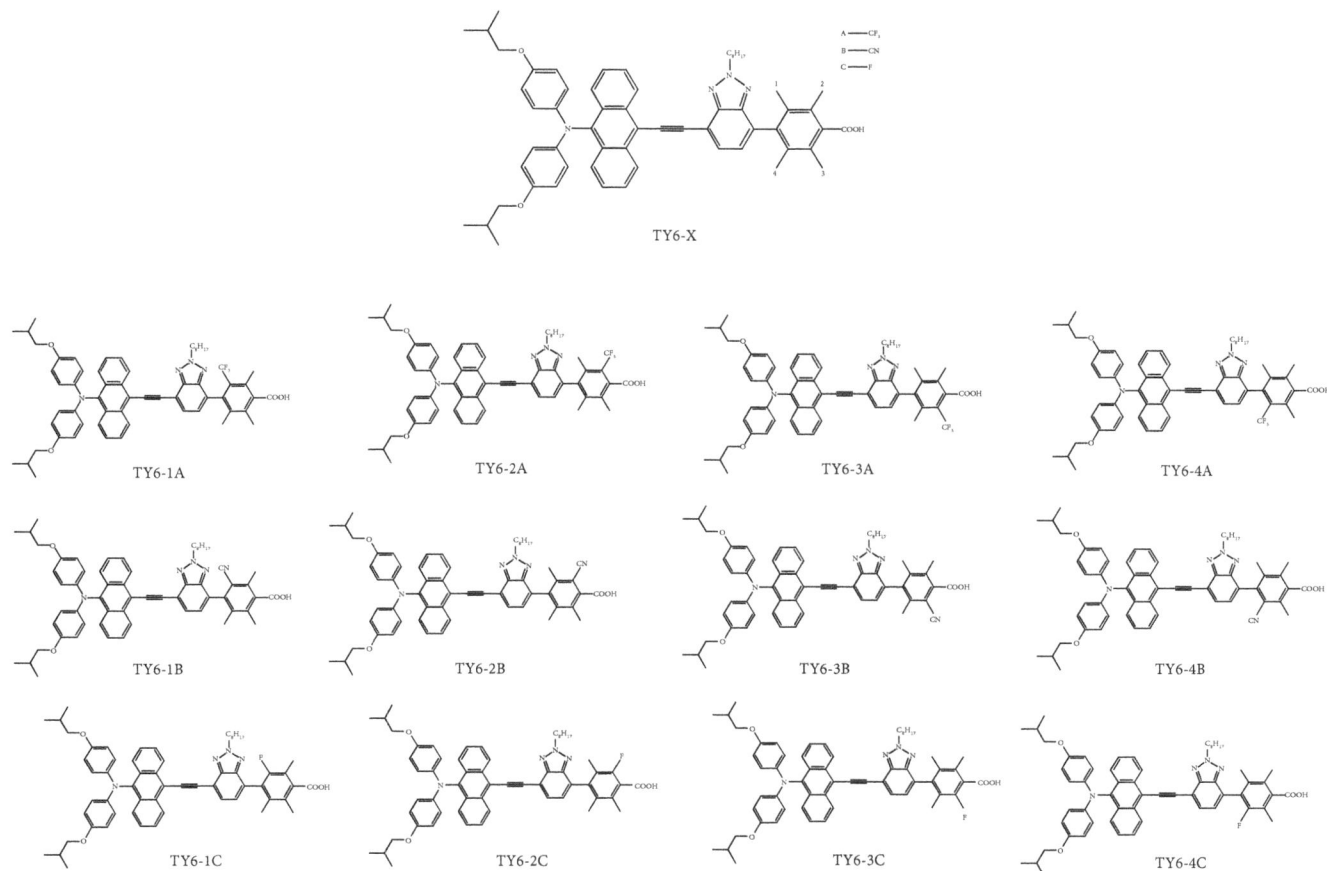

FIGURE 8: Chemical structures of TY6-X and the structure of each designed molecule.

TABLE 10: Selected bond lengths (Å) and dihedral angles (°) of TY6 and TY6-X.

		TY6	TY6-1A	TY6-2A	TY6-3A	TY6-4A	TY6-1B	TY6-2B
Dihedral angle	C1-C2-N3-C4	35.20	35.30	35.24	36.69	34.88	35.38	35.63
	C2-N3-C4-C5	69.60	69.45	69.21	71.38	70.71	69.31	69.10
	C6-C7-C8-C9	29.70	65.82	26.82	29.65	49.99	55.01	26.9
Bond length	C3-C4	1.429	1.429	1.429	1.431	1.430	1.429	1.428
	C7-C8	1.478	1.489	1.477	1.477	1.486	1.480	1.476
	C10-C11	1.483	1.487	1.493	1.496	1.486	1.487	1.486
		TY6-3B	TY6-4B	TY6-1C	TY6-2C	TY6-3C	TY6-4C	
Dihedral angle	C1-C2-N3-C4	35.42	35.17	34.6	31.38	35.29	31.12	
	C2-N3-C4-C5	68.97	69.46	70.69	70.07	69.28	70.65	
	C6-C7-C8-C9	27.65	38.33	49.15	26.70	28.43	30.95	
Bond length	C3-C4	1.428	1.429	1.429	1.430	1.429	1.431	
	C7-C8	1.476	1.477	1.478	1.477	1.477	1.477	
	C10-C11	1.491	1.486	1.486	1.483	1.486	1.485	

the –CF$_3$ group, and least influence corresponding to the –F group. The electronegativity of the 3 groups increased gradually, which was the same as that of the LUMO. Therefore, the greater the electronegativity of the inserted EWG is, the higher the energy of the LUMO is.

Table 12 lists the oscillator strength and transition energy of the design molecule. The results show a significant change in the absorption properties of the design molecules. For example, compared with TY6 (481 nm), the maximum absorption peak of the designed molecules has changed significantly. Related data is arranged in the following order: TY6-3B = TY6-2B > TY6-4B > TY6-2A > TY6-3C > TY6-1B > TY6 > TY6-1C > TY6-4A > TY6-2C = TY6-3A > TY6-1A > TY6-4C. Through the above sorting

TABLE 11: Frontier molecular orbital energies and energy gaps (eV) of TY6-X.

	TY6-1A	TY6-2A	TY6-3A	TY6-4A	TY6-1B	TY6-2B
H	−4.58	−4.64	−4.76	−4.60	−4.61	−4.67
L	−2.23	−2.43	−2.45	−2.31	−2.38	−2.54
Gap	2.35	2.21	2.31	2.29	2.23	2.13
	TY6-3B	TY6-4B	TY6-1C	TY6-2C	TY6-3C	TY6-4C
H	−4.67	−4.63	−4.58	−4.75	−4.62	−4.72
L	−2.54	−2.47	−2.27	−2.42	−2.38	−2.37
Gap	2.13	2.16	2.31	2.33	2.24	2.35

analysis, half of the designed molecules have a red-shifted absorption. Among them, the designed molecules inserted into the −CN group have an obvious red-shifted absorption, and the maximum value appears in TY6-3B/TY6-2B (492 nm), which is the insertion of the −CN group in acceptor position 3/2. As mentioned absove, it seems that there is no obvious regularity change for the insertion of different positions on the acceptor. For insertion of different EWGs, the displacement of the maximum absorption peak of the −CN group is the most obvious. This is the same characteristic as the low-energy gap, and low LUMO is caused by −CN. Similarly, compared with TY6 (0.8886), the designed molecules have greatly increased the oscillation intensity and are arranged in the following order: TY6-2C > TY6-4C > TY6-3A > TY6-2B > TY6-3B > TY6-2A > TY6-3C > TY6-4B > TY6 > TY6-1B > TY6-1C > TY6-4A > TY6-1A. As shown, 2/3 of the designed molecules are higher than the original molecules. The maximum value appears in TY6-2C (0.9726), which is the insertion of the −F group in acceptor position 2. The minimum value appears in TY6-1A (0.7379), which is the insertion of the −CF$_3$ group in acceptor position 4. It is also found that the designed molecules of the insertion group in acceptor position 1 are less than the original molecules, and the performance is poor. However, the designed molecules inserted at positions 2 and 3 of the acceptor are higher than the original molecules, which shows a better performance. This phenomenon is consistent with the previous conclusions that positions 2 and 3 are better than positions 1 and 4. Besides, the magnitude of oscillator strength is directly related to the light harvesting efficiency (LHE). It can be seen from Table 13 that the order of the LHE of the designed molecules is consistent with the oscillator strength (TY6, 0.871). What is important is that a higher LHE leads to a better PCE, and it comes to the following conclusion: Molecules with groups inserted at positions 2 and 3 will have a better performance than those at positions 1 and 4.

Table 13 shows some of the chemical reaction parameters of the design molecules. Compared with the original molecule TY6, most of the designed molecules have significantly increased the values of chemical reaction parameters, and the effect is obvious. For EA, the designed molecules are arranged in the following order: TY6-3B = TY6-2B = TY6-3A TY6-4B > TY6-2C > TY6-2A > TY6-4C > TY6-1B > TY6-3C > TY6-

4A > TY6 (1.40 eV) > TY6-1A > TY6-1C. The maximum value of EA appears at TY6-3B/TY6-2B/TY6-3A (1.64 eV), which adds a −CN/− CF$_3$ group on position 3/2 of the acceptor; the minimum value of EA occurs at TY6-1C (1.35 eV), which adds an −F group on position 1 of the acceptor. For ω^+, the designed molecules are arranged in the following order: TY6-3A > TY6-3B = TY6-2B > TY6-2C = TY6-4B > TY6-2A > TY6-4C > TY6-1B > TY6-3C > TY6-4A > TY6 (1.43 eV) > TY6-1A > TY6-1C. The maximum value of ω^+ appears at TY6-3A (1.77 eV), which adds a −CF$_3$ group on position 3 of the acceptor; the minimum value of ω^+ occurs at TY6-1C (1.37 eV), which adds an −F group on position 1 of the acceptor. For ω, the designed molecules are arranged in the following order: TY6-3B = TY6-2B > TY6-3A > TY6-4B > TY6-2C > TY6-2A > TY6-4C > TY6-1B > TY6 3C > TY6-4A > TY6 (2.89 eV) > TY6-1A > TY6-1C. The maximum value of ω appears at TY6-3B/TY6-2B (3.31 eV), which adds a −CN group on position 3/2 of the acceptor; the minimum value of ω occurs at TY6-1C (2.83 eV), which adds an −F group on position 1 of the acceptor. As mentioned above, we can get some general conclusions as follows. First, for the three kinds of chemical reactivity parameters, the order of the designed molecules is almost consistent, and the order of EA and ω is exactly the same. Second, the maximum value of the chemical reactivity parameters is found in TY6-3X (position 3 of the acceptor); the corresponding minimum value appears at TY6-1C, and its value is smaller than the original molecular TY6. As a whole, the molecular properties designed at positions 2 and 3 of the acceptor are better than those designed at position 1 and 4 of the acceptor. Finally, all the molecules inserted by the −CN group showed better performance than the original molecule TY6 did. Furthermore, the introduction of the −CN group not only can effectively improve the electronic transmission performance and the ability to receive electrons but also can improve the stability of energy.

The total static first hyperpolarizability of the designed molecules is listed in Table 13. Compared with TY6 (747115), the designed molecules have obvious changes and are arranged in the following order: TY6-3B > TY6-1B > TY6-2B > TY6-3A > TY6-2A > TY6-1C > TY6-4B > TY6-2C > TY6-3C > TY6-4C > TY6 > TY6-1A > TY6-4A. The maximum value of β_{tot} appears at TY6-3B (1068163), which adds a −CN group on position 3 of the acceptor; the minimum value of β_{tot} occurs at TY6-4A (529611), which adds a −CF$_3$ group on position 4 of the acceptor. Through the above analysis, it can be concluded that, for different positions of the acceptor, the molecules at acceptor positions 2 and 3 are designed to be better than the acceptors designed at positions 1 and 4; for different groups, the designed molecules inserted by the −CN group have the highest β_{tot}, while the dye inserted by the −CF$_3$ group has the worst performance. Therefore, by inserting the −CN group, the molecules can have more obvious ICT. The lifetime (t) of the designed molecules is listed in Table 13. Compared to the original molecule TY6 (3.91 ns), the majority of the molecules behaved better and are arranged in the following order: TY6-1A > TY6-4A > TY6-1C > TY6-1B > TY6-4B > TY6-3C > TY6-3B > TY6-2A = TY6 > TY6-2B > TY6-3A > TY6-4C > TY6-2C.

TABLE 12: Calculated transition energies (E) and oscillator strengths (f) of TY6-X.

Dye	State	Contribution Mo	E (eV)	Absorption peak λ (nm)	Strength f
TY6-1A	1	0.64911H-L	2.61	474	0.7379
	2	0.64184H-1-L	3.19	389	0.3080
	3	0.46081H-L + 1	3.72	333	0.1848
TY6-2A	1	0.62146H-L	2.55	487	0.9095
	2	0.60609H-1-L	3.11	399	0.3775
	3	0.49844H-L + 1	3.56	349	0.1175
TY6-3A	1	0.63567H-L	2.61	476	0.9494
	2	0.60880H-1-L	3.16	392	0.3024
	3	0.51245H-L + 1	3.61	344	0.1524
TY6-4A	1	0.63494H-L	2.59	478	0.7723
	2	0.62841H-1-L	3.14	395	0.3537
	3	0.48229H-L + 1	3.66	339	0.1645
TY6-1B	1	0.61488H-L	2.57	482	0.8056
	2	0.60560H-1-L	3.14	395	0.3335
	3	0.46057H-L + 1	3.59	345	0.1073
TY6-2B	1	0.59970H-L	2.52	492	0.9340
	2	0.58364H-1-L	3.09	402	0.3819
	3	0.46337H-L + 1	3.51	353	0.0761
TY6-3B	1	0.60186H-L	2.52	492	0.9248
	2	0.58604H-1-L	3.09	401	0.3789
	3	0.46237H-L + 1	3.51	353	0.0761
TY6-4B	1	0.60783H-L	2.54	489	0.8887
	2	0.59154H-1-L	3.10	400	0.3533
	3	0.48002H-L + 1	3.52	352	0.0757
TY6-1C	1	0.64038H-L	2.59	479	0.7841
	2	0.63212H-1-L	3.16	393	0.3601
	3	0.49671H-L + 1	3.66	338	0.1861
TY6-2C	1	0.63810H-L	2.61	476	0.9726
	2	0.60898H-1-L	3.17	391	0.3018
	3	0.51590H-L + 1	3.60	344	0.1599
TY6-3C	1	0.63042H-L	2.56	485	0.8984
	2	0.61512H-1-L	3.12	397	0.3732
	3	0.50881H-L + 1	3.58	346	0.1387
TY6-4C	1	0.64020H-L	2.62	473	0.9611
	2	0.61318H-1-L	3.17	391	0.3012
	3	0.50465H-L + 1	3.62	342	0.1880

The maximum value of t appears at TY6-1A (4.57 ns), which adds a –CF$_3$ group on position 3 of the acceptor; the minimum value of t occurs at TY6-2C (3.48 ns), which adds a –F group on position 2 of the acceptor. Through the above analysis, we can get some conclusions: Molecules designed at position 1 of the acceptor have a longer lifetime. The lifetime of the designed molecules inserted at position 2 of the acceptor is lower than that of the original molecule (TY6), which is contrary to the conclusion that positions 2 and 3 are better than positions 1 and 4. However, there is no significant difference between the different groups.

4. Conclusion

In this study, the DFT and TD-DFT methods were used to calculate the properties of the ground and excited states for MS3, TY3, and TY6. The calculated results show that TY6 has the highest HOMO energy compared to MS3 and TY3, resulting in the smallest energy gap. Smaller energy gaps favor the red-shifted absorption; so, TY6 has the most pronounced red-shifted absorption and the highest molar extinction coefficient, which results in more effective absorption of sunlight. TY6 not only has the largest EA and the lowest IP but also has lower hole/electron reorganization

TABLE 13: Chemical reactivity parameters (eV), light harvesting efficiencies (LHE), hyperpolarizabilities (β_{tot}), and lifetime (t).

	EA	ω^+	ω	LHE	β_{tot}	τ (ns)
TY6-1A	1.37	1.40	2.86	0.817	589,038	4.57
TY6-2A	1.55	1.63	3.15	0.877	909,491	3.91
TY6-3A	1.64	1.77	3.30	0.888	963,150	3.57
TY6-4A	1.43	1.47	2.95	0.831	529,611	4.43
TY6-1B	1.50	1.56	3.07	0.844	1,030,142	4.33
TY6-2B	1.64	1.75	3.31	0.884	967,198	3.88
TY6-3B	1.64	1.75	3.31	0.881	1,068,163	3.92
TY6-4B	1.59	1.68	3.22	0.871	818,100	4.02
TY6-1C	1.35	1.37	2.83	0.836	833,292	4.39
TY6-2C	1.58	1.68	3.20	0.893	815,616	3.48
TY6-3C	1.48	1.53	3.03	0.874	803,191	3.93
TY6-4C	1.52	1.60	3.09	0.891	755,814	3.49

energies for higher hole and electron transfer. Through orbital and NBO analysis, the D-A-π-A (TY6) structure enhances the donoring electron ability and acceptor ability of receiving electrons. At the same time, TY6 also has the highest β and the most obvious ICT. Through ESP analysis, we can find that TY6 has better reactivity. TY6 has the lowest h and the highest LHE, ΔG^{inject}, $\Delta G_{\text{dye}}^{\text{regen}}$, and ω^+, which leads to higher J_{SC}. A higher μ_{normal} results in a higher V_{OC}. This is consistent with the experimental results that TY6 has the highest J_{SC} and V_{OC}.

The calculated results for the design molecules show that the overall effect is improved to some extent. For different groups, by introducing the –CN group, the LUMO energy and energy gap can be reduced, leading to an obvious red-shifted absorption. Chemical reactivity parameters and β were significantly improved. For the different positions of the molecular acceptors, the positions away from the acceptors (positions 2 and 3) performed better than the other positions did. Especially, the oscillator strength led to a significant increase in LHE. Therefore, it will be of some reference to molecular design.

Conflicts of Interest

The authors declare that there is no conflict of interest regarding the publication of this article.

Acknowledgments

This work was supported by the Fundamental Research Funds for the Central Universities (2572018BC24), the China Postdoctoral Science Foundation (2016 M590270), the Heilongjiang Postdoctoral Science Foundation (Grant LBH-Z15002), the National Natural Science Foundation of China (Grant no. 11404055), and the college students' innovation project of the Northeast Forestry University (201709000001).

References

[1] S. Zhang, X. Yang, Y. Numata, and L. Han, "Highly efficient dye-sensitized solar cells: progress and future challenges," *Energy & Environmental Science*, vol. 6, no. 5, pp. 1443–1464, 2013.

[2] M. Grätzel, "Photoelectrochemical cells," *Nature*, vol. 414, no. 6861, pp. 338–344, 2001.

[3] M. Grätzel, "Recent advances in sensitized mesoscopic solar cells," *Accounts of Chemical Research*, vol. 42, no. 11, pp. 1788–1798, 2009.

[4] S. Ahmad, E. Guillén, L. Kavan, M. Grätzel, and M. K. Nazeeruddin, "Metal free sensitizer and catalyst for dye sensitized solar cells," *Energy & Environmental Science*, vol. 6, no. 12, pp. 3439–3466, 2013.

[5] B. O'Regan and M. Grätzel, "A low-cost, high-efficiency solar cell based on dye-sensitized colloidal TiO2 films," *Nature*, vol. 353, no. 6346, pp. 737–740, 1991.

[6] J. Wu, Z. Lan, J. Lin et al., "Electrolytes in dye-sensitized solar cells," *Chemical Reviews*, vol. 115, no. 5, pp. 2136–2173, 2015.

[7] V. Sugathan, E. John, and K. Sudhakar, "Recent improvements in dye sensitized solar cells: a review," *Renewable and Sustainable Energy Reviews*, vol. 52, pp. 54–64, 2015.

[8] N. Robertson, "Optimizing dyes for dye-sensitized solar cells," *Angewandte Chemie International Edition*, vol. 45, no. 15, pp. 2338–2345, 2006.

[9] Y.-S. Yen, H. H. Chou, Y. C. Chen, C. Y. Hsu, and J. T. Lin, "Recent developments in molecule-based organic materials for dye-sensitized solar cells," *Journal of Materials Chemistry*, vol. 22, no. 18, pp. 8734–8747, 2012.

[10] A. Yella, C. L. Mai, S. M. Zakeeruddin et al., "Molecular engineering of push–pull porphyrin dyes for highly efficient dye-sensitized solar cells: the role of benzene spacers," *Angewandte Chemie*, vol. 53, no. 11, pp. 2973–2977, 2014.

[11] A. Mishra, M. K. R. Fischer, and P. Bäuerle, "Metal-free organic dyes for dye-sensitized solar cells: from structure: property relationships to design rules," *Angewandte Chemie International Edition*, vol. 48, no. 14, pp. 2474–2499, 2009.

[12] S.-W. Wang, C. C. Chou, F. C. Hu et al., "Panchromatic Ru(ii) sensitizers bearing single thiocyanate for high efficiency dye sensitized solar cells," *Journal of Materials Chemistry A*, vol. 2, no. 41, pp. 17618–17627, 2014.

[13] M. K. Nazeeruddin, A. Kay, I. Rodicio et al., "Conversion of light to electricity by cis-X2bis(2,2'-bipyridyl-4,4'-dicarboxylate)ruthenium(II) charge-transfer sensitizers (X = Cl-, Br-, I-, CN-, and SCN-) on nanocrystalline titanium dioxide electrodes," *Journal of the American Chemical Society*, vol. 115, no. 14, pp. 6382–6390, 1993.

[14] T. Bessho, S. . M. Zakeeruddin, C. Y. Yeh, E. . W. G. Diau, and M. Grätzel, "Highly efficient mesoscopic dye-sensitized solar cells based on donor–acceptor-substituted porphyrins," *Angewandte Chemie International Edition*, vol. 49, no. 37, pp. 6646–6649, 2010.

[15] T. Higashino and H. Imahori, "Porphyrins as excellent dyes for dye-sensitized solar cells: recent developments and

insights," *Dalton Transactions*, vol. 44, no. 2, pp. 448–463, 2015.

[16] S. Collavini, S. F. Völker, and J. L. Delgado, "Understanding the outstanding power conversion efficiency of perovskite-based solar cells," *Angewandte Chemie International Edition*, vol. 54, no. 34, pp. 9757–9759, 2015.

[17] Y. Zhao and K. Zhu, "Organic-inorganic hybrid lead halide perovskites for optoelectronic and electronic applications," *Chemical Society Reviews*, vol. 45, no. 3, pp. 655–689, 2016.

[18] M. A. Green, A. Ho-Baillie, and H. J. Snaith, "The emergence of perovskite solar cells," *Nature Photonics*, vol. 8, no. 7, pp. 506–514, 2014.

[19] Y. Rong, L. Liu, A. Mei, X. Li, and H. Han, "Beyond efficiency: the challenge of stability in mesoscopic perovskite solar cells," *Advanced Energy Materials*, vol. 5, no. 20, article 1501066, 2015.

[20] A. Yella, C. L. Mai, S. M. Zakeeruddin et al., "Molecular engineering of push–pull porphyrin dyes for highly efficient dye-sensitized solar cells: the role of benzene spacers," *Angewandte Chemie*, vol. 126, no. 11, pp. 3017–3021, 2014.

[21] Z. Yao, M. Zhang, H. Wu, L. Yang, R. Li, and P. Wang, "Donor/acceptor indenoperylene dye for highly efficient organic dye-sensitized solar cells," *Journal of the American Chemical Society*, vol. 137, no. 11, pp. 3799–3802, 2015.

[22] M. Liang and J. Chen, "Arylamine organic dyes for dye-sensitized solar cells," *Chemical Society Reviews*, vol. 42, no. 8, pp. 3453–3488, 2013.

[23] Y. S. Tingare, M. T. Shen, C. Su et al., "Novel oxindole based sensitizers: synthesis and application in dye-sensitized solar cells," *Organic Letters*, vol. 15, no. 17, pp. 4292–4295, 2013.

[24] Y. Wu and W. Zhu, "Organic sensitizers from D–π–A to D–A–π–A: effect of the internal electron-withdrawing units on molecular absorption, energy levels and photovoltaic performances," *Chemical Society Reviews*, vol. 42, no. 5, pp. 2039–2058, 2013.

[25] Y. Wu, W. H. Zhu, S. M. Zakeeruddin, and M. Grätzel, "Insight into D–A–π–a structured sensitizers: a promising route to highly efficient and stable dye-sensitized solar cells," *ACS Applied Materials & Interfaces*, vol. 7, no. 18, pp. 9307–9318, 2015.

[26] K. Kakiage, Y. Aoyama, T. Yano, K. Oya, J. I. Fujisawa, and M. Hanaya, "Highly-efficient dye-sensitized solar cells with collaborative sensitization by silyl-anchor and carboxy-anchor dyes," *Chemical Communications*, vol. 51, no. 88, pp. 15894–15897, 2015.

[27] P. Bazylewski, K. H. Kim, D. H. Choi, and G. S. Chang, "Self-ordering properties of functionalized Acenes for annealing-free organic thin film transistors," *The Journal of Physical Chemistry B*, vol. 117, no. 36, pp. 10658–10664, 2013.

[28] H. Cha, D. S. Chung, S. Y. Bae et al., "Complementary absorbing star-shaped small molecules for the preparation of ternary cascade energy structures in organic photovoltaic cells," *Advanced Functional Materials*, vol. 23, no. 12, pp. 1556–1565, 2013.

[29] P. Zhang, W. Dou, Z. Ju et al., "A 9,9'-bianthracene-cored molecule enjoying twisted intramolecular charge transfer to enhance radiative-excitons generation for highly efficient deep-blue OLEDs," *Organic Electronics*, vol. 14, no. 3, pp. 915–925, 2013.

[30] C. Teng, X. Yang, C. Yang et al., "Molecular design of anthracene-bridged metal-free organic dyes for efficient dye-sensitized solar cells," *The Journal of Physical Chemistry C*, vol. 114, no. 19, pp. 9101–9110, 2010.

[31] I. T. Choi, B. S. You, Y. K. Eom et al., "Triarylamine-based dual-function coadsorbents with extended π-conjugation aryl linkers for organic dye-sensitized solar cells," *Organic Electronics*, vol. 15, no. 11, pp. 3316–3326, 2014.

[32] V. Mallam, H. Elbohy, Q. Qiao, and B. A. Logue, "Investigation of novel anthracene-bridged carbazoles as sensitizers and Co-sensitizers for dye-sensitized solar cells," *International Journal of Energy Research*, vol. 39, no. 10, pp. 1335–1344, 2015.

[33] C.-L. Wang, P. T. Lin, Y. F. Wang et al., "Cost-effective anthryl dyes for dye-sensitized cells under one sun and dim light," *The Journal of Physical Chemistry C*, vol. 119, no. 43, pp. 24282–24289, 2015.

[34] Y.-S. Yen, Y. C. Chen, H. H. Chou, S. T. Huang, and J. T. Lin, "Novel organic sensitizers containing 2,6-difunctionalized anthracene unit for dye sensitized solar cells," *Polymer*, vol. 4, no. 3, pp. 1443–1461, 2012.

[35] C.-L. Mai, T. Moehl, Y. Kim et al., "Acetylene-bridged dyes with high open circuit potential for dye-sensitized solar cells," *RSC Advances*, vol. 4, no. 66, pp. 35251–35257, 2014.

[36] Y. S. Tingare, N. S.'. Vinh, H. H. Chou et al., "New acetylene-bridged 9,10-conjugated anthracene sensitizers: application in outdoor and indoor dye-sensitized solar cells," *Advanced Energy Materials*, vol. 7, no. 18, article 1700032, 2017.

[37] C. Lee, W. Yang, and R. G. Parr, "Development of the Colle-Salvetti correlation-energy formula into a functional of the electron density," *Physical Review B*, vol. 37, no. 2, pp. 785–789, 1988.

[38] P. J. Stephens, F. J. Devlin, C. F. Chabalowski, and M. J. Frisch, "Ab initio calculation of vibrational absorption and circular dichroism spectra using density functional force fields," *The Journal of Physical Chemistry*, vol. 98, no. 45, pp. 11623–11627, 1994.

[39] B. J. Lynch, P. L. Fast, M. Harris, and D. G. Truhlar, "Adiabatic connection for kinetics," *The Journal of Physical Chemistry A*, vol. 104, no. 21, pp. 4811–4815, 2000.

[40] A. D. Becke, "Density-functional thermochemistry. III. The role of exact exchange," *The Journal of Chemical Physics*, vol. 98, no. 7, pp. 5648–5652, 1993.

[41] M. Cossi, V. Barone, B. Mennucci, and J. Tomasi, "Ab initio study of ionic solutions by a polarizable continuum dielectric model," *Chemical Physics Letters*, vol. 286, no. 3-4, pp. 253–260, 1998.

[42] R. E. Stratmann, G. E. Scuseria, and M. J. Frisch, "An efficient implementation of time-dependent density-functional theory for the calculation of excitation energies of large molecules," *The Journal of Chemical Physics*, vol. 109, no. 19, pp. 8218–8224, 1998.

[43] T. Yanai, D. P. Tew, and N. C. Handy, "A new hybrid exchange–correlation functional using the Coulomb-attenuating method (CAM-B3LYP)," *Chemical Physics Letters*, vol. 393, no. 1–3, pp. 51–57, 2004.

[44] J. P. Foster and F. Weinhold, "Natural hybrid orbitals," *Journal of the American Chemical Society*, vol. 102, no. 24, pp. 7211–7218, 1980.

[45] M. J. Gwt, H. B. Frisch, G. E. Schlegel et al., *Gaussian 09, Revision A.01*, Gaussian, Inc, Wallingford, 2009.

[46] Y. Mo, Z. Lin, W. Wu, and Q. Zhang, "Bond-distorted orbitals and effects of hybridization and resonance on C– C bond

lengths," *The Journal of Physical Chemistry*, vol. 100, no. 28, pp. 11569–11572, 1996.

[47] M. Kręglewski, "The geometry and inversion-internal rotation potential function of methylamine," *Journal of Molecular Spectroscopy*, vol. 133, no. 1, pp. 10–21, 1989.

[48] L. X. Wang, Y. Liu, X. L. Tuo, N. Song, and X. G. Wang, "Effect of H+ and NH4+ on the N-NO2 bond dissociation energy of HMX," *Acta Physico-Chimica Sinica*, vol. 23, no. 10, pp. 1560–1564, 2007.

[49] Y. Li, C. Sun, D. Qi, P. Song, and F. Ma, "Effects of different functional groups on the optical and charge transport properties of copolymers for polymer solar cells," *RSC Advances*, vol. 6, no. 66, pp. 61809–61820, 2016.

[50] P. Song, Y. Li, F. Ma, T. Pullerits, and M. Sun, "External electric field-dependent photoinduced charge transfer in a donor–acceptor system for an organic solar cell," *The Journal of Physical Chemistry C*, vol. 117, no. 31, pp. 15879–15889, 2013.

[51] C. G. Zhan, J. A. Nichols, and D. A. Dixon, "Ionization potential, electron affinity, electronegativity, hardness, and electron excitation energy: molecular properties from density functional theory orbital energies," *The Journal of Physical Chemistry A*, vol. 107, no. 20, pp. 4184–4195, 2003.

[52] G. Zhang and C. B. Musgrave, "Comparison of DFT methods for molecular orbital eigenvalue calculations," *The Journal of Physical Chemistry A*, vol. 111, no. 8, pp. 1554–1561, 2007.

[53] C. Sun, Y. Li, P. Song, and F. Ma, "An experimental and theoretical investigation of the electronic structures and photoelectrical properties of ethyl red and carminic acid for DSSC application," *Materials*, vol. 9, no. 10, p. 813, 2016.

[54] L. L. Estrella, M. P. Balanay, and D. H. Kim, "The effect of donor group rigidification on the electronic and optical properties of arylamine-based metal-free dyes for dye-sensitized solar cells: a computational study," *The Journal of Physical Chemistry. A*, vol. 120, no. 29, pp. 5917–5927, 2016.

[55] J. Martínez, "Local reactivity descriptors from degenerate frontier molecular orbitals," *Chemical Physics Letters*, vol. 478, no. 4–6, pp. 310–322, 2009.

[56] R. G. Parr and R. G. Pearson, "Absolute hardness: companion parameter to absolute electronegativity," *Journal of the American Chemical Society*, vol. 105, no. 26, pp. 7512–7516, 1983.

[57] J. L. Gázquez, A. Cedillo, and A. Vela, "Electrodonating and electroaccepting powers," *The Journal of Physical Chemistry A*, vol. 111, no. 10, pp. 1966–1970, 2007.

[58] R. G. Parr, L. v. Szentpály, and S. Liu, "Electrophilicity index," *Journal of the American Chemical Society*, vol. 121, no. 9, pp. 1922–1924, 1999.

[59] S. Kushwaha and L. Bahadur, "Enhancement of power conversion efficiency of dye-sensitized solar cells by co-sensitization of phloxine B and bromophenol blue dyes on ZnO photoanode," *Journal of Luminescence*, vol. 161, pp. 426–430, 2015.

[60] J. Preat, D. Jacquemin, C. Michaux, and E. A. Perpète, "Improvement of the efficiency of thiophene-bridged compounds for dye-sensitized solar cells," *Chemical Physics*, vol. 376, no. 1–3, pp. 56–68, 2010.

[61] J. Preat, C. Michaux, D. Jacquemin, and E. A. Perpète, "Enhanced efficiency of organic dye-sensitized solar cells: triphenylamine derivatives," *The Journal of Physical Chemistry C*, vol. 113, no. 38, pp. 16821–16833, 2009.

[62] J. B. Asbury, Y. Q. Wang, E. Hao, H. N. Ghosh, and T. Lian, "Evidences of hot excited state electron injection from sensitizer molecules to TiO2 nanocrystalline thin films," *Research on Chemical Intermediates*, vol. 27, no. 4-5, pp. 393–406, 2001.

[63] M. Salazar-Villanueva, A. Cruz-López, A. A. Zaldívar-Cadena, A. Tovar-Corona, M. L. Guevara-Romero, and O. Vazquez-Cuchillo, "Effect of the electronic state of Ti on M-doped TiO2 nanoparticles (M=Zn, Ga or Ge) with high photocatalytic activities: a experimental and DFT molecular study," *Materials Science in Semiconductor Processing*, vol. 58, pp. 8–14, 2017.

[64] T. Daeneke, A. J. Mozer, Y. Uemura et al., "Dye regeneration kinetics in dye-sensitized solar cells," *Journal of the American Chemical Society*, vol. 134, no. 41, pp. 16925–16928, 2012.

[65] G. Boschloo and A. Hagfeldt, "Characteristics of the iodide/triiodide redox mediator in dye-sensitized solar cells," *Accounts of Chemical Research*, vol. 42, no. 11, pp. 1819–1826, 2009.

[66] Z. Yang, Y. Liu, C. Liu, C. Lin, and C. Shao, "TDDFT screening auxiliary withdrawing group and design the novel D-A-π-A organic dyes based on indoline dye for highly efficient dye-sensitized solar cells," *Spectrochimica Acta Part A: Molecular and Biomolecular Spectroscopy*, vol. 167, pp. 127–133, 2016.

[67] M. Li, L. Kou, L. Diao et al., "Theoretical study of WS-9-based organic sensitizers for unusual Vis/NIR absorption and highly efficient dye-sensitized solar cells," *The Journal of Physical Chemistry C*, vol. 119, no. 18, pp. 9782–9790, 2015.

[68] Z. Ning, Y. Fu, and H. Tian, "Improvement of dye-sensitized solar cells: what we know and what we need to know," *Energy & Environmental Science*, vol. 3, no. 9, p. 1170, 2010.

[69] J. Zhang, Y.-H. Kan, H.-B. Li, Y. Geng, Y. Wu, and Z.-M. Su, "How to design proper π-spacer order of the D-π-A dyes for DSSCs? A density functional response," *Dyes and Pigments*, vol. 95, no. 2, pp. 313–321, 2012.

[70] S. R. Marder, D. N. Beratan, and L.-T. Cheng, "Approaches for optimizing the first electronic hyperpolarizability of conjugated organic molecules," *Science*, vol. 252, no. 5002, pp. 103–106, 1991.

[71] M. Sun, Y. Ding, and H. Xu, "Direct visual evidence for quinoidal charge delocalization in poly-p-phenylene cation radical," *The Journal of Physical Chemistry B*, vol. 111, no. 46, pp. 13266–13270, 2007.

[72] Y. Li, S. Wang, Y. Lv, Y. Li, and Q. Wang, "Insight into optoelectronic property by modifying optical layers with multipolar and multi-branched structures," *Journal of Materials Science: Materials in Electronics*, vol. 28, no. 2, pp. 1489–1500, 2017.

[73] D. Matthews, P. Infelta, and M. Grätzel, "Calculation of the photocurrent-potential characteristic for regenerative, sensitized semiconductor electrodes," *Solar Energy Materials and Solar Cells*, vol. 44, no. 2, pp. 119–155, 1996.

[74] W.-Q. Deng, L. Sun, J. D. Huang, S. Chai, S. H. Wen, and K. L. Han, "Quantitative prediction of charge mobilities of π-stacked systems by first-principles simulation," *Nature Protocols*, vol. 10, no. 4, pp. 632–642, 2015.

[75] J. S. Murray and P. Politzer, "The electrostatic potential: an overview," *Wiley Interdisciplinary Reviews: Computational Molecular Science*, vol. 1, no. 2, pp. 153–163, 2011.

11

Novel Imidazole Substituted Bodipy-Based Organic Sensitizers in Dye-Sensitized Solar Cells

Mao Mao[ID],[1] Xiao-Lin Zhang[ID],[1] and Guo-Hua Wu[2]

[1]School of Atmospheric Physics, Nanjing University of Information Science & Technology, Nanjing 210044, China
[2]College of Science and Technology, Nihon University, 1-18-14 Kanda Surugadai, Chiyoda-ku, Tokyo 101-8308, Japan

Correspondence should be addressed to Mao Mao; mmao@nuist.edu.cn and Xiao-Lin Zhang; xlnzhang@nuist.edu.cn

Academic Editor: P. Davide Cozzoli

A comparative study on the photophysical, electrochemical properties and photovoltaic performances of pure imidazole dyes containing varying linker groups is done. Two new organic dyes containing 4,5-bis(4-methoxyphenyl)-1H-imidazole (BPI) unit as an electron donor, boron dipyrromethene (Bodipy) chromophore as a conjugate bridge, cyanoacetic acid as an electron acceptor, and phenylene (BPI-P) or thienyl (BPI-T) as a additional linker have been synthesized for fabricating dye-sensitized solar cells (DSSCs). A reference dye (DPI-T) with 6,9-dimethoxy-1H-phenanthro[9,10-d]imidazole as the donor has also been synthesized for comparison. The overall conversion efficiencies of 0.18%, 0.32%, and 1.28% were obtained for DSSCs based on BPI-P, BPI-T, and DPI-T, respectively. DPI-T was found to be more efficient than BPI-P and BPI-T because of its enhanced light harvesting efficiency and better coplanar geometry of the electronic structure.

1. Introduction

Global energy consumption and environmental pollution are leading to the increasing demand for viable renewable energy sources [1–4]. Dye-sensitized solar cells (DSSCs) are attractive solar energy conversion devices because of their peculiarity in terms of low-cost production, relative ease of fabrication, environmental friendliness, and stability, which paves the way to novel applications [5–9]. Besides the electrolyte and semiconductor, the performance of DSSCs is vitally influenced by the structural engineering of the sensitizing dyes, which absorb sunlight to generate electrons and transmit generated electrons [10–15]. Enormous studies have been focusing on searching novel fully organic dyes, providing advantages in terms of the molecularly tailored design flexibility and raw material abundance comparable to their inorganic counterparts. Most metal-free organic sensitizers with high efficiency contain dye molecules with linearly shaped structures comprising a strong D-π-A dipole with an electron donor (D), a π-bridge, and an electron acceptor (A), which owns photoinduced intramolecular charge transfer natures [16, 17].

Imidazole, an archetypical heterocyclic molecule, is well known in the field of medicine as a drug with anticancer, antibacterial, and antifungal activities and as an antioxidant [18–21]. Recently, researchers have already introduced imidazole into organic light-emitting diode [22–26] and DSSC application [27–31] because of its attractive properties. According to previous studies, introducing electron donors in the 4,5-site and an electron acceptor in the 2-site of imidazole is propitious to form conjugated dipolar sensitizers as well as strengthen their light-harvesting capability. Moreover, as a result of weakening positive charge density at the donor group by electronic delocalization of the two substituents in the 4,5-site of the imidazolyl ring, charge recombination after electron injection may be retarded [27, 30]. Nevertheless, light sensitivity of imidazole photosensitizers in near-infrared spectra is limited. Boron dipyrromethene (Bodipy), known as the "little sister of porphyrins," is a unique chromophore with an intensive absorption profile in the visible/near-infrared region and could be conveniently and flexibly tuned by chemical modification at five different points of the Bodipy core [32]. Extensive efforts have been devoted to the development of Bodipy-modified dyes with a

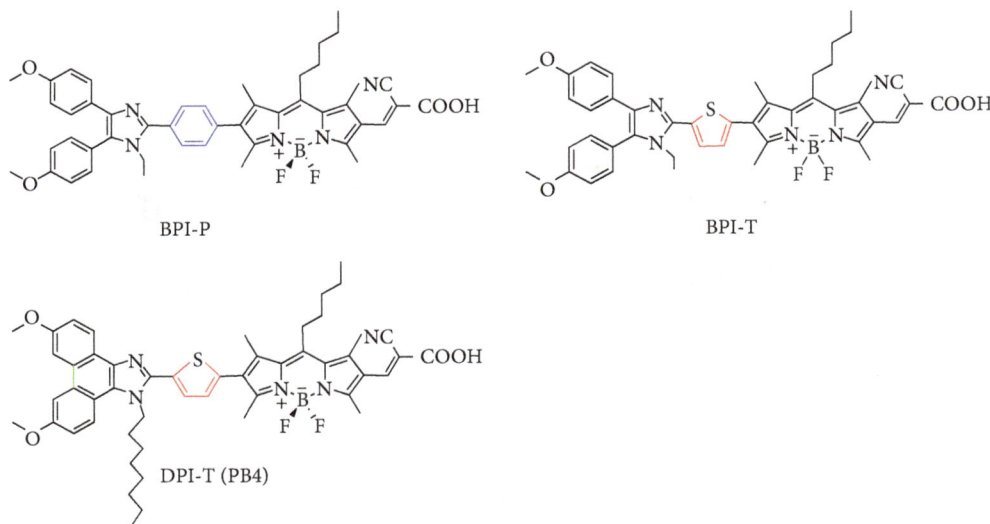

FIGURE 1: Molecular structures of the three dyes.

D-π-A system to increase the light-harvesting properties [33–36]. With the aforementioned in mind, synthesis of some properties of new 4,5-di-substituted imidazole which is conjugated with Bodipy for broadening the spectral range could be worthwhile.

In the present work, two new organic sensitizers containing 4,5-bis(4-methoxyphenyl)-1H-imidazole (BPI) and Bodipy as an electron donor and a conjugated bridge, respectively, along with different additional linker groups (phenylene for BPI-P and thienyl for BPI-T) incorporated into the bridge to further expand the absorption spectra and improve the solubility were synthesized and applied in DSSCs. 6,9-Dimethoxy-1H-phenanthro[9,10-d]imidazole (DPI) is structurally analogous to the BPI unit except for an additional single C–C bond that links two phenyl rings. For the purpose of comparison, dye DPI-T [37] with DPI as the electron donor, Bodipy as the conjugated bridge, and thienyl unit as the additional linker was prepared. The corresponding molecular structures are shown in Figure 1. The structural and electronic properties of these sensitizers are also investigated using density functional theory (DFT) calculations.

2. Experimental Section

2.1. Materials. All reagents and materials in the experiments were purchased from Sigma-Aldrich and used without further purification. Solvents for measurements of spectroscopy are high-performance liquid chromatography (HPLC) grade. The starting materials (4-(1-ethyl-4,5-bis(4-methoxyphenyl)-1H-imidazol-2-yl)phenyl)boronic acid (3a), (5-(1-ethyl-4,5-bis(4-methoxyphenyl)-1H-imidazol-2-yl)thiophen-2-yl)boronic acid (3b), and DPI-T were prepared according to reported literatures [27, 37].

2.2. Characterization and Measurement. The NMR spectra were obtained using a Bruker AV spectrometer operating at 300 MHz for ^{1}H NMR and 75 MHz for ^{13}C NMR. MALDI-

TOF MS spectra were recorded on the Thermo LTQ Orbitrap mass spectrometer. UV-Vis absorption spectrum was performed on a Shimadzu UV/Vis-2450 spectrometer.

2.3. Synthesis and Characterization of New Compounds. The synthetic routes of BPI-P and BPI-T are depicted in Figure 2. We note that the synthesis routes of BPI-P and BPI-T are similar with the only difference in the linker unit. The synthetic routes start from a Suzuki coupling reaction of 3a and 3b with 4 to afford the important precursor aldehyde intermediates 5a and 5b in favorable yields, respectively. Afterwards, the target product dyes BPI-P and BPI-T were obtained through a Knoevenagel condensation reaction between 5 and cyanoacetic acid in the presence of piperidine. The structures of the two compounds were characterized using spectroscopy.

2.3.1. Synthesis and Characterization of 5a and 5b. Under N_2, compound 3 (0.6 mmol) was reacted with 4 (0.5 mmol) by Suzuki coupling reaction using Pd(PPh$_3$)$_4$ (50 mg) and 2 M K_2CO_3 (3 mL) aqueous solution as catalysis in the mixture solution of THF (15 mL) and toluene (15 mL) at 90°C for 12 h. After cooling, water (50 mL) was added and the reaction mixture was extracted with CH_2Cl_2 (100 mL). The combined organic layer was washed with brine, dried over anhydrous $MgSO_4$, and evaporated under reduced pressure. The crude product was purified by column chromatography on silica gel using CH_2Cl_2 and EtOAc as eluent.

For 5a: red solid, yield: 80%. $R_f = 0.15$ (CH_2Cl_2/EtOAc 100 : 1). Mp 134-135°C. ^{1}H NMR (300 MHz, CDCl$_3$): δ 10.14 (s, 1H), 7.83 (d, $J = 7.7$ Hz, 2H), 7.50 (d, $J = 8.0$ Hz, 2H), 7.36–7.34 (m, 4H), 7.03 (d, $J = 8.5$ Hz, 2H), 7.77 (d, $J = 8.6$ Hz, 2H), 4.02–4.00 (m, 2H), 3.90 (s, 3H), 3.76 (s, 3H), 3.14 (m, 2H), 2.81–2.78 (m, 6H), 2.55 (s, 3H), 2.42 (s, 3H), 1.69 (m, 2H), 1.52–1.47 (m, 2H), 1.44–1.37 (m, 2H), 1.10 (t, $J = 7.1$ Hz, 3H), 0.94 (t, $J = 7.1$ Hz, 3H). ^{13}C NMR (75 MHz, CDCl$_3$): δ 186.1, 159.9, 158.5, 158.2, 155.8, 149.4, 146.3, 140.2, 139.9, 137.9, 136.0, 134.0, 133.1, 132.4, 131.2,

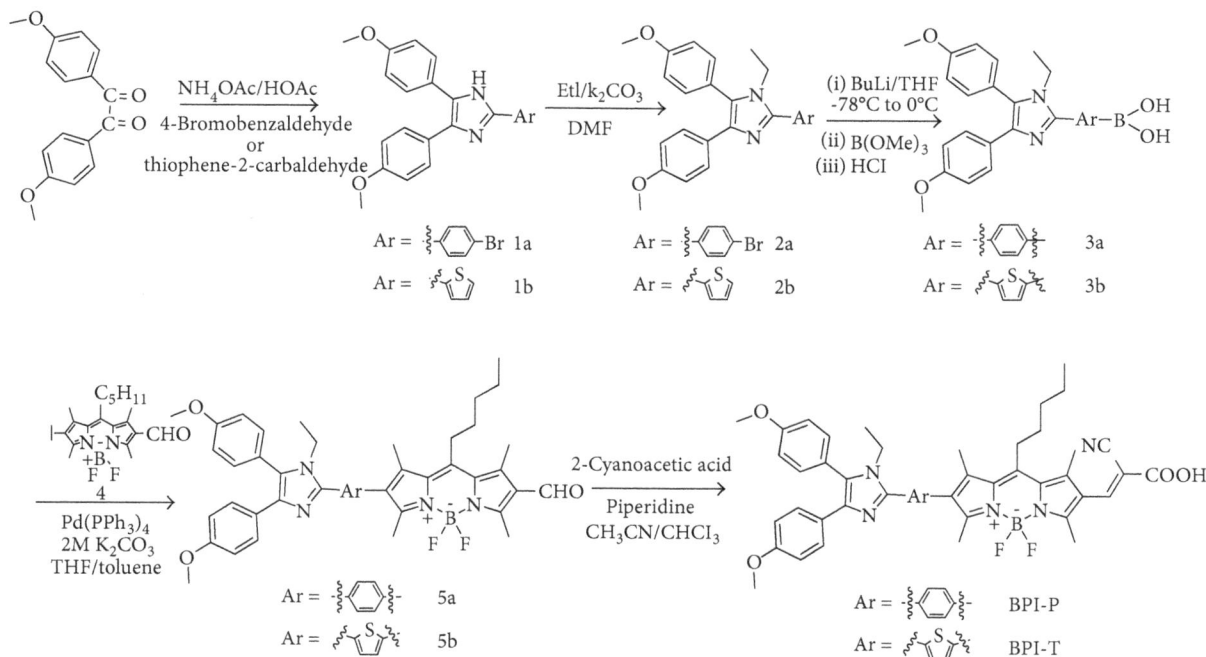

FIGURE 2: Synthetic route to the dyes BPI-P and BPI-T.

130.5, 130.4, 129.3, 129.1, 128.7, 128.6, 128.4, 127.8, 127.5, 126.1, 123.6, 114.6, 113.5, 55.3, 55.2, 39.6, 32.5, 31.6, 28.9, 22.5, 16.4, 14.9, 14.0, 13.8, 13.0, 12.7. IR (KBr, cm^{-1}): 1667 (s), 1539 (s), 1495 (s), 1323 (m), 1247 (m), 1177 (m), 1073 (m), 1024 (m). MALDI-TOF MS calcd for ([M + H]$^+$) $C_{44}H_{48}BF_2N_4O_3$: 729.3782, found: 729.3780.

For 5b: dark solid, yield: 89%. $R_f = 0.25$ (CH$_2$Cl$_2$/EtOAc 100 : 1). Mp 183–185°C. ^{1}H NMR (300 MHz, CDCl$_3$): δ 10.15 (s, 1H), 7.48–7.46 (m, 3H), 7.32 (d, $J = 8.4$ Hz, 2H), 7.03 (d, $J = 8.4$ Hz, 2H), 6.98 (m, 1H), 6.77 (d, $J = 8.4$ Hz, 2H), 4.08–4.06 (m, 2H), 3.90 (s, 3H), 3.77 (s, 3H), 3.15 (m, 2H), 2.82–2.79 (m, 6H), 2.56 (s, 3H); 2.52 (s, 3H), 1.46–1.41 (m, 6H), 0.96 (t, $J = 6.9$ Hz, 3H), 0.88–0.86 (t, $J = 7.1$ Hz, 3H). ^{13}C NMR (75 MH$_Z$, CDCl$_3$): δ 186.1, 160.0, 158.6, 158.3, 156.3, 149.9, 141.5, 140.7, 140.1, 138.1, 134.9, 134.2, 133.7, 132.4, 130.6, 129.1, 128.4, 127.9, 127.4, 127.2, 126.5, 126.3, 125.9, 123.1, 114.6, 113.6, 55.3, 55.1, 39.6, 32.5, 31.6, 28.9, 22.5, 16.3, 15.1, 14.8, 13.9, 13.1, 12.8. IR (KBr, cm^{-1}): 1681 (m), 1538 (s), 1505 (s), 1323 (m), 1248 (m), 1175 (m), 1063 (m). MALDI-TOF MS calcd for ([M + H]$^+$) $C_{42}H_{46}BF_2N_4O_3S$: 735.3346, found: 735.3343.

2.3.2. Synthesis and Characterization of BPI-P and BPI-T.

Compound 5 (0.20 mmol) and 2-cyanoacetic acid (0.60 mmol) were mixed with 20 mL MeCN and 20 mL CHCl$_3$. The mixture was heated to reflux for 24 h in the presence of a few drops of piperidine. After cooling to room temperature, solvents were removed by rotary evaporation. The residue was extracted with CH$_2$Cl$_2$, washed with brine, and dried over anhydrous Na$_2$SO$_4$. The solvent was evaporated under reduced pressure, and the crude product was purified by column chromatography on silica gel with CH$_2$Cl$_2$ and MeOH as eluent to afford target dyes.

For BPI-P: red solid, yield: 87%. $R_f = 0.45$ (CH$_2$Cl$_2$/CH$_3$OH 3 : 1). Mp 192-193°C. ^{1}H NMR (300 MHz, CDCl$_3$): δ 8.09 (s, 1H), 7.86 (m, 2H), 7.51 (d, $J = 8.5$ Hz, 2H), 7.34–7.26 (m, 4H), 7.02 (m, 2H), 6.80 (d, $J = 7.5$ Hz, 2H), 4.03 (m, 2H), 3.88 (s, 3H), 3.77 (s, 3H), 2.98 (m, 2H), 2.48 (s, 6H), 2.34 (s, 6H), 1.48–1.36 (m, 6H), 1.10 (t, $J = 7.1$ Hz, 3H), 0.89 (m, $J = 6.9$ Hz, 3H). IR (KBr, cm^{-1}): 1724 (m), 1614 (m), 1538 (s), 1390 (m), 1248 (m), 1015 (s). MALDI-TOF MS calcd for ([M + H]$^+$) $C_{47}H_{49}BF_2N_5O_4$: 796.3846, found: 796.3840.

For BPI-T: dark solid, yield: 87%. $R_f = 0.50$ (CH$_2$Cl$_2$/CH$_3$OH 3 : 1). Mp 203-205°C. ^{1}H NMR (300 MHz, CDCl$_3$): δ 8.12 (s, 1H), 7.72 (m, 1H), 7.49 (d, $J = 7.5$ Hz, 2H), 7.34 (d, $J = 7.5$ Hz, 2H), 7.05–7.03 (m, 3H), 6.82–6.80 (d, $J = 7.5$ Hz, 2H), 4.13 (m, 2H), 3.89 (s, 3H), 3.77 (s, 3H), 3.01 (m, 2H), 2.59–2.55 (m, 6H), 2.46 (s, 3H), 2.38 (s, 3H), 1.64 (m, 2H), 1.48–1.40 (m, 4H), 1.27 (t, $J = 7.1$ Hz, 3H), 0.93 (t, $J = 7.1$ Hz, 3H). IR (KBr, cm^{-1}): 1715 (m), 1612 (m), 1537 (s), 1249 (m), 1202 (s), 1012 (s). MALDI-TOF MS calcd for ([M + H]$^+$) $C_{45}H_{47}BF_2N_5O_4S$: 802.3404, found: 802.3404.

2.4. Fabrication and Characterization of the DSSC Devices.

The nanocrystalline titanium dioxide (TiO$_2$) electrode and the platinum (Pt) cathode were prepared according to our previous report [37]. Nanocrystalline electrodes are about 14.5 μm thick [determined by a profilometer (XP-2, AMBIOS Technology Inc.)]. The TiO$_2$ electrodes were immersed in a dry CHCl$_3$ solution containing 0.3 mM dye sensitizer for 12 h at the room temperature to ensure complete dye uptake and were then rinsed with anhydrous CHCl$_3$ and EtOH to remove the unbound dye. The dye-coated TiO$_2$ films were placed under vacuum for further drying and used as the photo-anode in the DSSCs. A thermally Pt counter

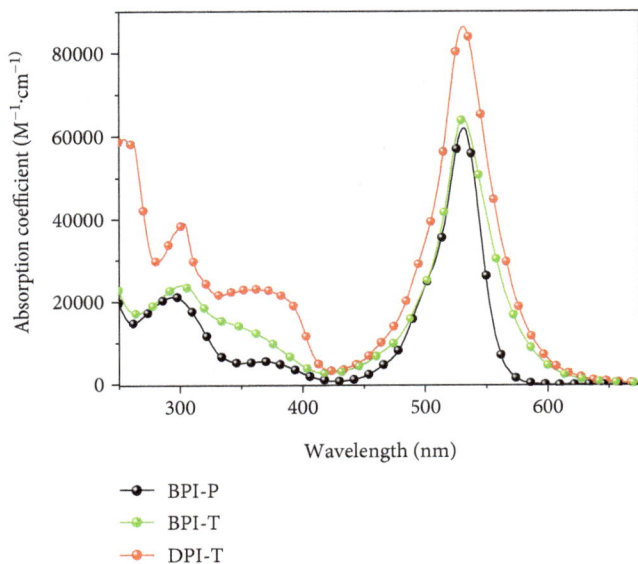

FIGURE 3: Absorption spectra of the dyes BPI-P, BPI-T, and DPI-T in $CHCl_3$ (conc. $= 1.0 \times 10^{-5}$ M).

electrode was then combined with the dye-adsorbed TiO_2 electrode by use of a hot-melt Surlyn film as a spacer to produce a sandwich-type electrochemical cell, and then an electrolyte was injected into the seam between two electrodes. An acetonitrile (CH_3CN) solution containing 1,2-dimethyl-3-n-propylimidazolium iodide (0.6 M), LiI (0.1 M), I_2 (0.05 M), and tert-butylpyridine (TBP, 0.5 M) was used as an electrolyte. The opening was sealed with Araldite glue after filling the electrolyte. To prevent inflated photocurrents arising from stray light, a black metal mask surrounding the active area was a testing cell during all measurements. The active geometrical area of the DSSCs was $0.25\,cm^2$ ($0.5\,cm \times 0.5\,cm$). The cell parameters were obtained under incident light with an intensity of $100\,mW \cdot cm^{-2}$ using a Keithley 2420 3A source meter controlled by Test-point software under a solar simulator (solar AAA simulator, oriel USA, calibrated with a standard crystalline silicon solar).

3. Results and Discussion

3.1. Photophysical Properties. The UV-Vis absorption spectra of BPI-P, BPI-T, and DPI-T in $CHCl_3$ solution are depicted in Figure 3, and the characteristic parameters are listed in Table 1. The absorption spectra of all sensitizers show two distinct bands in the range of 250–675 nm. The shorter wavelength region (250–400 nm) is assigned to the aromatic $\pi-\pi^*$ electron transition of BPI or DPI with an additional linker, whereas the longer wavelength region with high-intensity absorption (420–650 nm) corresponds to the Bodipy unit with some intramolecular charge transfer transition. The λ_{max} (absorption maximum wavelength) of BPI-P with molar absorption coefficient (ε) of $62,000\,M^{-1} \cdot cm^{-1}$ and BPI-T with ε of $64,000\,M^1 \cdot cm^{-1}$ are both at about 531 nm, due to the similar chemical structures except for the additional linker substituents. Remarkably, resulting from its increased donor

property and smaller steric hindrance between the thiophene and BPI unit (vide infra), the thienyl-containing dye (BPI-T) shows an apparent broadening especially for longer wavelength when compared to the phenylene-containing dye (BPI-P). Generally, the better conjugation in the thiophene-based analogue would cause some small bathochromic shift of the band maximum, not only its broadening. The $CHCl_3$ solution is medium-polarity solvent, and aggregation of the studied dyes is very probable in it. To verify this, their absorption spectra have been measured in some high-polarity solvent, such as CH_3CN solution. However, the UV/Vis absorption spectra of BPI-P and BPI-T in CH_3CN which exhibit simultaneously blue shifts of ca. 11 nm compared to those in the medium-polarity solvent.

Additionally, the reference dye DPI-T has a relatively red-shifted absorption maximum wavelength and greater molar absorption coefficient than BPI-T. This is presumably caused by two reasons: on the one hand, the torsion angle between the DPI and thiophene is smaller than BPI and thiophene (vide infra), leading to favorable charge transfer from the donor unit to the acceptor unit; on the other hand, the electron-donating ability of DPI is stronger than that of BPI due to the fact that four aromatic rings in the former are more propitious to conjugate extension. Based on the above analysis, the order of the light-harvesting ability is DPI-T > BPI-T > BPI-P.

Upon adsorption onto the nanocrystalline TiO_2 films (Figure 4), the λ_{max} values for BPI-P, BPI-T, and DPI-T are 505 nm, 527 nm, and 528 nm, respectively. The absorption spectra of all three sensitizers on the TiO_2 films exhibited significantly hypsochromic shift and a broader full width at half maximum (fwhm) than absorption spectra measured in $CHCl_3$ due to strong interactions between the dyes and TiO_2 surfaces. Meanwhile, both in solutions and on TiO_2 film, the fwhm values of sensitizers with additional thiophene chromophore were broader in the region of 400–700 nm than those in the phenylene group, representing an advantageous spectral property for light harvesting and thus increasing the photocurrent response region.

3.2. Electrochemical Properties. In order to evaluate the feasibility of electron transfer from the excited dye molecule to the conduction band (CB) of TiO_2 and the regeneration of oxidized dyes, the cyclic voltammetry (CV) has been performed in CH_2Cl_2 solutions with 0.1 M tetrabutylammonium hexafluorophosphate ($TBAPF_6$) as supporting electrolyte and a three-electrode configuration consisting of a glass carbon as working electrode, an auxiliary Pt wire electrode, and a saturated calomel electrode (SCE) reference electrode. All potentials reported are calibrated with Fc/Fc^+ as an external reference. The relevant data are compiled in Table 1.

The E_{0-0} values were estimated from the onset point of UV-vis spectra. As compared to BPI-P, both BPI-T and DPI-T have narrow E_{0-0} values, a trend consistent with the wider absorption ranges illustrated in Figures 3 and 4, indicating the additional linker with the thiophene moiety which reduces energy gap energies. The excited state potentials (E_{ox}^*), which correspond to the lowest unoccupied molecular orbital (LUMO) levels, were estimated by

TABLE 1: Photophysical and electrochemical data of three dyes.

Dye	λ_{max} (nm)[a]	$\varepsilon \times 10^{-4}$ ($M^{-1}\cdot cm^{-1}$)[a]	λ_{max} (nm)[b]	fwhm (nm)[c]	E_{ox}^{onset} (V)[d]	$E_{0,0}$ (eV)[e]	E_{ox}^* (V)[f]
BPI-P	531	6.20	505	38 (92)	1.26	2.22 (560 nm)	−0.96
BPI-T	531	6.40	527	47 (>175)	1.19	2.11 (588 nm)	−0.92
DPI-T	532	8.02	528	49 (170)	1.20	2.11 (589 nm)	−0.91

[a]Absorption maximum in $CHCl_3$ solution. [b]Absorption maximum on TiO_2 was obtained through measuring the dyes adsorbed on 6 μm TiO_2 nanoparticle films. [c]Values in $CHCl_3$ solutions (on TiO_2 film). [d]E_{ox}^{onset} was measured in CH_2Cl_2 with 0.1 M TBAPF$_6$ as the electrolyte. [e]$E_{0,0}$ was estimated from the onset point of the absorption spectra. [f]E_{ox}^* was estimated by subtracting $E_{0,0}$ from the E_{ox}^{onset}.

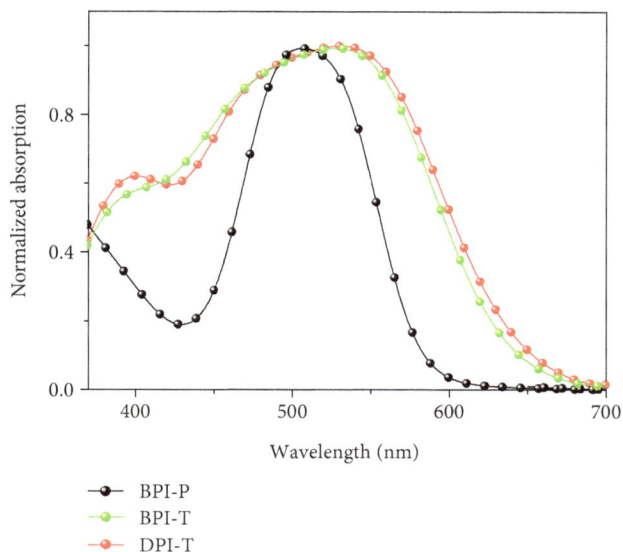

FIGURE 4: Normalized absorption spectra for BPI-P, BPI-T, and DPI-T immobilized on TiO_2.

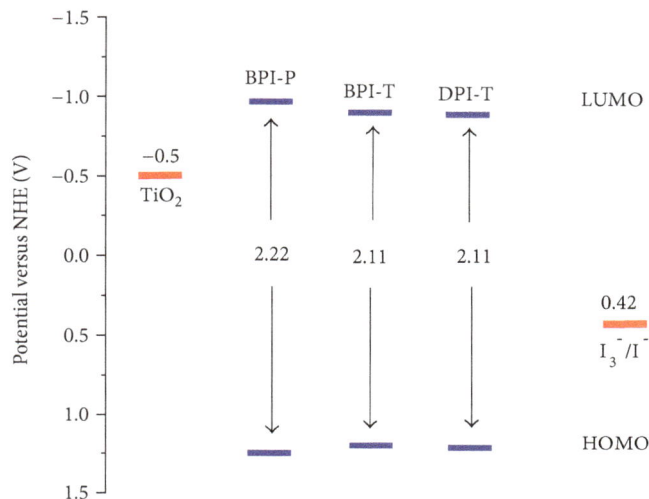

FIGURE 5: Cyclic voltammogram of the dyes BPI-P, BPI-T, and DPI-T in CH_2Cl_2 solutions.

virtue of the following equation: $E_{ox}^* = E_{ox}^{onset} - E_{0,0}$, where E_{ox}^{onset} stands for the onset potential of oxidation of the ground state of a dye dissolved in CH_2Cl_2. As shown in Figure 5, all E_{ox}^* values are more negative than the CB edge of TiO_2 (−0.5 V versus NHE), ensuring an efficient electron injection process from the excited state of the dyes into the TiO_2 electrode. In contrast, the more positive oxidation potential of E_{ox}^{onset} corresponding to the highest occupied molecular orbital (HOMO) level of all photosensitizers compared with that of the I_3^-/I^- redox couples (0.42 V versus NHE) indicates that regeneration of the dyes is thermodynamically feasible. All dyes can be used as sensitizers for feasible electron transfer in DSSCs.

3.3. Computational Study.
Density functional theory (DFT) calculations using the Gaussian 09 [38] software (B3LYP/6-31 G) were conducted in order to gain in-depth insight into the optimized geometrical configuration (Figure 6) and electron distribution of the frontier orbitals of the three molecules (Figure 7). Time-dependent density functional theory (TD-DFT) was then utilized at the same theoretical level to find vertical transitions (Table 2).

In the structure of BPI-P, the dihedral angle between the imidazole core and the adjacent phenyl ring is 41.9°, while in the case of BPI-T, the dihedral angle between imidazole and the neighboring 'thienyl ring is 19.0°. Encouragingly, when the BPI was replaced with DPI to construct DPI-T, the torsion angle calculated between the imidazole core and thiophene linker gets smaller (9.7 °). Based on the calculated molecular geometry, a more distinctly molecular planar can be observed for DPI-T, which may naturally increase the effective conjugation and strengthen the charge transfer from the donor to the acceptor causing a broadening absorption spectrum compared to BPI-P and BPI-T.

From Table 2, the relative trend of UV-Vis absorption spectra based on the TD-DFT calculations is well consistent with our experimental values. The pathways for excitation and electron injection process can be learned by investigating and analyzing the computational results. For example, the low-energy band located in the range of 420–675 nm can be assigned to HOMO/LUMO, HOMO−1/LUMO, and HOMO−2/LUMO transitions. From Figure 7, the electron density of the HOMOs of all dyes is located primarily at the electron-rich BPI or DPI moiety and extended to the conjugated linker group, while the HOMO−1s and LUMOs are concentrated at electron withdrawing unit of cyanoacetic acid and its nearby Bodipy bridge. It is interesting to note that HOMO−2 of thienyl-containing dyes is localized on the whole molecular skeleton and phenylene-containing dye is a donor part. The maximum oscillator strength (f) for lower-energy transitions is higher for DPI-T (0.75) compared to that of BPI-T (0.58), respectively. The

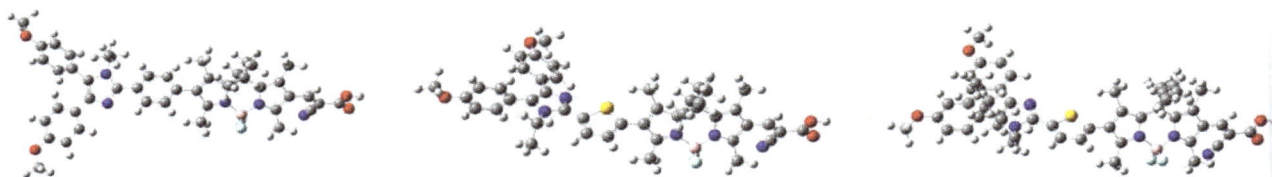

Figure 6: Optimized molecular structures of molecules BPI-P (left), BPI-T (middle), and DPI-T (right).

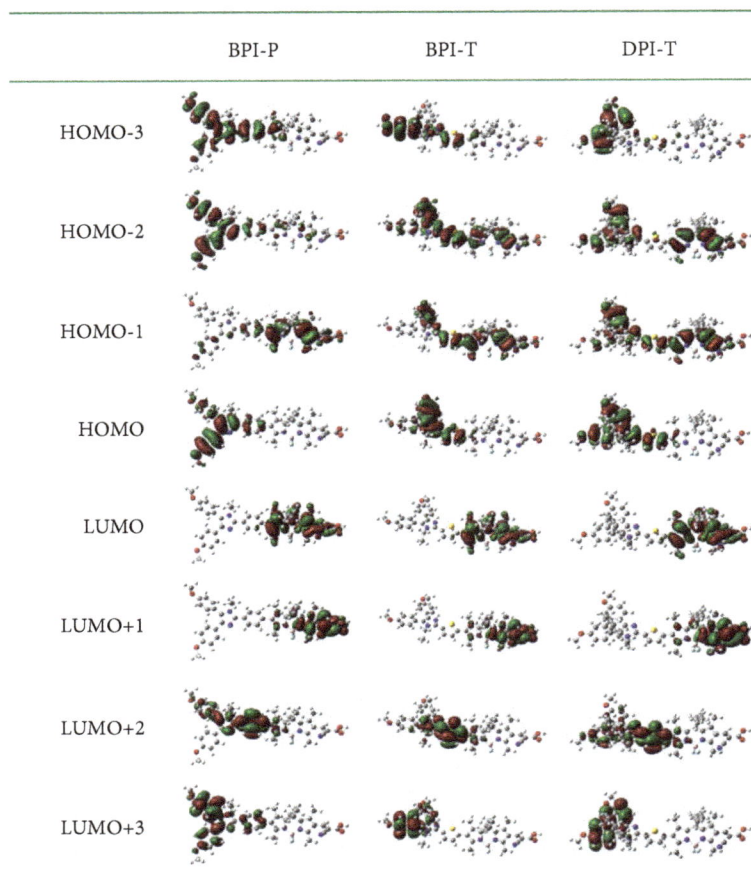

Figure 7: Frontier molecular orbitals of the dyes calculated with DFT on a B3LYP/6-31G (d) level.

results were consistent with the molar absorption coefficient of the dyes in solution. As shown in Table 2, the excited electrons could be successfully transferred from the HOMO or HOMO–1 or HOMO–2 to LUMO upon photoexcitation, facilitating the charge separation in the molecular and ensuring subsequent electron injection into the conduction band of the TiO$_2$ semiconductor.

Overall, the results of DFT and TD-DFT calculations give a better understanding of electronic structures and optical properties and the nature of transition of synthesized molecules. Previous relevant studies have reported that a disturbance to charge delivery, through either a distortion of the molecular geometry or intramolecular aggregation, may reduce charge migration rate [39]. Better coplanar geometry and hydrophobic long chains

(n-C$_8$H$_{17}$) in DPI-T might be favorable to reduce intermolecular aggregation and dark current.

3.4. Photovoltaic Performance of DSSCs Based on the Bodipy Dyes.
The typical photocurrent-voltage (J-V) curves of the devices fabricated with BPI-P, BPI-T, and DPI-T as sensitizers under the illumination of AM 1.5G (100 mW·cm^{-2}) are shown in Figure 8. The photovoltaic conversion efficiency (η) of the DSSCs is calculated from short-circuit current density (J_{sc}), open-circuit voltage (V_{oc}), fill factor (ff), and intensity of the incident light (P) according to the following equation: η (%) = ($J_{sc} \times V_{oc} \times$ ff/P) $\times$ 100. The detailed photovoltaic parameters are summarized in Table 3.

The overall η for phenylene-conjugated derivative (BPI-P) and thiophene-linked analogue (BPI-T) are 0.18%

TABLE 2: Calculated excitation energy characteristics of BPI-T and DPI-T.

Molecule	Wavelength (nm)	f^a	Energy level (eV)	Composition (%)[b,c]
BPI-T	690	0.04	1.80	H→L (70%)
	481	0.35	2.58	H−1→L (64%)
	444	0.58	2.79	H−2→L (57%)
	377	0.13	3.28	H−5→L (61%)
	354	0.24	3.50	H→L + 2 (64%)
	302	0.12	4.11	H−3→L + 1 (59%)
DPI-T	735	0.06	1.69	H→L (70%)
	481	0.34	2.58	H−1→L (64%)
	454	0.75	2.73	H−2→L (67%)
	388	0.45	3.19	H→L + 2 (45%)
	301	0.13	4.11	H−4→L + 1 (52%)

[a]The oscillator strength of a transition. [b]The composition means contribution of each transition for excitation energies. [c]H→HOMO; L→LUMO.

TABLE 3: Photovoltaic performance of DSSCs based on the Bodipy series of dyes.

Dye	J_{sc} (mA/cm²)	V_{oc} (V)	ff	η (%)
BPI-P	0.76	0.45	0.54	0.18
BPI-T	1.04	0.52	0.60	0.32
DPI-T	3.18	0.58	0.69	1.28

[a]Light source: 100 mW/cm², AM 1.5G simulated solar light; working area: 0.25 cm²; thickness: 14.5 um; dye bath: CHCl₃ solution (0.3 mM); and electrolyte: 0.05 M I₂ + 0.1 M LiI + 0.6 M DMPImI + 0.5 M TBP in CH₃CN solution.

twisted nonplanar geometry, which leads to poor orbital overlap. As a result, the benzene ring cannot be of benefit to the conjugated system and unflavored the electron injection. It is confirmed that the introduction of phenylene lowered the extinction coefficient not only in the ultraviolet regions but also in the visible regions, compared with the thiophene-linked dye (Figure 3). Hence, the photocurrent dropped due to the inefficient light harvest. The thiophene-linked analogues possess a high planar configuration, which can lead to an increased intermolecular π–π^* interaction. Compared to the devices using BPI-T and BPI-P with the BPI as the electron donor, the device based on DPI-T has dramatically improved η of 1.28%. Thus, the order of photovoltaic performance is DPI-T > BPI-T > BPI-P, consistent with their light-harvesting ability.

It is found that DPI-T shows a higher photocurrent than BPI-T due to the excellent conjugation system with relatively better electron-donating ability and smaller steric hindrance, which produces a broader absorption spectrum and a higher extinction coefficient. Previous reports have revealed that the long hydrophobic chain might effectively suppress the electron recombination and reduce the interaction between the dye molecules [40, 41], and in this study, DPI-T with octyl chain in the 1-site of the imidazolyl ring exhibited higher V_{oc} than the sensitizers with ethyl chain in the same substitution site.

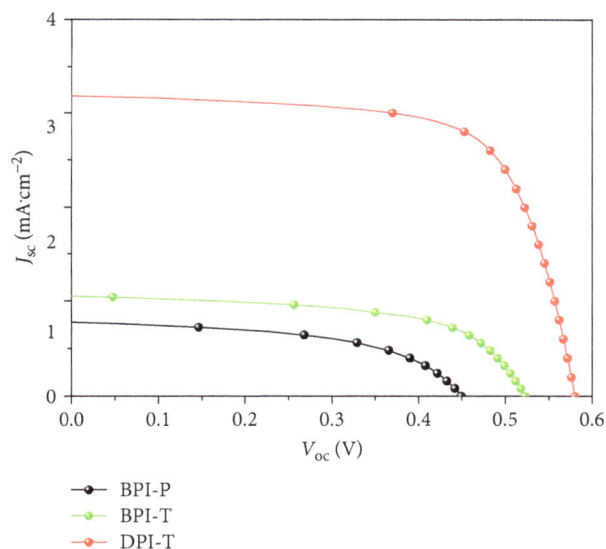

FIGURE 8: The photocurrent density-voltage curve for DSSCs based on BPI-P, BPI-T, and DPI-T under AM 1.5 full simulated sunlight irradiation.

(J_{sc} = 0.76 mA/cm², V_{oc} = 0.45 V, and ff = 0.54) and 0.32% (J_{sc} = 1.04 mA/cm², V_{oc} = 0.52 V, and ff = 0.60), respectively. Under the same conditions, the solar cell based on reference dye DPI-T generated an efficiency of 1.28%, with a J_{sc} of 3.18 mA/cm², a V_{oc} of 0.58 V, and an ff of 0.69. Despite their relatively low overall conversion efficiencies, the dyes show amusing structural dependence in their DSSC performance. Both V_{oc} and J_{sc} increase with the thiophene units. The difference in performance between the BPI-P and BPI-T probably stems from the difference in the coplanarity of the aromatic segment that links the donor and π bridge unit. According to the DFT study, the phenylene derivative possesses a more

4. Conclusions

In summary, two new 4,5-bis(4-methoxyphenyl)-1H-imidazole (BPI) organic dyes BPI-P and BPI-T containing the same Bodipy-conjugated bridges and different additional linkers (phenylene/thiophene) were designed and synthesized. Their optical DFT calculations and photovoltaic properties were systematically investigated because of their promising potential as efficient photosensitizers in DSSCs. The BPI-P- and BPI-T-based cell generated the overall η of 0.18% and 0.32%, respectively, whereas the reference dye DPI-T-based cell gave the efficiency of 1.28%. Dye DPI-T was found to be more efficient than both BPI-P and BPI-T and is attributed to its relatively outstanding light-harvesting efficiency as a result of smaller steric hindrance between the thiophene and DPI unit and better extension of π-conjugated system with four rings on the donor.

Conflicts of Interest

The authors declare that there is no conflict of interests regarding the publication of this paper.

Acknowledgments

This work was supported by grants from the Natural Science Foundation of China (nos. 21406189 and 41505127).

References

[1] X. L. Zhang and M. Mao, "Brown haze types due to aerosol pollution at Hefei in the summer and fall," *Chemosphere*, vol. 119, pp. 1153–1162, 2015.

[2] X. L. Zhang, M. Mao, M. J. Berg, and W. B. Sun, "Insight into winter time aerosol characteristics over Beijing," *Journal of Atmospheric and Solar-Terrestrial Physics*, vol. 121, pp. 63–71, 2014.

[3] X. L. Zhang, M. Mao, Y. Yin, and B. Wang, "Absorption enhancement of aged black carbon aerosols affected by their microphysics: a numerical investigation," *Journal of Quantitative Spectroscopy & Radiative Transfer*, vol. 202, pp. 90–97, 2017.

[4] X. L. Zhang, M. Mao, Y. Yin, and B. Wang, "Numerical investigation on absorption enhancement of black carbon aerosols partially coated with non-absorbing organics," *Journal of Geophysical Research: Atmospheres*, vol. 123, 2018.

[5] B. O'Regan and M. Grätzel, "A low-cost, high-efficiency solar cell based on dye-sensitized colloidal TiO₂ films," *Nature*, vol. 353, no. 6346, pp. 737–740, 1991.

[6] A. Hagfeldt, G. Boschloo, L. Sun, L. Kloo, and H. Pettersson, "Dye-sensitized solar cells," *Chemical Reviews*, vol. 110, no. 11, pp. 6595–6663, 2010.

[7] Z. Ning, Y. Fu, and H. Tian, "Improvement of dye-sensitized solar cells: what we know and what we need to know," *Energy & Environmental Science*, vol. 3, no. 9, pp. 1170–1181, 2010.

[8] H. S. Jung and J. K. Lee, "Dye sensitized solar cells for economically viable photovoltaic systems," *Journal of Physical Chemistry Letters*, vol. 4, no. 10, pp. 1682–1693, 2013.

[9] M. Urbani, M. Grätzel, M. K. Nazeeruddin, and T. Torres, "Meso-substituted porphyrins for dye-sensitized solar cells," *Chemical Reviews*, vol. 114, no. 24, pp. 12330–12396, 2014.

[10] M. Grätzel, "Photoelectrochemical cells," *Nature*, vol. 414, no. 6861, pp. 338–344, 2001.

[11] A. Listorti, B. O'Regan, and J. R. Durrant, "Electron transfer dynamics in dye-sensitized solar cells," *Chemistry of Materials*, vol. 23, no. 15, pp. 3381–3399, 2011.

[12] J. N. Clifford, E. Martínez-Ferrero, A. Viterisi, and E. Palomares, "Sensitizer molecular structure-device efficiency relationship in dye sensitized solar cells," *Chemical Society Reviews*, vol. 40, no. 3, pp. 1635–1646, 2011.

[13] M. Grätzel, R. A. J. Janssen, D. B. Mitzi, and E. H. Sargent, "Materials interface engineering for solution-processed photovoltaics," *Nature*, vol. 488, no. 7411, pp. 304–312, 2012.

[14] D. Joly, L. Pellejà, S. Narbey et al., "Metal-free organic sensitizers with narrow absorption in the visible for solar cells exceeding 10% efficiency," *Energy & Environmental Science*, vol. 8, no. 7, pp. 2010–2018, 2015.

[15] Y. Wu and W. Zhu, "Organic sensitizers from D–π–A to D–A–π–A: effect of the internal electron-withdrawing units on molecular absorption, energy levels and photovoltaic performances," *Chemical Society Reviews*, vol. 42, no. 5, pp. 2039–2058, 2013.

[16] R. Grisorio, L. De Marco, R. Agosta et al., "Enhancing dye-sensitized solar cell performances by molecular engineering: highly efficient π-extended organic sensitizers," *ChemSusChem*, vol. 7, no. 9, pp. 2659–2669, 2014.

[17] F. Zhang, K. J. Jiang, J. H. Huang et al., "A novel compact DPP dye with enhanced light harvesting and charge transfer properties for highly efficient DSCs," *Journal of Materials Chemistry A*, vol. 1, no. 15, pp. 4858–4863, 2013.

[18] N. Nagarajan, G. Vanitha, D. A. Ananth, A. Rameshkumar, T. Sivasudha, and R. Renganathan, "Bioimaging, antibacterial and antifungal properties of imidazole-pyridine fluorophores: synthesis, characterization and solvatochromism," *Journal of Photochemistry and Photobiology B: Biology*, vol. 127, pp. 212–222, 2013.

[19] F. Bellina, S. Cauteruccio, S. Montib, and R. Rossia, "Novel imidazole-based combretastatin A-4 analogues: evaluation of their in vitro antitumor activity and molecular modeling study of their binding to the colchicine site of tubulin," *Bioorganic & Medicinal Chemistry Letters*, vol. 16, no. 22, pp. 5757–5762, 2006.

[20] A. Melaiye, Z. Sun, K. Hindi et al., "Silver(I)–imidazole cyclophane gem-diol complexes encapsulated by electrospun tecophilic nanofibers: formation of nanosilver particles and antimicrobial activity," *Journal of the American Chemical Society*, vol. 127, no. 7, pp. 2285–2291, 2005.

[21] B. F. Abdel-Wahab, G. E. A. Awad, and F. A. Badria, "Synthesis, antimicrobial, antioxidant, anti-hemolytic and cytotoxic evaluation of new imidazole-based heterocycles," *European Journal of Medicinal Chemistry*, vol. 46, no. 5, pp. 1505–1511, 2011.

[22] Y. Zhang, S. L. Lai, Q. X. Tong et al., "High efficiency non-doped deep-blue organic light emitting devices based on imidazole-π-triphenylamine derivatives," *Chemistry of Materials*, vol. 24, no. 1, pp. 61–70, 2011.

[23] H. Huang, X. Yang, B. Pan et al., "Benzimidazole–carbazole-based bipolar hosts for high efficiency blue and white electrophosphorescence applications," *Journal of Materials Chemistry*, vol. 22, no. 26, pp. 13223–13230, 2012.

[24] X. Yang, S. Zheng, R. Bottger et al., "Efficient fluorescent deep-blue and hybrid white emitting devices based on carbazole/benzimidazole compound," *Journal of Physical Chemistry C*, vol. 115, no. 29, pp. 14347–14352, 2011.

[25] S. Park, J. E. Kwon, S. H. Kim et al., "A white-light-emitting molecule: frustrated energy transfer between constituent emitting centers," *Journal of the American Chemical Society*, vol. 131, no. 39, pp. 14043–14049, 2009.

[26] N. Nagarajan, A. Prakash, G. Velmurugan et al., "Synthesis, characterization and electroluminescence behaviour of π-conjugated imidazole-isoquinoline derivatives," *Dyes and Pigments*, vol. 102, pp. 180–188, 2014.

[27] M. Velusamy, Y. Hsu, J. T. Lin, C. W. Chang, and C. P. Hsu, "1-Alkyl-1*H*-imidazole-based dipolar organic compounds for dye-sensitized solar cells," *Chemistry - An Asian Journal*, vol. 5, no. 1, pp. 87–96, 2010.

[28] N. Nagarajan, G. Velmurugan, P. Venuvanalingam, and R. Renganathan, "Tunable single and dual emission behavior of imidazole fluorophores based on D-π-A architecture," *Journal of Photochemistry and Photobiology A: Chemistry*, vol. 284, pp. 36–48, 2014.

[29] R. K. Aulakh, S. Sandhu, S. Kumar, A. Mahajan, R. K. Bedi, and S. Kumar, "Designing and synthesis of imidazole based hole transporting material for solid state dye sensitized solar cells," *Synthetic Metals*, vol. 205, pp. 92–97, 2015.

[30] Z. Wan, L. Zhou, C. Jia, X. Chen, Z. Li, and X. Yao, "Comparative study on photovoltaic properties of imidazole-based dyes containing varying electron acceptors in dye-sensitized solar cells," *Synthetic Metals*, vol. 196, pp. 193–198, 2014.

[31] D. Karthik, V. Kumar, K. J. Thomas, C. T. Li, and K. C. Ho, "Synthesis and characterization of thieno[3,4- d]imidazole-based organic sensitizers for photoelectrochemical cells," *Dyes and Pigments*, vol. 129, pp. 60–70, 2016.

[32] A. Loudet and K. Burgess, "BODIPY dyes and their derivatives: syntheses and spectroscopic properties," *Chemical Reviews*, vol. 107, no. 11, pp. 4891–4932, 2007.

[33] S. Kolemen, O. A. Bozdemir, Y. Cakmak et al., "Optimization of distyryl-Bodipy chromophores for efficient panchromatic sensitization in dye sensitized solar cells," *Chemical Science*, vol. 2, no. 5, pp. 949–954, 2011.

[34] M. Mao, X. L. Zhang, X. Q. Fang et al., "Highly efficient light-harvesting boradiazaindacene sensitizers for dye-sensitized solar cells featuring phenothiazine donor antenna," *Journal of Power Source*, vol. 268, pp. 965–976, 2014.

[35] C. Qin, A. Mirloup, N. Leclerc et al., "Molecular engineering of new thienyl-Bodipy dyes for highly efficient panchromatic sensitized solar cells," *Advanced Energy Materials*, vol. 4, no. 11, article 1400085, 2014.

[36] M. Mao and Q. H. Song, "The structure-property relationships of D-π-A BODIPY dyes for dye-sensitized solar cells," *The Chemical Record*, vol. 16, no. 2, pp. 719–733, 2016.

[37] M. Mao, J. B. Wang, Z. F. Xiao, S. Y. Dai, and Q. H. Song, "New 2,6-modified BODIPY sensitizers for dye-sensitized solar cells," *Dyes and Pigments*, vol. 94, no. 2, pp. 224–232, 2012.

[38] M. J. Frisch, G. W. Trucks, H. B. Schlegel et al., *Gaussian 09, Revision E.01*, Gaussian, Inc, Wallingford, CT, USA, 2009.

[39] R. Y. Huang, Y. H. Chiu, Y. H. Chang et al., "Influence of a D-π-A system through a linked unit of double and triple bonds in a triarylene bridge for dye-sensitised solar cells," *New Journal of Chemistry*, vol. 41, no. 16, pp. 8016–8025, 2017.

[40] Z. Wan, C. Jia, Y. Wang, and X. Yao, "Dithiafulvenyl–triphenylamine organic dyes with alkyl chains for efficient coadsorbent-free dye-sensitized solar cells," *RSC Advances*, vol. 5, no. 63, pp. 50813–50820, 2015.

[41] B. Nagarajan, S. Kushwaha, R. Elumalai, and S. Mandal, "Novel ethynyl-pyrene substituted phenothiazine based metal free organic dyes in DSSC with 12% conversion efficiency," *Journal of Materials Chemistry A*, vol. 5, no. 21, pp. 10289–10300, 2017.

12

Analytical Approach to Circulating Current Mitigation in Hexagram Converter-Based Grid-Connected Photovoltaic Systems Using Multiwinding Coupled Inductors

Abdullrahman A. Al-Shamma'a[1,2] **Abdullah M. Noman**[1,2] **Khaled E. Addoweesh,**[1]
Ayman A. Alabduljabbar,[3] **and A. I. Alolah**[1]

[1]*Department of Electrical Engineering, College of Engineering, King Saud University, Riyadh 11421, Saudi Arabia*
[2]*Department of Mechatronics Engineering, College of Engineering, Taiz University, Taiz, Yemen*
[3]*King Abdulaziz City for Science and Technology (KACST), P.O. Box 6086, Riyadh 11442, Saudi Arabia*

Correspondence should be addressed to Abdullrahman A. Al-Shamma'a; ashammaa@ksu.edu.sa

Academic Editor: Francesco Riganti-Fulginei

The hexagram multilevel converter (HMC) is composed of six conventional two-level voltage source converters (VSCs), where each VSC module is connected to a string of PV arrays. The VSC modules are connected through inductors, which are essential to minimize the circulating current. Selecting inductors with suitable inductance is no simple process, where the inductance value should be large to minimize the circulating current as well as small to reduce an extra voltage drop. This paper analyzes the utilization of a multiwinding (e.g., two, three, and six windings) coupled inductor to interconnect the six VSC modules instead of six single inductors, to minimize the circulating current inside the HMC. Then, a theoretical relationship between the total impedance to the circulating current, the number of coupled inductor windings, and the magnetizing inductance is derived. Owing to the coupled inductors, the impedance on the circulating current path is a multiple of six times the magnetizing inductance, whereas the terminal voltage is slightly affected by the leakage inductance. The HMC is controlled to work under variable solar radiation, providing active power to the grid. Additional functions such as DSTATCOM, during daytime, are also demonstrated. The controller performance is found to be satisfactory for both active and reactive power supplies.

1. Introduction

Recently, photovoltaic (PV) energy systems have gained more attention as distributed generation units, as they offer low cost of generation closer to that of conventional plants, as well as less maintenance and no grid noise [1, 2]. Moreover, PV systems can solve multiple typical problems present in conventional AC power systems. However, PV systems present frequency and voltage fluctuations when islanding operation occurs. Therefore, PV plants should be integrated with the power system in order to maintain the overall frequency and voltage at a stable condition. Several studies suggest interconnection methods of PV systems to the grid through voltage source converters (VSCs), because they provide versatile functions enhancing capabilities of the system

[3–5]. The main purpose of the VSC is to connect the PV plant to the grid while guaranteeing power quality (PQ) standards. However, the high-frequency switching of VSCs introduces additional harmonic components to the system, hence creating PQ problems if not implemented accurately [6].

Most of the available VSCs for PV systems are traditional two-level three-phase VSCs with low power capacity [7]. In the literature, researchers have suggested numerous multilevel topologies for grid-connected PV plants [8–10]. Multilevel VSCs are considered more attractive than traditional two-level VSCs, as they improve the output voltage quality and reduce electromagnetic interference, voltage stress on IGBTs, and common-mode voltage. Moreover, multilevel VSCs operate at a low switching frequency and hence increase system efficiency [10]. Consequently, multilevel

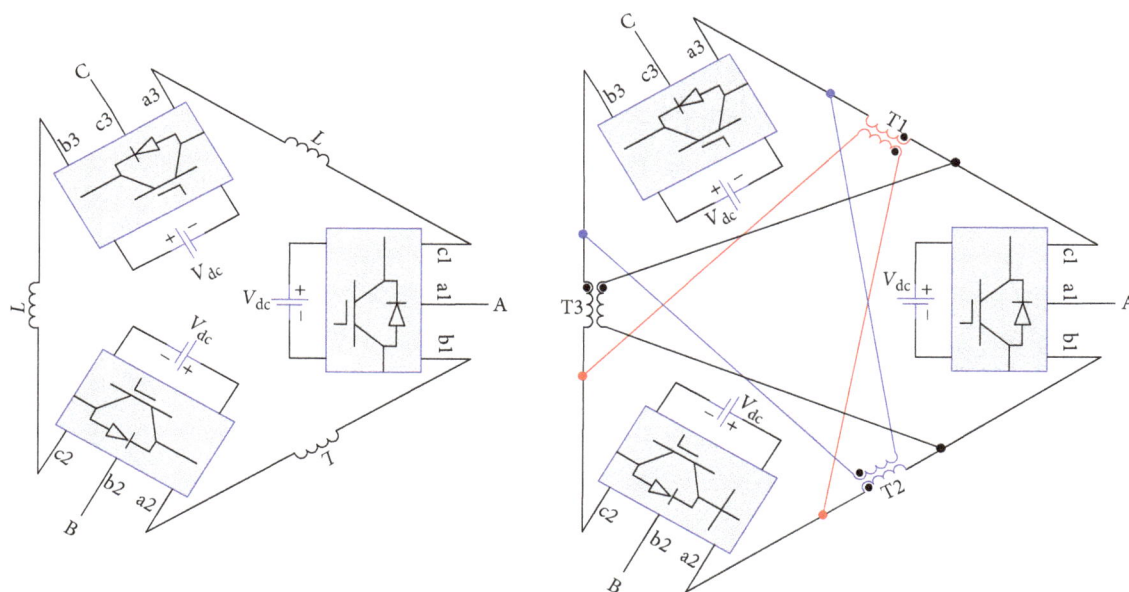

FIGURE 1: Topologies of cascade two-level converters.

VSCs have been widely used in chemical, oil, and different kinds of plants, as well as in power plants, transmission systems, and PQ compensators [11].

Current research is mainly focused on three specific multilevel VSC topologies, namely, the neutral-point clamped (NPC) [12], flying capacitor (FC) [13] and the Cascaded H-bridge (CHB) [14] topologies. The NPC VSC requires a high number of clamping diodes to increase the number of voltage levels, which can cause problems related to switches with different ratings, high-voltage rating in blocking diodes, and capacitor voltage imbalance. However, the FC VSC consists of a large number of capacitors, which leads to complications in regulating capacitor voltages. Finally, the CHB VSC provides isolated DC sources, hence being appropriate for use in PV systems [15]. In addition, other advantages of this topology include its modular structure and the reduced number of components compared to the other multilevel converters (i.e., NPC and FC). Therefore, the CHB VSC can reach the same number of voltage levels with a simpler assembly and maintenance. However, this topology presents three main drawbacks: (i) high overall component count; (ii) high energy storage requirement because the instantaneous power related to each H-bridge varies at twice the fundamental frequency, given its single-phase modular structure; and (iii) difficulty to control the voltage across DC-link capacitors.

The concept of interconnecting three traditional three-phase VSCs to produce a multilevel converter was first proposed in [16] and applied to medium-voltage variable-speed drives. In [17], three three-phase two-level VSCs are interconnected using three single-phase transformers with a 1:1 turn ratio. These interconnected transformers increase the output voltage and suppress the circulating current inside the converter. The power capacity of the overall converter is three times the capacity of each interconnected converter, whereas the volt-ampere rating of each intermediate transformer is equal to that of each interconnected converter. In [18], another topology known as the hexagram multilevel converter (HMC) is proposed, which combines six two-level converters by using six inductors. This topology shares many advantages of the CHB but uses fewer switches and reduces the size of the DC-link capacitor [19]. The advantages of the proposed topology can be summarized as follows: (1) only six standard three-phase VSCs are necessary to generate multilevel output voltage; (2) each VSC module is balanced in operation, equally loaded, and supplies 1/6 output power; (3) modular construction which facilitates system maintenance and spare management; (4) only six isolated DC links with no voltage unbalance problem; (5) low number of power electronics switches and low DC energy storage requirement; and (6) the output transformer contributes to higher output voltage. Figures 1 and 2 are schematic diagrams illustrating multilevel topologies based on cascaded two-level VSCs according to previous research.

However, selecting an inductor with suitable inductance is no simple process, where the inductance value should be adequate to minimize the circulating current as well as to reduce an extra voltage drop on the inductors that affects the terminal voltage. This difficulty can be circumvented by using a multiwinding coupled inductor. Therefore, this paper analyzes the utilization of a multiwinding (e.g., two, three, and six windings) coupled inductor to interconnect the six VSCs instead of six single inductors. Consequently, both goals minimize the circulating current and the minimal effects to the output voltage can be accomplished instantaneously. Then, an analytical model to calculate the total inductance imposed to the circulating current path is derived. The equivalent circuit model of the HMC is detailed in the abc reference frame and then transformed into the orthogonal dq0 reference frame. Moreover, to extract the maximum power from the PV arrays, a control algorithm for maximum power point tracking is also presented.

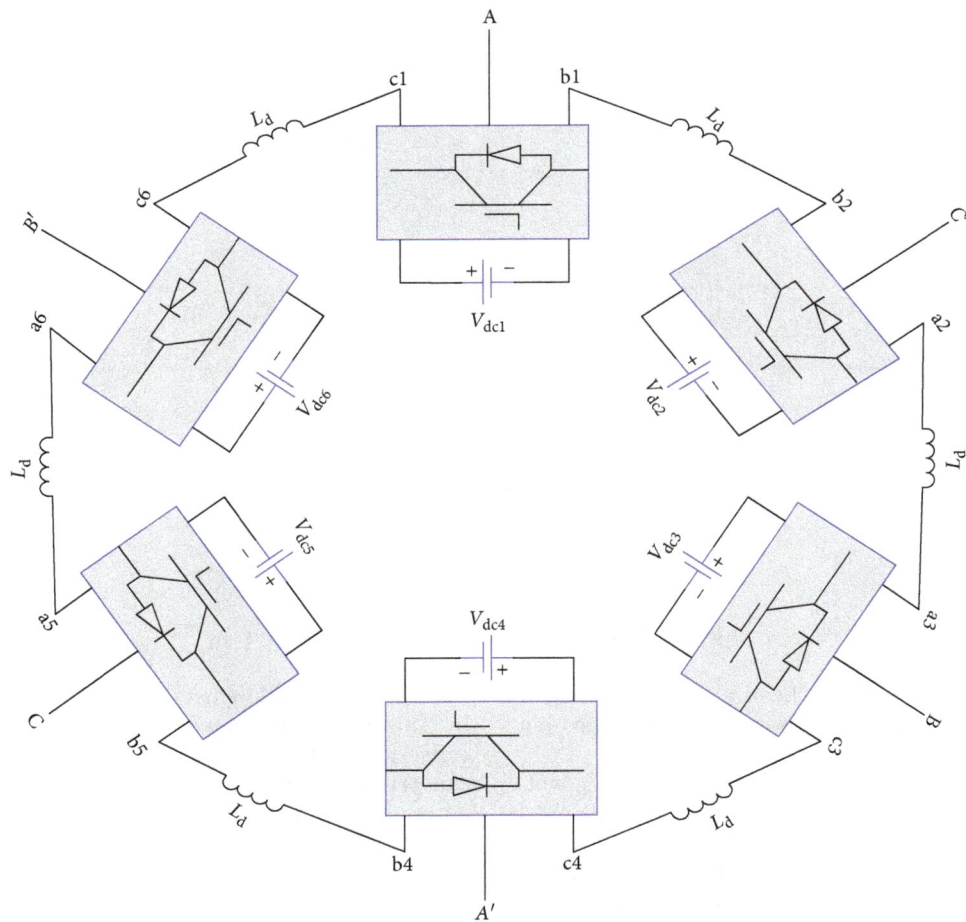

FIGURE 2: HMC electrical diagram.

The paper is organized as follows: Section 2 describes the HMC. Section 3 presents the mathematical model for a grid-connected PV system. Section 4 demonstrates the benefits of the multiwinding coupled inductor inside the HMC. Section 5 describes the control system and modulation strategy. Section 6 shows the simulation results followed by the corresponding discussion. Finally, the conclusions achieved from the present work are summarized in Section 7.

2. System Description

The proposed configuration of the three-phase grid-connected PV plants is shown in Figure 3. This configuration is composed of six traditional three-phase two-level VSC modules, which have a hexagonal interconnection to produce higher voltage levels as shown in Figure 2. The circulating current in the obtained loop can be suppressed using six inductors, which present a small impedance at the switching frequency. Each VSC module supplies 1/6 of the converter output power. In addition, each module has one of its AC terminals (i.e., a, b, or c) designated as converter output. The three-phase AC output terminals of the HMC are labeled as

follows: A (AC terminal a1 of module 1), B (AC terminal b3 of module 3), and C (AC terminal c5 of module 5). The remaining three-phase AC output terminals are labeled as follows: A' (AC terminal a4 of module 4), B' (AC terminal b6 of module 6), and C' (AC terminal c2 of module 2). The other two AC terminals of each module are, respectively, connected to an adjacent module through an inductor. For instance, AC terminal b1 of module 1 is coupled to AC terminal b2 of module 2 through an inductor L_d; AC terminal c1 of module 1 is coupled to AC terminal c6 of module 6 through another inductor. The remaining modules consist of similar connections, as shown in Figure 2.

As shown in Figure 3, the PV system based on the HMC has six output terminals. Therefore, when linked to the three-phase electrical grid, an open-end winding (OEW) transformer is necessary to provide these six terminals. The secondary windings of the OEW transformer are connected differentially. In the proposed converter, AC terminals A-A', B-B', and C-C' are used to provide phases A, B, and C, respectively. High-voltage windings are arranged in a star configuration and coupled directly to the three-phase grid. Each VSC module is supplied with a separate PV string to produce two-level individual

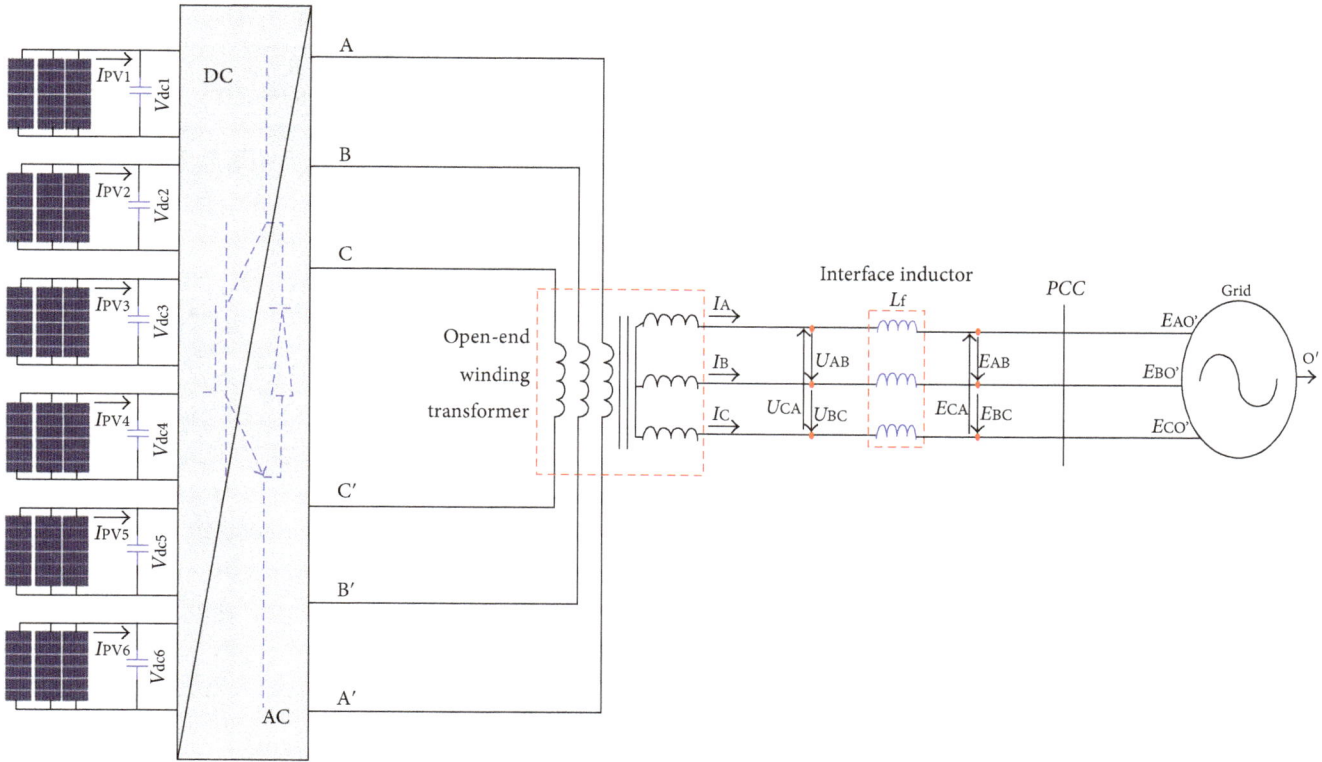

FIGURE 3: Complete HMC for a grid-connected PV system.

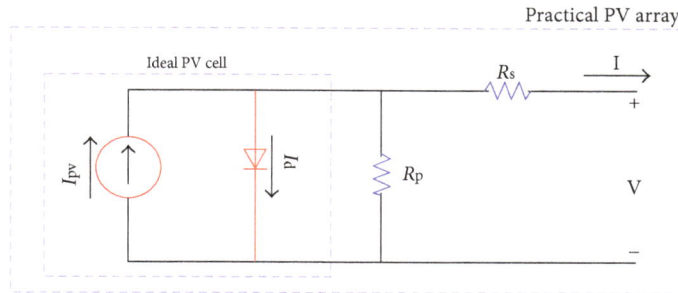

FIGURE 4: Equivalent model of a PV array.

outputs. The proposed converter topology presents several advantages, such as the use of fewer power electronic switches, diodes, and storage capacitors when compared with other topologies, making it suitable for renewable energy and other applications.

3. System Modeling

3.1. PV System Modeling and MPPT Control. Figure 4 shows the equivalent model of a PV array. The PV array is composed of several series-parallel connected solar cells. The basic equation of the PV array is given by [20].

$$I = I_{\mathrm{pv}} - I_0 \left[\exp\left(\frac{V + R_s I}{a V_t} \right) - 1 \right] - \frac{V + R_s I}{R_p}, \quad (1)$$

where $I_{\mathrm{pv}} = N_{\mathrm{p}} I_{\mathrm{pv,cell}}$ is the PV current of the array, $I_{\mathrm{pv,cell}}$ is the current generated by incident light (directly proportional to Sun irradiation), N_{p} is the number of cells connected in parallel, $I_0 = N_{\mathrm{p}} I_{0,\mathrm{cell}}$ is the saturation current of the array, $I_{0,\mathrm{cell}}$ is the reverse saturation or leakage current of the diode, $V_{\mathrm{t}} = N_{\mathrm{s}} kT/q$ is the thermal voltage of the array, N_{s} is the number of cells connected in series, q is the electron charge, k is the Boltzmann constant, T is the temperature of the p-n junction, a is the diode ideality constant, R_{s} is the equivalent series resistance of the array, and R_{p} is the equivalent parallel resistance. Table 1 shows the nominal parameters of the KC200GT PV array. Figure 5 shows the I-V and P-V curves obtained using (1) of the PV array under variable solar radiation.

TABLE 1: Parameters of the KC200GT PV array at 25°C, AM1.5, and 1000 W/m².

I_{mp} (A)	V_{mp} (V)	P_{max} (W)	I_{sc} (A)	V_{oc} (V)	K_v (V/K)	K_I (A/K)	N_{ss} (No.)	R_p (Ω)	R_s (Ω)	$I_{o,n}$ (A)	a
7.61	26.3	200.143	8.21	32.9	−0.123	0.0032	54	415.405	0.221	9.825×10^{-8}	1.3

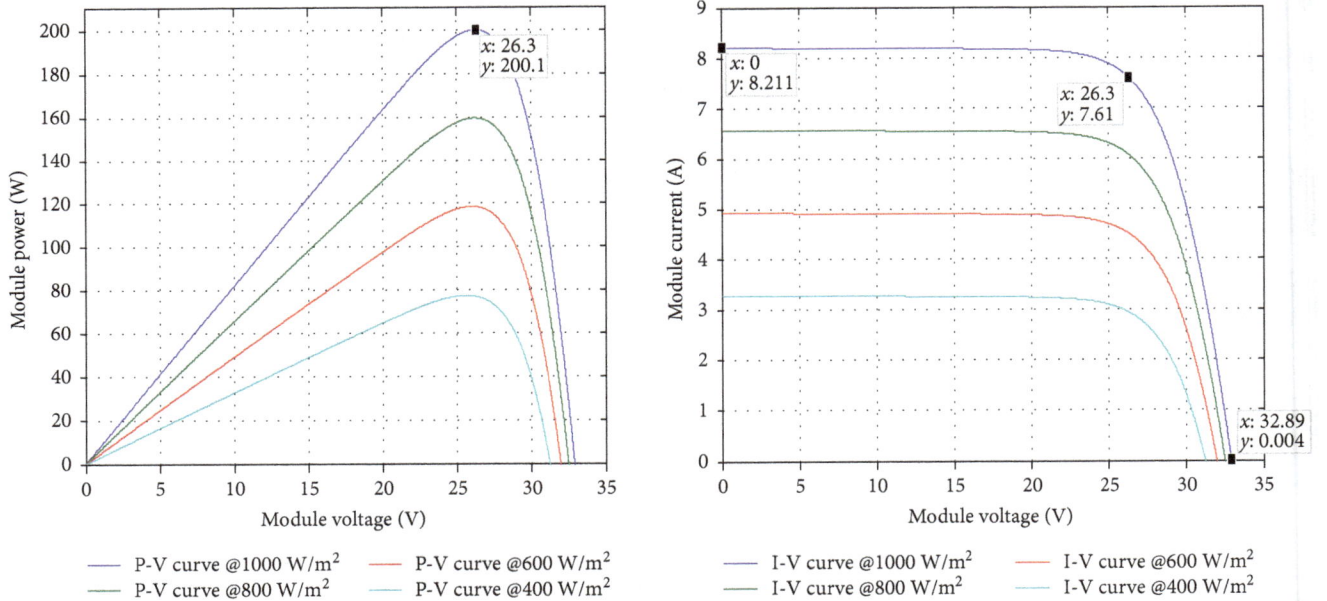

FIGURE 5: I-V and P-V curves of the KC200GT PV array.

3.2. Maximum Power Point Tracking.

The proposed system consists of 12 PV modules distributed as six pairs of series-connected modules and coupled directly to a single DC link (see Figure 3). Therefore, the power rating of the proposed converter is 2.4 kW. The controller is aimed to regulate the voltage of the PV-module pair at 52.6 V, to guarantee maximum power transfer to the grid.

Next, the aim of the MPPT control algorithm used in this paper is to ensure that under any solar radiation and temperature conditions, the maximum power is extracted from the PV modules. This is achieved by matching the PV-array maximum power point to the corresponding operating voltage and current of the converters. The perturb and observe (P&O) algorithm is a commonly used MPPT technique because it is easy to implement [21]. The operating principle of the P&O algorithm is shown in Figure 6. This algorithm measures the PV plant voltage and current, then it varies the operating voltage and compares the power received between the two voltage values. After each perturbation, the algorithm compares the output power from the PV before and after the perturbation. The direction of a new perturbation depends upon the output power: the perturbation will follow the same direction if higher power is measured when the output voltage varies and the opposite direction otherwise. These procedures are repeated continuously, and the reference voltage is generated and fed to the converter controller.

The numerical illustration of the P&O algorithm [2] is given as follows:

$$\frac{dP_{pv}(k)}{dV_{pv}(k)} = \frac{P_{pv}(k) - P_{pv}(k-1)}{V_{pv}(k) - V_{pv}(k-1)}. \tag{2}$$

Here, $P_{pv}(k)$ and $P_{pv}(k-1)$ stand for current power and previous measured power, while $V_{pv}(k)$ and $V_{pv}(k-1)$ stand for current PV voltage and previous one.

3.3. Voltage and Current Analysis.

Under the symmetrical operation condition, the fundamental components of the six VSC modules are the same. The corresponding open-circuit voltage of the HMC is presented in Figure 7.

3.3.1. Current Relations.

The phase currents of each VSC modules fulfill the following expressions:

$$\begin{bmatrix} i_{a1} + i_{b1} + i_{c1} \\ i_{a2} + i_{b2} + i_{c2} \\ i_{a3} + i_{b3} + i_{c3} \\ i_{a4} + i_{b4} + i_{c4} \\ i_{a5} + i_{b5} + i_{c5} \\ i_{a6} + i_{b6} + i_{c6} \end{bmatrix} = 0. \tag{3}$$

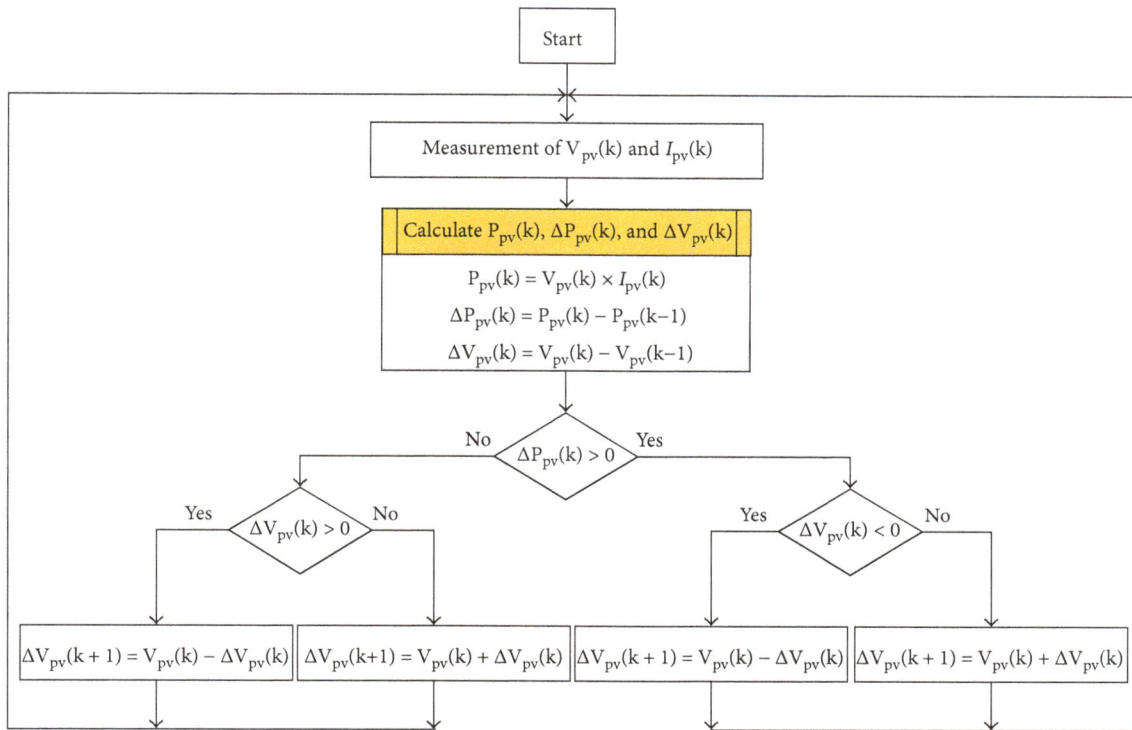

FIGURE 6: Flowchart for P&O MPPT method.

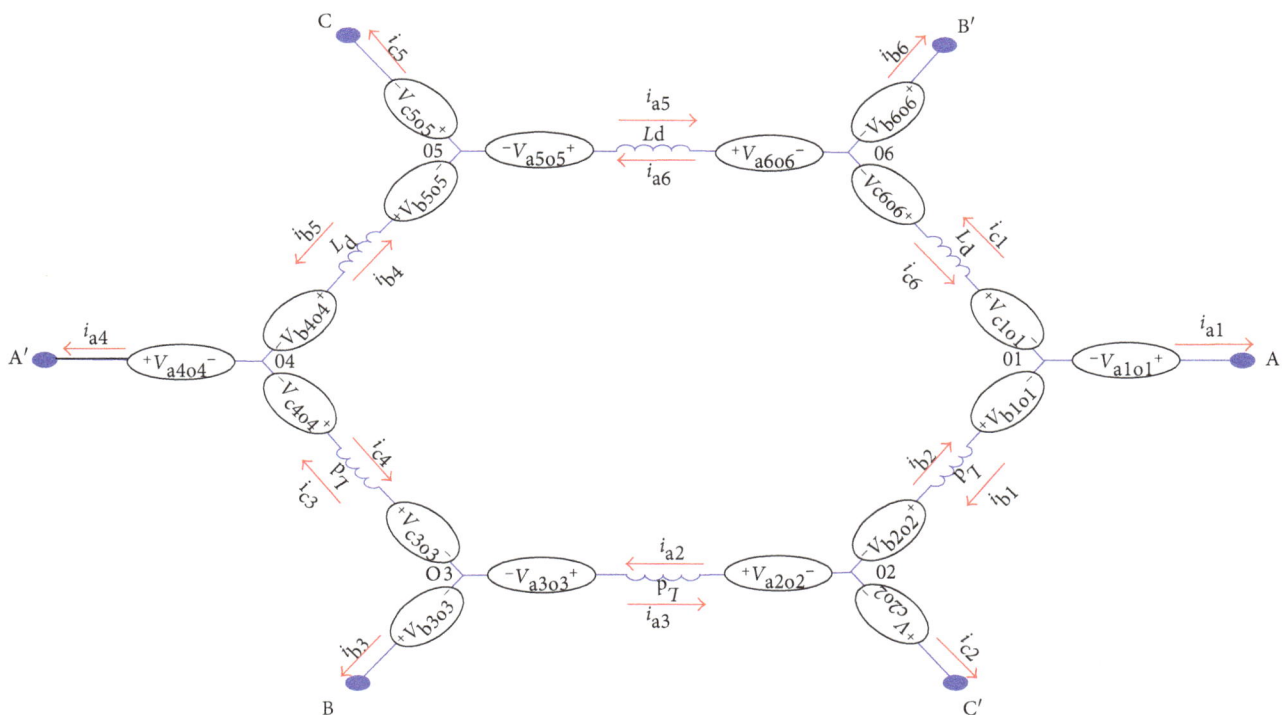

FIGURE 7: Phasor voltage of the HMC.

Assuming that the intermediate coupled inductors have large magnetizing inductance, the circulating currents are suppressed to a low value and can be ignored, hence

$$i_{b1} + i_{a2} + i_{c3} + i_{b4} + i_{a5} + i_{c6} = 0. \tag{4}$$

Because every pair of the six VSC modules is coupled, the current within the proposed converter has the following expressions:

$$\begin{bmatrix} i_{b1} \\ i_{a2} \\ i_{c3} \\ i_{b4} \\ i_{a5} \\ i_{c6} \end{bmatrix} = - \begin{bmatrix} i_{b2} \\ i_{a3} \\ i_{c4} \\ i_{b5} \\ i_{a6} \\ i_{c1} \end{bmatrix}. \tag{5}$$

Next, suppose that the HMC is linked to a three-phase grid. Then, the output currents will satisfy the following equation:

$$\begin{bmatrix} I_A \\ I_B \\ I_C \end{bmatrix} = \begin{bmatrix} i_{a1} \\ i_{b3} \\ i_{c5} \end{bmatrix} = - \begin{bmatrix} i_{a4} \\ i_{b6} \\ i_{c2} \end{bmatrix} = \begin{bmatrix} I\lfloor 0 \\ I\lfloor -120 \\ I\lfloor +120 \end{bmatrix}. \tag{6}$$

Using (2), (3), and (4), the output current of each VSC module can be shown to be

$$\begin{bmatrix} i_{a1} \\ i_{b1} \\ i_{c1} \end{bmatrix} = \begin{bmatrix} i_{a3} \\ i_{b3} \\ i_{c3} \end{bmatrix} = \begin{bmatrix} i_{a5} \\ i_{b5} \\ i_{c5} \end{bmatrix} = - \begin{bmatrix} i_{a2} \\ i_{b2} \\ i_{c2} \end{bmatrix} = - \begin{bmatrix} i_{a4} \\ i_{b4} \\ i_{c4} \end{bmatrix} = - \begin{bmatrix} i_{a6} \\ i_{b6} \\ i_{c6} \end{bmatrix}$$
$$= \begin{bmatrix} I_A \\ I_B \\ I_C \end{bmatrix} = \begin{bmatrix} I\lfloor 0 \\ I\lfloor -120 \\ I\lfloor +120 \end{bmatrix}, \tag{7}$$

where $[I_A, I_B, I_C]^T$ and I are the converter output phase currents and their RMS values, respectively. It can be concluded from (5) that each VSC module within the HMC will have an identical current under the symmetrical operation condition.

3.3.2. Voltage Relations.
Under the symmetrical operation condition, the fundamental component (RMS) of the phase voltages at the VSC modules can be described as

$$\begin{bmatrix} V_{a1o1} \\ V_{b1o1} \\ V_{c1o1} \end{bmatrix} = \begin{bmatrix} V_{a3o3} \\ V_{b3o3} \\ V_{c3o3} \end{bmatrix} = \begin{bmatrix} V_{a5o5} \\ V_{b5o5} \\ V_{c5o5} \end{bmatrix} = - \begin{bmatrix} V_{a2o2} \\ V_{b2o2} \\ V_{c2o2} \end{bmatrix}$$
$$= - \begin{bmatrix} V_{a4o4} \\ V_{b4o4} \\ V_{c4o4} \end{bmatrix} = - \begin{bmatrix} V_{a6o6} \\ V_{b6o6} \\ V_{c6o6} \end{bmatrix} = \begin{bmatrix} V\lfloor 0 \\ V\lfloor -120 \\ V\lfloor +120 \end{bmatrix}, \tag{8}$$

$$V = \frac{1}{2\sqrt{2}} V_{dc} m_a, \tag{9}$$

where V, V_{dc}, and m_a are the RMS values of each VSC-module phase voltage, the DC-link voltage of each VSC module, and the amplitude modulation index, respectively; subscripts o1, o2, o3, o4, o5, and o6 represent the virtual neutral points of each VSC.

The output voltages of the proposed converter are written as

$$\begin{bmatrix} V_{AA'} \\ V_{BB'} \\ V_{CC'} \end{bmatrix} = \begin{bmatrix} V_{a1a4} \\ V_{b3b6} \\ V_{c2c5} \end{bmatrix}$$
$$= \begin{bmatrix} V_{a1o1} - V_{a4o4} + V_{c4o4} - V_{c3o3} + V_{a3o3} - V_{a2o2} + V_{b2o2} - V_{b1o1} \\ V_{b3o3} - V_{b6o6} + V_{a6o6} - V_{a5o5} + V_{b5o5} - V_{b4o4} + V_{c4o4} - V_{c3o3} \\ V_{c5o5} - V_{c2o2} + V_{b2o2} - V_{b1o1} + V_{c1o1} - V_{c6o6} + V_{a6o6} - V_{a5o5} \end{bmatrix}$$
$$- 2Z_d \begin{bmatrix} i_{a1} \\ i_{b3} \\ i_{c5} \end{bmatrix}, \tag{10}$$

where impedance $Z_d = j\omega L_d$.

Thus, using (5), (6), (7), and (8), the net output voltages of the HMC under the symmetrical operation condition are given by

$$\begin{bmatrix} V_{AA'} \\ V_{BB'} \\ V_{CC'} \end{bmatrix} = \begin{bmatrix} V_{a1a4} \\ V_{b3b6} \\ V_{c2c5} \end{bmatrix} = \begin{bmatrix} 6V\lfloor 0 \\ 6V\lfloor -120 \\ 6V\lfloor +120 \end{bmatrix} - 2Z_d \begin{bmatrix} I_A \\ I_B \\ I_C \end{bmatrix}. \tag{11}$$

Consequently, (9) demonstrates that the three-phase voltage of the HMC is approximately six times the phase voltage of each VSC module. In other words, the voltage stress is reduced by a factor of six.

3.4. Equivalent Circuit Model.
Based on Figure 3, the line voltages of the HMC are expressed by

$$
\begin{bmatrix} U_{AB} \\ U_{BC} \\ U_{CA} \end{bmatrix} = \begin{bmatrix} V_{AA'} - V_{BB'} \\ V_{BB'} - V_{CC'} \\ V_{CC'} - V_{AA'} \end{bmatrix} = \begin{bmatrix} 6\sqrt{3}V\lfloor-30 \\ 6\sqrt{3}V\lfloor-150 \\ 6\sqrt{3}V\lfloor+90 \end{bmatrix}
$$

$$
- 2j\omega L_d \begin{bmatrix} I_A - I_B \\ I_B - I_C \\ I_C - I_A \end{bmatrix} = \begin{bmatrix} E_{AB} \\ E_{BC} \\ E_{CA} \end{bmatrix} \qquad (12)
$$

$$
+ j\omega L_f \begin{bmatrix} I_A \\ I_B \\ I_C \end{bmatrix} - j\omega L_f \begin{bmatrix} I_B \\ I_C \\ I_A \end{bmatrix},
$$

where $[E_{AB}, E_{BC}, E_{CA}]^T$ are the RMS values of the line-to-line voltages in the grid, $[U_{AB}, U_{BC}, U_{CA}]^T$ and $[I_A, I_B, I_C]^T$ are the RMS values of line voltages and the RMS values of the output current, respectively, and L_f is the AC-side filtering inductance.

When the conversion powers between the VSC modules within the HMC are balanced, the six DC-link voltages are equal to V_{DCav}, hence

$$
V_{DCav} = V_{dc1} = V_{dc2} = V_{dc3} = V_{dc4} = V_{dc5}
$$
$$
= V_{dc6} = \frac{V_{dc1} + V_{dc2} + V_{dc3} + V_{dc4} + V_{dc5} + V_{dc6}}{6}.
$$
$$(13)$$

Moreover, the equivalent DC-link voltage can be expressed as

$$
U_{DC\_eq} = 6V_{DC\_av}. \qquad (14)
$$

If each VSC is driven using the same PWM, the relationship between the DC-link voltages and the phase voltages can be expressed as

$$
V = \frac{1}{2\sqrt{2}} U_{DC\_eq} m_a. \qquad (15)
$$

From (12) and (14), the HMC can be modeled by

$$
\begin{bmatrix} \overline{U_{AB}} \\ \overline{U_{BC}} \\ \overline{U_{CA}} \end{bmatrix} = \begin{bmatrix} E_{AB} \\ E_{BC} \\ E_{CA} \end{bmatrix} + j\omega(L_f + 2L_d) \begin{bmatrix} I_A \\ I_B \\ I_C \end{bmatrix} - j\omega(L_f + 2L_d) \begin{bmatrix} I_B \\ I_C \\ I_A \end{bmatrix},
$$
$$(16)$$

where

$$
\begin{bmatrix} \overline{U_{AB}} \\ \overline{U_{BC}} \\ \overline{U_{CA}} \end{bmatrix} = \begin{bmatrix} 6\sqrt{3}V\lfloor-30 \\ 6\sqrt{3}V\lfloor-150 \\ 6\sqrt{3}V\lfloor+90 \end{bmatrix}. \qquad (17)
$$

Using (14) and (17), the HMC can be modeled as a conventional three-phase two-level VSC, as shown in Figure 8.

Figure 8 shows the equivalent circuit model of the HMC. The equivalent circuit is composed of an open-circuit voltage source linked in series to the HMC output impedance. The open-circuit voltage of the HMC is shown in Figure 7. As shown in Figure 8, the equivalent interface inductor of the HMC is $2L_d + L_f$. This feature is advantageous, because the filtering inductor L_f could be minimalized or even removed. However, this can be inopportune when the line transformer has sufficient inductance for filtering the output current. In this case, essentially if the HMC supplies reactive power, the output impedance causes a voltage drop, decreasing the power capability. Thus, it might be better to minimize the interconnected total inductances. However, the inductances are essential to suppress the circulating current within the HMC. The aforementioned decision shows that the interconnected inductance value selection is not a simple process, where the inductance should be large enough to minimize the circulating current but small enough to avoid an extra voltage drop. In the following section, this feature is investigated using multiwinding coupled inductors.

4. The Role of the Coupled Inductors

The function of the multiwinding coupled inductors within the HMC is investigated in this section. The HMC performance analysis can be demonstrated based on the function of the coupled inductors. The six VSC modules are interconnected to each other, creating a hexagon as explained in Section 3.4. However, instead of using six inductors, multiwinding coupled inductors are used to interconnect the VSC modules.

4.1. HMC with Three Two-Winding Coupled Inductors. Two inductors with an equal number of turns are coupled together; that is, the input to one side will produce an output on both, as shown in Figure 9. Since the turn ratio of the coupled inductor is approximately $1:1$, the self-inductances of the primary and secondary windings are the same (e.g., $L_p = L_s = L_B$). Applying the voltage relationships of coupled inductors, the following equations are given:

$$
v_{b1b2} = L_{11} \frac{di_{b1}}{dt} + L_{12} \frac{di_{b5}}{dt}, \qquad (18)
$$

and

$$
v_{b5b4} = L_{22} \frac{di_{b5}}{dt} + L_{21} \frac{di_{b1}}{dt}, \qquad (19)
$$

where $L_{22} = L_{11} = L_B + L_m$, $L_m = L_{21} = L_{12}$, $i_{b1} = i_{b1} + i_{cir}$, and $i_{b5} = -i_{b1} + i_{cir}$, and where i_x $(x = a, b, c)$ is the line current, L_B is the leakage inductance of the inductor windings, which is expected to be identical for all windings, and L_m is the magnetizing inductance of the coupled inductor.

It is obvious in Figure 10 that the voltage drop on one of the coupled inductors is

$$
v = v_{b1b2} + v_{b5b4}. \qquad (20)
$$

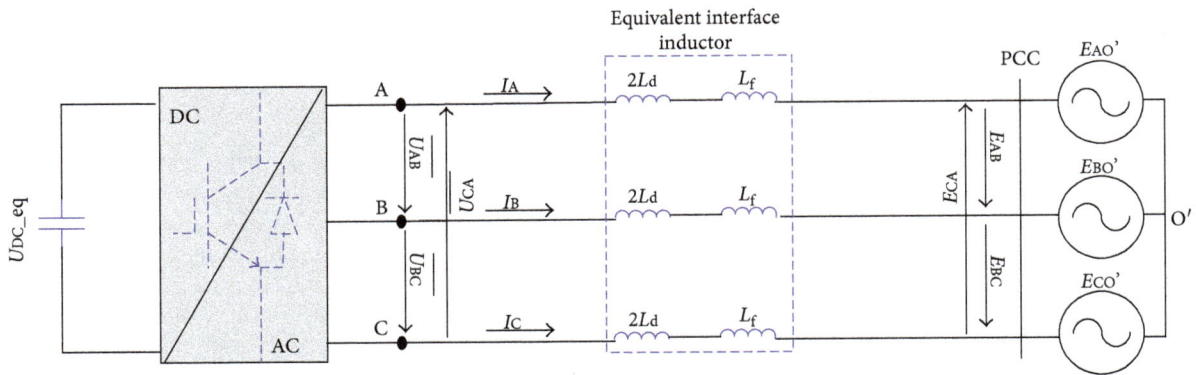

FIGURE 8: Equivalent circuit model of HMC.

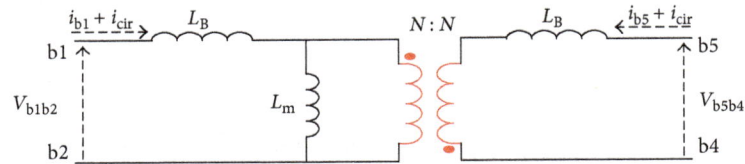

FIGURE 9: Equivalent circuit of the two-winding coupled inductor.

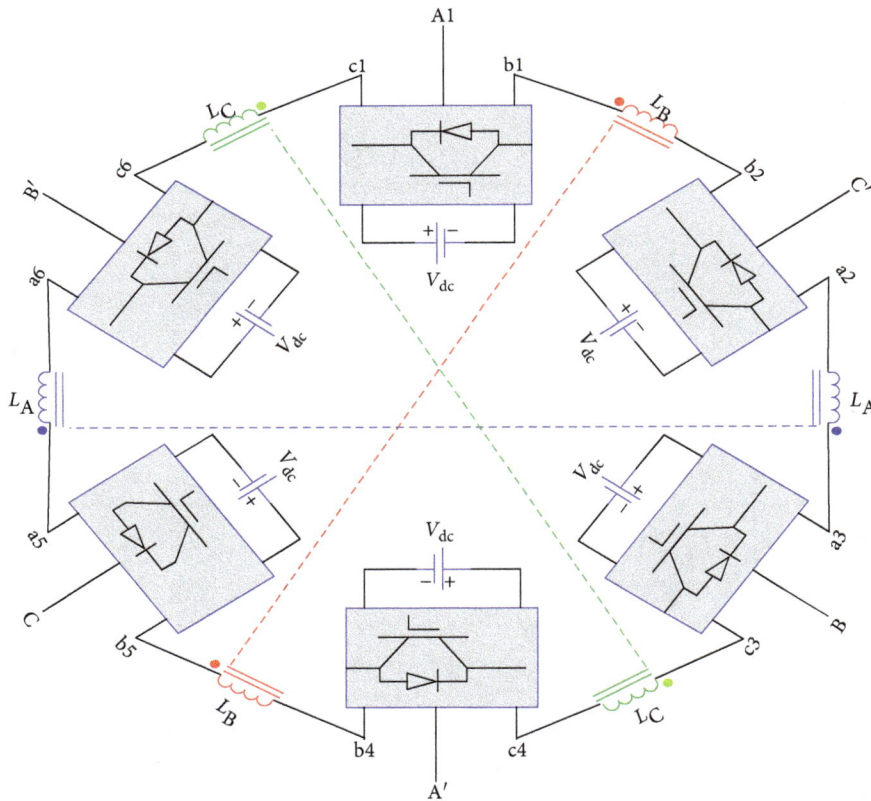

FIGURE 10: The HMC with three two-winding coupled inductors.

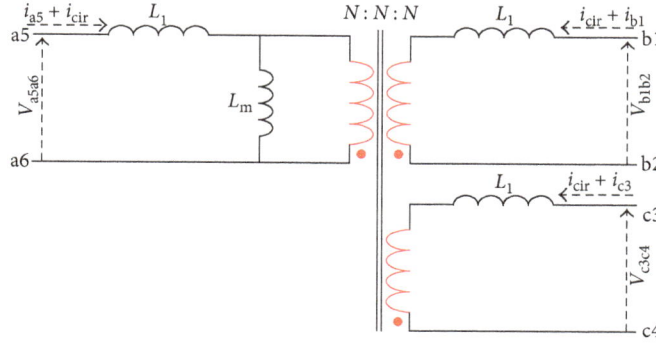

FIGURE 11: Equivalent circuit of the three-winding coupled inductor.

Using (18) and (19), the voltage across the coupled inductor can be expressed as

$$v = (L_B + L_m)\frac{d(i_{b1} + i_{cir})}{dt} + L_m\frac{d(-i_{b1} + i_{cir})}{dt}$$
$$+ (L_B + L_m)\frac{d(-i_{b1} + i_{cir})}{dt} + L_m\frac{d(i_{b1} + i_{cir})}{dt}, \quad (21)$$
$$v = (2L_B + 4L_m)\frac{d(i_{cir})}{dt}.$$

Therefore, the total voltage drop across the three coupled inductors can be expressed as

$$v_{total} = 3(2L_B + 4L_m)\frac{d(i_{cir})}{dt} = (6L_B + 12L_m)\frac{d(i_{cir})}{dt}. \quad (22)$$

4.2. HMC with Two Three-Winding Coupled Inductors.

Three inductors with an equal number of turns are coupled together as shown in Figure 11. Writing the voltage equations of one coupled inductor with winding inductance Z1, the following equations are given:

$$v_{a5a6} = L_a\frac{di_a}{dt} + L_{ab}\frac{di_b}{dt} + L_{ac}\frac{di_c}{dt},$$
$$v_{b1b2} = L_b\frac{di_b}{dt} + L_{ba}\frac{di_a}{dt} + L_{bc}\frac{di_c}{dt}, \quad (23)$$
$$v_{c3c4} = L_c\frac{di_c}{dt} + L_{ca}\frac{di_a}{dt} + L_{cb}\frac{di_b}{dt},$$

where

$$L_a = L_b = L_c = L + L_m, L_m = L_{ab} = L_{ba} = L_{bc} = L_{cb} = L_{ca} = L_{ac}, \quad (24)$$

and

$$i_a = i_{a5} + i_{cir},$$
$$i_b = i_{b1} + i_{cir}, \quad (25)$$
$$i_c = i_{c3} + i_{cir},$$

where $i_x\ (x = a, b, c)$ is the line current, L is the leakage inductance of the inductor windings, and L_m is the magnetizing

inductance. It is obvious in Figure 12 that the voltage drop on the coupled inductor is

$$v = v_{a5a6} + v_{b1b2} + v_{c3c4}. \quad (26)$$

Using (23), the voltage drop across one of the coupled inductor can be expressed as

$$v = (L + L_m)\frac{d(i_{a5} + i_{cir})}{dt} + L_m\frac{d(i_{b1} + i_{cir})}{dt} + L_m\frac{d(i_{c3} + i_{cir})}{dt}$$
$$+ (L + L_m)\frac{d(i_{b1} + i_{cir})}{dt} + L_m\frac{d(i_{a5} + i_{cir})}{dt} + L_m\frac{d(i_{c3} + i_{cir})}{dt}$$
$$+ (L + L_m)\frac{d(i_{c3} + i_{cir})}{dt} + L_m\frac{d(i_{a5} + i_{cir})}{dt} + L_m\frac{d(i_{b1} + i_{cir})}{dt},$$
$$v = (L + L_m)\frac{d(i_{a5} + i_{b1} + i_{c3})}{dt} + 3(L + L_m)\frac{d(i_{cir})}{dt}$$
$$+ 2L_m\frac{d(i_{a5} + i_{b1} + i_{c3})}{dt} + 6L_m\frac{d(i_{cir})}{dt}.$$

$$(27)$$

The voltage drop across the coupled inductor windings is

$$v = (L + 3L_m)\frac{d(i_{a5} + i_{b1} + i_{c3})}{dt} + (3L + 9L_m)\frac{d(i_{cir})}{dt}. \quad (28)$$

Under balanced conditions,

$$i_{a5} + i_{b1} + i_{c3} = i_a + i_b + i_c = 0. \quad (29)$$

Accordingly, by applying (28) and (29), the voltage drop on the coupled inductor windings is

$$v = (3L + 9L_m)\frac{d(i_{cir})}{dt}. \quad (30)$$

The total voltage drop across the two coupled inductors can be expressed as

$$v_{total} = (6L_B + 18L_m)\frac{d(i_{cir})}{dt}. \quad (31)$$

4.3. HMC with One Six-Winding Coupled Inductor.

Instead of using six inductances, one six-winding coupled inductor can be used to interconnect the VSC modules. Since the turn ratio of the coupled inductor is approximately 1 : 1, the self-inductances of all the windings are the same as shown in

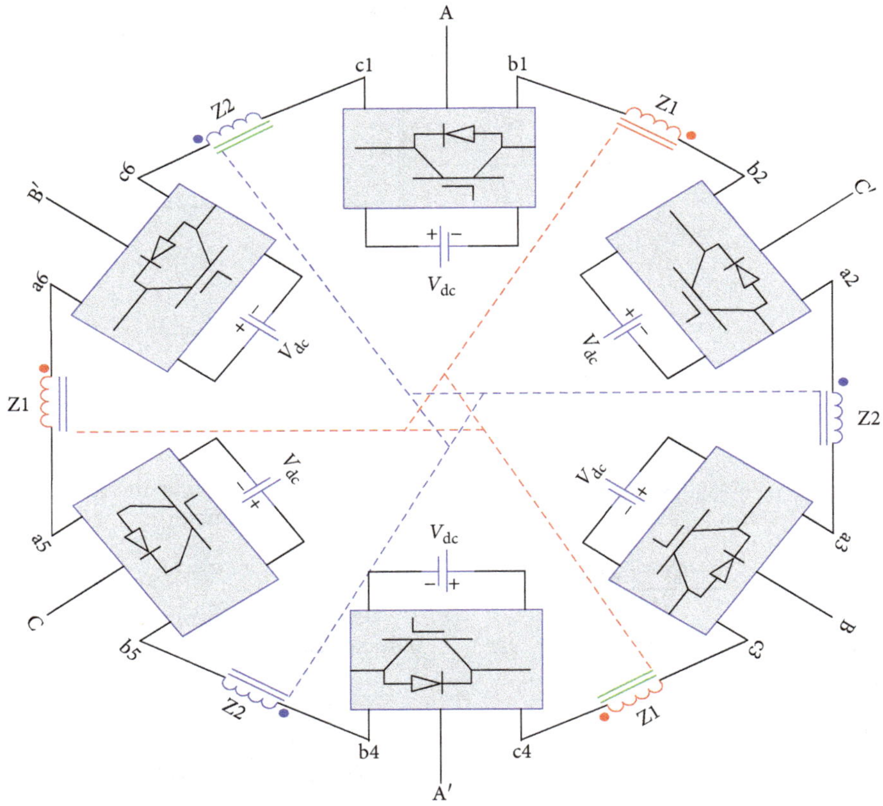

FIGURE 12: The HMC with two three-winding coupled inductors.

Figure 13. The HMC with one six-winding coupled inductor is shown in Figure 14. Applying the voltage equations of the coupled inductor, the following equations are given:

$$v_{a5a6} = L_a \frac{di_{a5}}{dt} + L_m \frac{di_{b1} + di_{c3} + di_{b5} + di_{a3} + di_{c1}}{dt},$$

$$v_{b1b2} = L_b \frac{di_{b1}}{dt} + L_m \frac{di_{a5} + di_{c3} + di_{b5} + di_{a3} + di_{c1}}{dt},$$

$$v_{c3c4} = L_c \frac{di_{c3}}{dt} + L_m \frac{di_{a5} + di_{b1} + di_{b5} + di_{a3} + di_{c1}}{dt},$$

$$v_{a3a2} = L_a \frac{di_{a3}}{dt} + L_m \frac{di_{b1} + di_{c3} + di_{b5} + di_{a5} + di_{c1}}{dt},$$ (32)

$$v_{b5b4} = L_b \frac{di_{b5}}{dt} + L_m \frac{di_{a5} + di_{c3} + di_{b1} + di_{a3} + di_{c1}}{dt},$$

$$v_{c1c6} = L_c \frac{di_{c1}}{dt} + L_m \frac{di_{a5} + di_{b1} + di_{b5} + di_{a3} + di_{c3}}{dt}.$$

Since,

$$i_a = i_{a5} = i_{a3} = i_{a1} = -i_{a2} = -i_{a4} = -i_{a6},$$

$$i_b = i_{b5} = i_{b3} = i_{b1} = -i_{b2} = -i_{b4} = -i_{b6},$$ (33)

$$i_c = i_{c5} = i_{c3} = i_{c1} = -i_{c2} = -i_{c4} = -i_{c6}.$$

The voltage drop equations across each coupled inductor can be expressed as

$$v_{a5a6} = v_{a3a2} = L_a \frac{di_a}{dt} + L_m \frac{di_a + 2di_b + 2di_c}{dt},$$

$$v_{b1b2} = v_{b5b4} = L_b \frac{di_b}{dt} + L_m \frac{di_b + 2di_a + 2di_c}{dt},$$ (34)

$$v_{c3c4} = v_{c1c6} = L_c \frac{di_b}{dt} + L_m \frac{di_c + 2di_a + 2di_b}{dt},$$

where

$$L_a = L_b = L_c = L + L_m,$$ (35)

and

$$i_a = i_a + i_{cir},$$

$$i_b = i_b + i_{cir},$$ (36)

$$i_c = i_c + i_{cir}.$$

Using (34), the voltage drop across the coupled inductor can be expressed as

$$v = v_{a5a6} + v_{b1b2} + v_{c3c4} + v_{a3a2} + v_{b5b4} + v_{c1c6},$$ (37)

$$v = 2v_{a5a6} + 2v_{b1b2} + 2v_{c3c4},$$ (38)

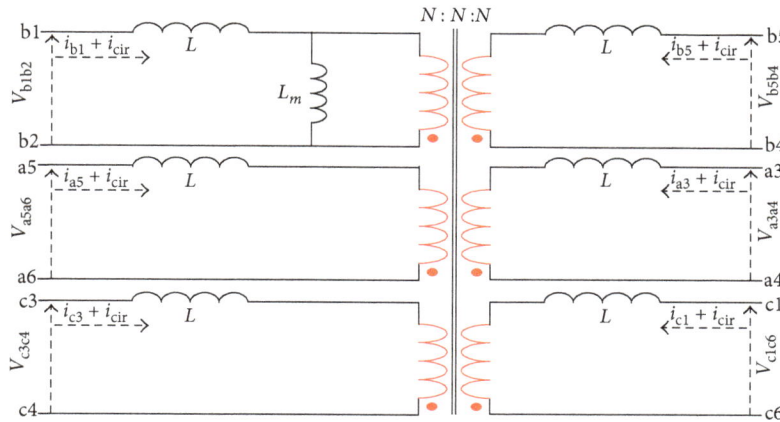

FIGURE 13: Equivalent circuit of the six-winding coupled inductor.

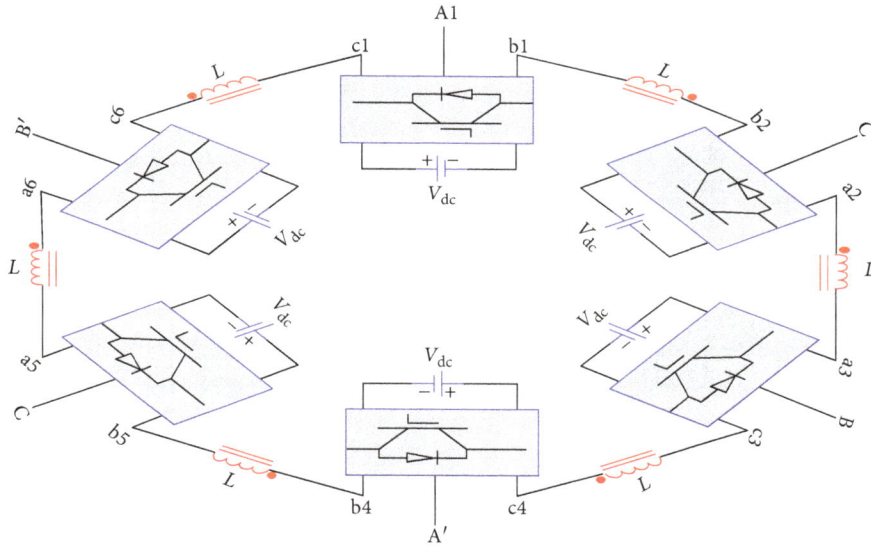

FIGURE 14: The HMC with one six-winding coupled inductor.

$$v = \left(6L + 36L_m\right)\frac{di_{cir}}{dt} + \left(2L + 12L_m\right)\frac{d(i_a + i_b + i_c)}{dt}. \quad (39)$$

Under balanced conditions,

$$i_a + i_b + i_c = 0. \quad (40)$$

Consequently, by applying (39) and (40), the voltage drop on the coupled inductor windings is

$$v = \left(6L + 36L_m\right)\frac{di_{cir}}{dt}. \quad (41)$$

It is considered that the magnetizing inductance is much higher than the leakage inductance. Therefore, neglecting the leakage inductance, the two-winding coupled inductor imposes twelve times the magnetizing inductance for the circulating current, while the impedance to the circulating current using three-winding coupled inductors is eighteen as much as the magnetizing inductance. Thanks to the six-winding coupled inductor, the impedance on the circulating current path is thirty six times the magnetizing inductance.

In a general way if k is the number of the coupled inductor windings, the impedance to the circulating current can be expressed as

$$Z_{cir} = 2\pi f \left(6kL_m\right). \quad (42)$$

5. Control Scheme and Modulation Strategy

5.1. Control Scheme. The main function of the controller is to generate reference currents such that the proposed converter only provides available active power from the DC links to the grid at the point of common coupling (PCC) [22, 23]. Using the equivalent circuit model presented in

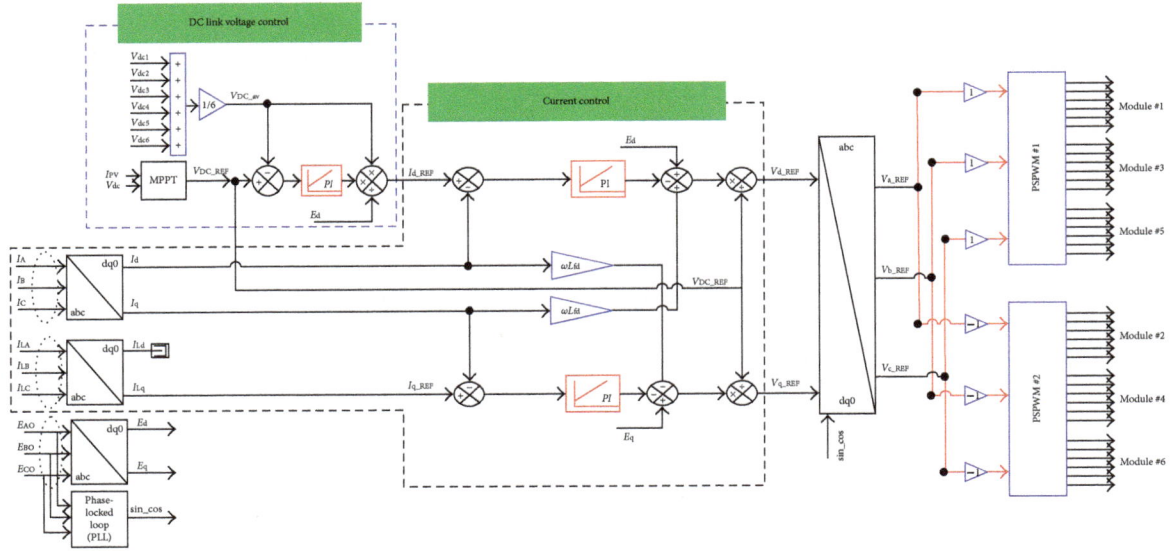

FIGURE 15: HMC control block diagram.

Figure 7 and applying Kirchhoff's voltage and current laws at the PCC, the following two equations in the abc frame can be obtained:

$$\begin{bmatrix} E_{AO'} \\ E_{BO'} \\ E_{CO'} \end{bmatrix} = \begin{bmatrix} \overline{U_{AO}} \\ \overline{U_{BO}} \\ \overline{U_{CO}} \end{bmatrix} + L_{fd}\frac{d}{dt}\begin{bmatrix} I_A \\ I_B \\ I_C \end{bmatrix} + U_{O'O}, \qquad (43)$$

$$C_{eq}\frac{dU_{DC\_av}}{dt} = [S_A\ S_B\ S_C]\begin{bmatrix} I_A \\ I_B \\ I_C \end{bmatrix}, \qquad (44)$$

where $L_{fd} = L_f + L_d$ is the equivalent interface inductor and S_A, S_B, and S_C represent the switching states of the equivalent circuit model under balanced conditions.

Assuming that the voltages are balanced and the zero-sequence component is zero, the voltage between the neutral virtual point of equivalent circuit model (O) and the grid neutral point (O') is given by

$$U_{O'O} = -\frac{(\overline{U_{AO}} + \overline{U_{BO}} + \overline{U_{CO}})}{3}, \qquad (45)$$

$$\begin{bmatrix} \overline{U_{AO}} \\ \overline{U_{BO}} \\ \overline{U_{CO}} \end{bmatrix} = U_{DC\_av}\begin{bmatrix} S_A \\ S_B \\ S_C \end{bmatrix}. \qquad (46)$$

Substituting (46) into (43) and (44), the following relation is obtained:

$$\frac{d}{dt}\begin{bmatrix} I_A \\ I_B \\ I_C \end{bmatrix} = \frac{1}{L_{fd}}\begin{bmatrix} E_{AO'} \\ E_{BO'} \\ E_{CO'} \end{bmatrix}$$
$$- \frac{U_{DC_{av}}}{L_{fd}}\left(\begin{bmatrix} S_A \\ S_B \\ S_C \end{bmatrix} - \frac{1}{3}[S_A\ \ S_B\ \ S_C]\begin{bmatrix} 1 \\ 1 \\ 1 \end{bmatrix}\right). \qquad (47)$$

The dynamic model in the abc frame of the HMC equivalent circuit is represented by (47). The switching state functions, d_i ($i =$ A, B, C), are defined as

$$\begin{bmatrix} d_A \\ d_B \\ d_C \end{bmatrix} = \left(\begin{bmatrix} S_A \\ S_B \\ S_C \end{bmatrix} - \frac{1}{3}[S_A\ \ S_B\ \ S_C]\begin{bmatrix} 1 \\ 1 \\ 1 \end{bmatrix}\right). \qquad (48)$$

The dynamic model of the equivalent circuit model in the abc frame is achieved by combining (47) and (48) in the following equation:

$$L_{fd}\frac{d}{dt}\begin{bmatrix} I_A \\ I_B \\ I_C \end{bmatrix} = \begin{bmatrix} E_{AO'} \\ E_{BO'} \\ E_{CO'} \end{bmatrix} - U_{DC_{av}}\begin{bmatrix} d_A \\ d_B \\ d_C \end{bmatrix}. \qquad (49)$$

The DC side differential equation can be written as

$$\frac{dU_{DC\_av}}{dt} = \frac{1}{C_{eq}}I_{dc} = [d_A\ \ d_B\ \ d_C]\begin{bmatrix} I_A \\ I_B \\ I_C \end{bmatrix}, \qquad (50)$$

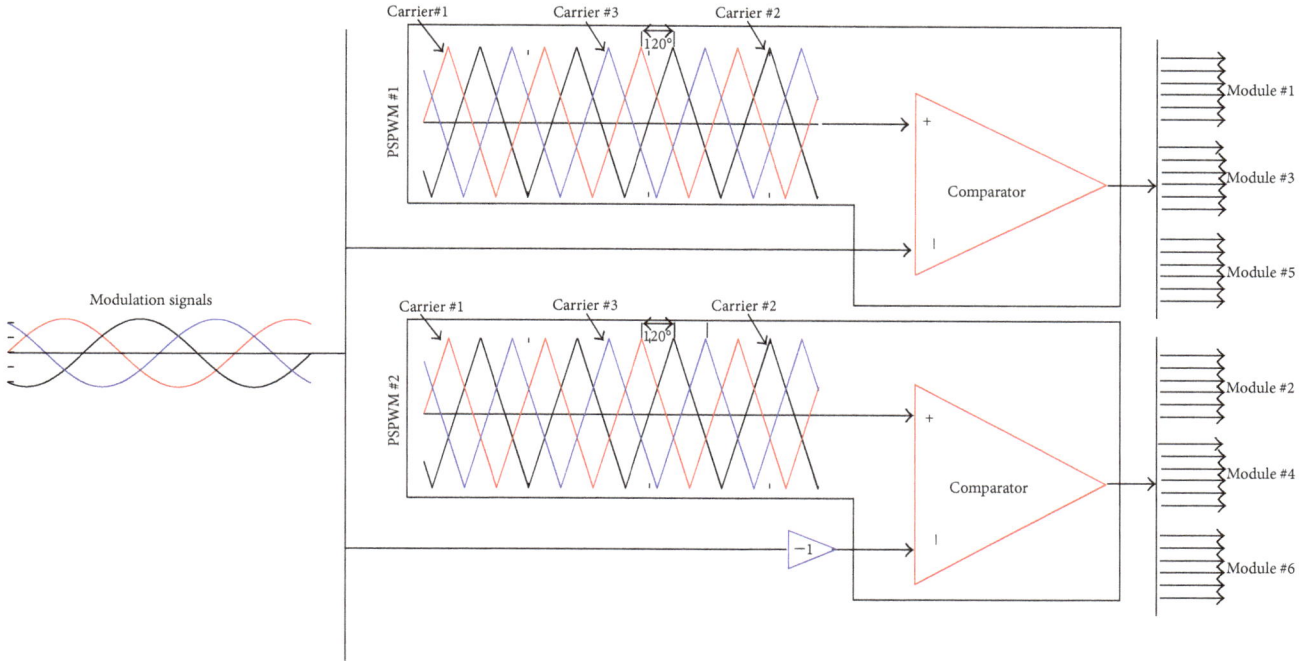

FIGURE 16: Proposed PS-PWM diagram.

$$\frac{dU_{DC\_av}}{dt} = \frac{1}{C_{eq}}(2d_A + d_B)I_A + \frac{1}{C_{eq}}(d_A + 2d_B)I_B. \quad (51)$$

It can be seen that the model represented by (49) and (51) is time varying. Thus, to facilitate the control algorithm implementation, the model can be expressed in the synchronous reference frame rotating at constant frequency ω. The corresponding conversion matrix is

$$C_{dq}^{abc} = \sqrt{\frac{2}{3}}\begin{bmatrix} \cos\theta & \cos\left(\theta - 2\frac{\pi}{3}\right) & \cos\left(\theta - 4\frac{\pi}{3}\right) \\ -\sin\theta & -\sin\left(\theta - 2\frac{\pi}{3}\right) & -\sin\left(\theta - 4\frac{\pi}{3}\right) \end{bmatrix}, \quad (52)$$

where $\theta = \omega t$.

Applying the coordinate transformation to (49) we obtain

$$L_{fd}\frac{d}{dt}\begin{bmatrix} I_d \\ I_q \end{bmatrix} = \begin{bmatrix} E_d \\ E_q \end{bmatrix} + L_f\omega\begin{bmatrix} I_q \\ -I_d \end{bmatrix} - U_{DC_{av}}\begin{bmatrix} d_d \\ d_q \end{bmatrix}. \quad (53)$$

Similarly, applying this transformation to (50) we obtain

$$C_{eq}\frac{dU_{DC\_av}}{dt} = d_dI_d + d_qI_q. \quad (54)$$

The obtained model, represented by (53) and (54), is nonlinear owing to the product between the state variables (i.e., I_d, I_q, and U_{DC_av}) and the inputs (i.e., d_d and d_q).

Figure 11 shows the control principle of the HMC. Because the proposed configuration is a three-wire system, only two phase currents are required to be measured.

TABLE 2: General data of the proposed system.

Parameter	Value
System parameters	
Nominal power	$P = 2.5\,\text{kVA}$
Phase voltage and frequency	$V_{ph} = 110\,\text{V (rms)}, f_{sys} = 50\,\text{Hz},$ $f_{swiching} = 2500\,\text{Hz}$
Output current	$I_{out} = 8.8\,\text{A (rms)}$
DC bus voltages	$V_{dc1} = V_{dc2} = V_{dc3} = V_{dc4}$ $= V_{dc5} = V_{dc6} = 52.6\,\text{V}$
Open-end winding transformers	$V_{pri}/V_{sec} = 1, f_{tran} = 50\,\text{Hz}$
Current controller parameters	$K_p = 55, K_i = 0.001$
Voltage controller parameters	$K_p = 3, K_i = 20$
Coupled inductor parameters	
RMS voltage, current	20 V (rms), 8.8 A (rms)
Magnetizing inductance	3.2 mH
Leakage inductance	150 μH

Currents I_A and I_B are measured and converted to the dq0 frame to obtain the corresponding currents I_d and I_q. Accordingly, (53) is rewritten as follows:

$$L_{fd}\begin{bmatrix} u_d \\ u_q \end{bmatrix} = \begin{bmatrix} E_d \\ E_q \end{bmatrix} + L_{fd}\omega\begin{bmatrix} I_q \\ -I_d \end{bmatrix} - U_{DC_{av}}\begin{bmatrix} d_d \\ d_q \end{bmatrix}, \quad (55)$$

where $\begin{bmatrix} u_d \\ u_q \end{bmatrix} = (d/dt)\begin{bmatrix} I_d \\ I_q \end{bmatrix}$.

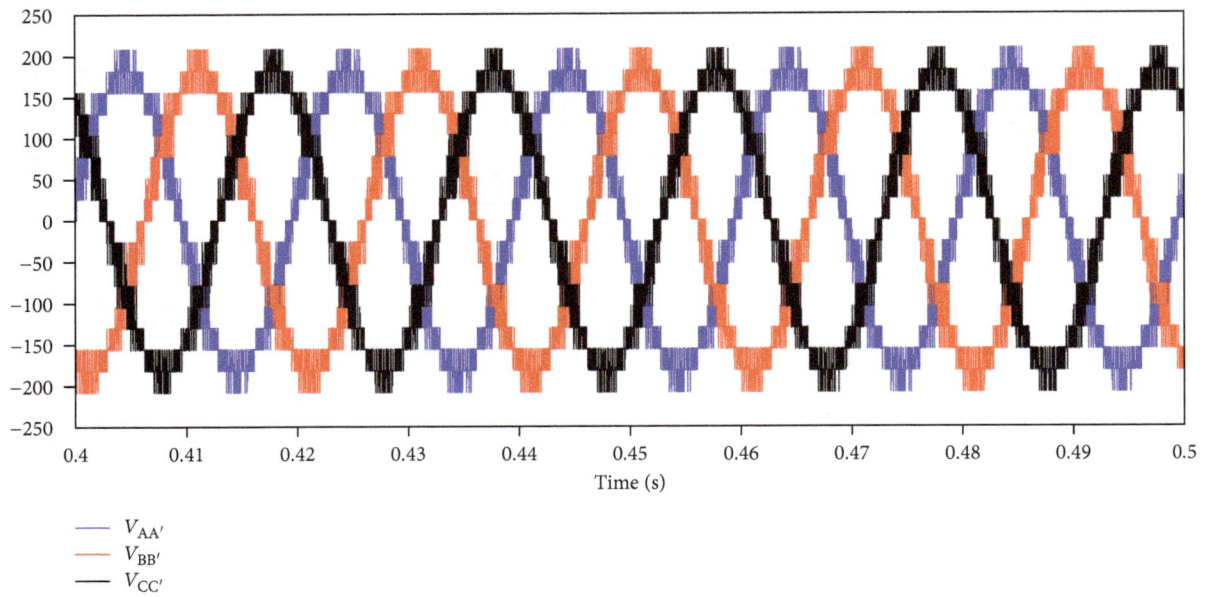

FIGURE 17: Three-phase output voltage of the HMC.

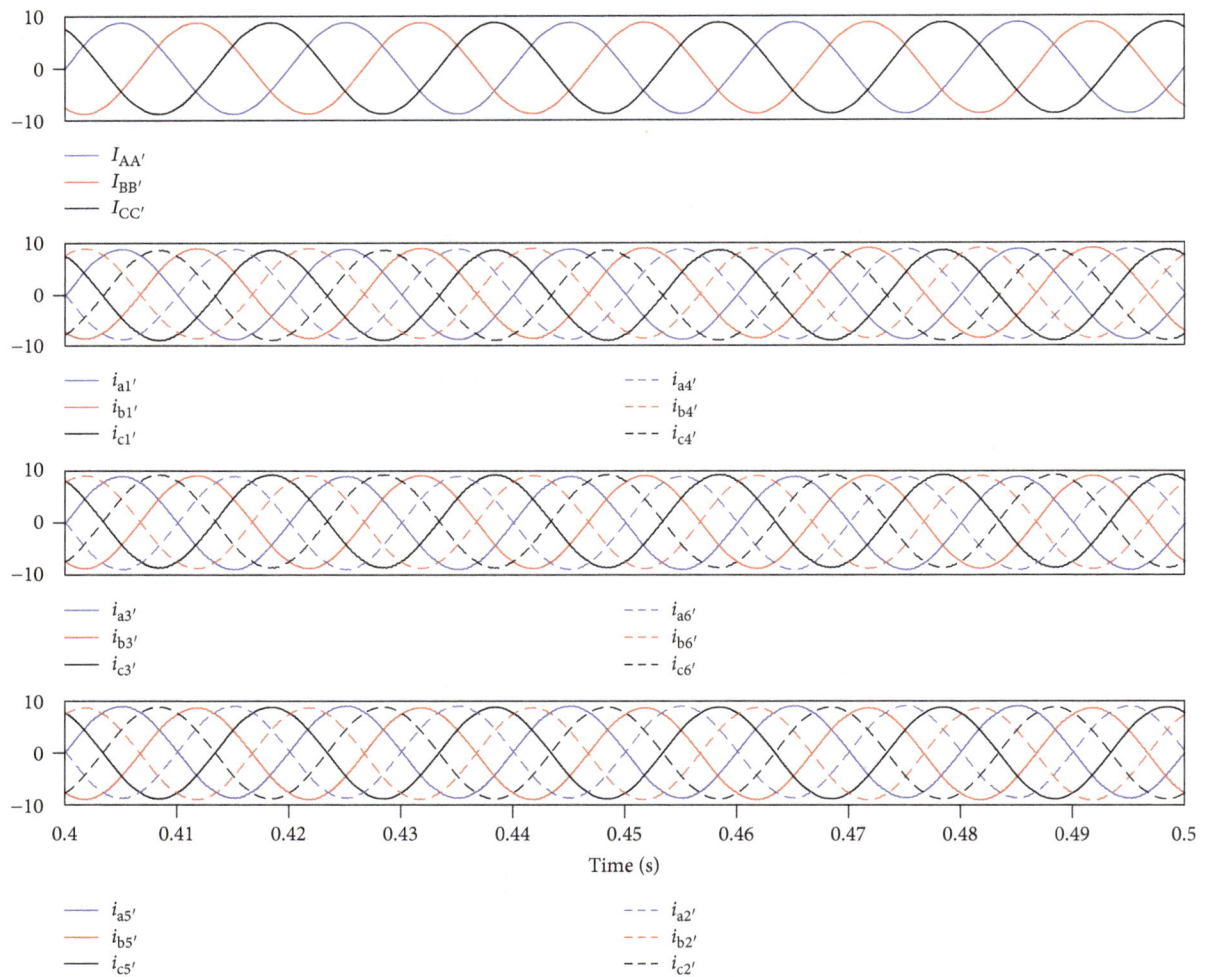

FIGURE 18: Currents inside the HMC.

Fundamental (50 Hz) = 184, THD = 13.57%

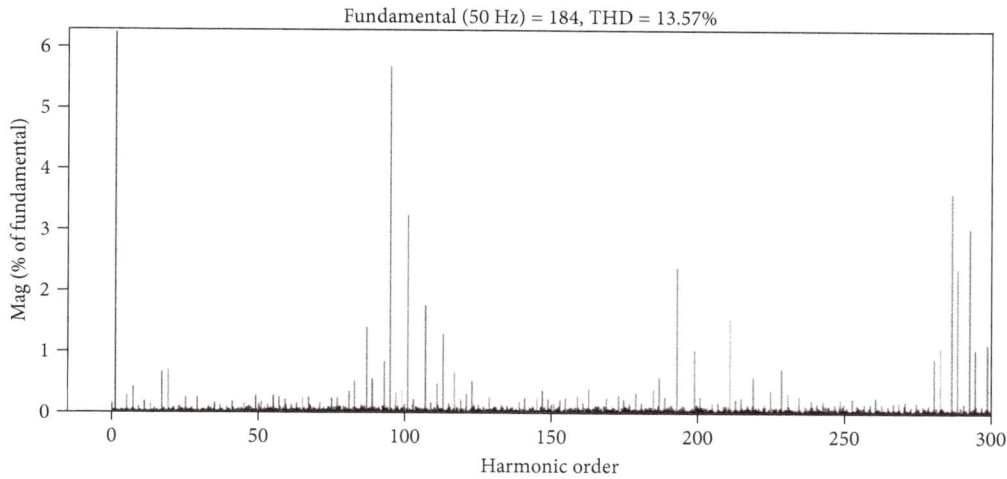

FIGURE 19: THD of the HMC output voltage.

Fundamental (50 Hz) = 8.8, THD = 0.68%

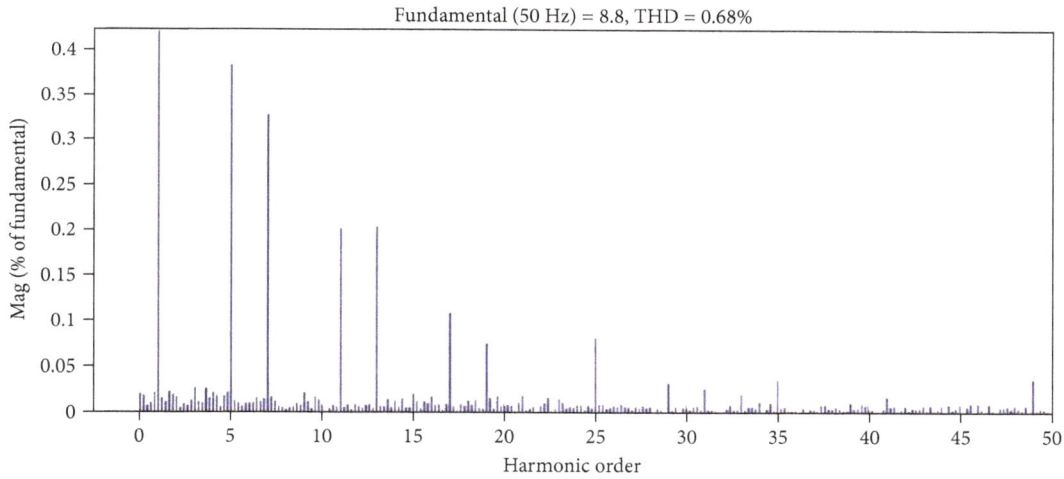

FIGURE 20: THD of the HMC output current.

Demonstrated that currents I_d and I_q can be controlled separately by acting on inputs u_d and u_q, respectively. Hence, the controller is designed using the following expressions:

$$u_d = k_p \tilde{i}_d + k_i \int \tilde{i}_d \, dt,$$

$$u_q = k_p \tilde{i}_q + k_i \int \tilde{i}_q \, dt, \tag{56}$$

where $\tilde{i}_d = I_{d\_REF} - I_d$ and $\tilde{i}_q = I_{q\_REF} - I_q$ are the current error signals, with I_{d_REF} and I_{q_REF} being the reference values for currents I_d and I_q, respectively.

Using (55), the current control law is given by the following expression:

$$\begin{bmatrix} d_d \\ d_q \end{bmatrix} = \frac{L_{fd}\omega}{U_{DC\_av}} \begin{bmatrix} I_q \\ -I_d \end{bmatrix} + \frac{1}{U_{DC\_av}} \begin{bmatrix} E_d \\ E_q \end{bmatrix} - \frac{L_{fd}}{U_{DC_{av}}} \begin{bmatrix} u_d \\ u_q \end{bmatrix}. \tag{57}$$

The d-axis reference current (I_{d_REF}) is produced using the DC-link voltage controller, while the q-axis reference current (I_{q_REF}) is taken as the load q-axis current (I_{Lq}). To aim for a unity power factor, the I_{q_REF} is set to zero (i.e., $i_{q\_REF} = 0$). The active power exchange between the DC links and the grid is proportional to the direct-axis current I_d and can be expressed as

$$P_{dc} = \frac{3}{2}\left(E_d I_d + E_q I_q\right) = \frac{3}{2} E_d I_d. \tag{58}$$

(58) it is shown that direct-axis current I_d is responsible for maintaining the DC-link voltages at a desired value. Thus, using (54), one deduces that

$$C_{eq} \frac{dU_{DC\_av}}{dt} = d_d I_d = u_{dc}. \tag{59}$$

Consequently, the active current is

$$I_d = \frac{u_{dc}}{d_d} = \frac{u_{dc} V_{DC\_av}}{d_d V_{DC\_av}}. \tag{60}$$

Assuming that the current loop is perfect and the HMC works under balanced conditions, the following expressions hold:

$$d_d V_{DC\_av} = E_d,$$

$$I_d = \frac{u_{dc}}{d_d} = \frac{u_{dc} V_{DC\_av}}{E_d}, \tag{61}$$

$$\begin{bmatrix} E_d \\ E_q \end{bmatrix} = \sqrt{\frac{3}{2}} \begin{bmatrix} \hat{V} \\ 0 \end{bmatrix},$$

where $\hat{V}$ and E_d are the RMS voltage and the direct-axis phase voltage at the PCC, respectively. Thus, the control effort of the DC link voltage loop is given by

$$I_{d\_REF} = \frac{u_{dc}}{d_d} = \sqrt{\frac{2}{3}} \frac{V_{DC\_av}}{\hat{V}} u_{dc}. \tag{62}$$

To control the DC-link voltage, a PI controller is used, which is expressed as

$$u_{dc} = k_{pdc} \widetilde{v_{dc}} + k_{idc} \int \widetilde{v_{dc}} \, dt, \tag{63}$$

where $\widetilde{v_{dc}} = V_{DC\_REF} - V_{DC\_av}$ is the DC-link voltage error, whereas V_{DC_REF} and V_{DC_av} are the DC-link reference and average voltages, respectively.

5.2. DC-Link Voltage Controller. With reference to Figure 3, the HMC is based on a symmetric configuration, having six converters with identical power capabilities that are supplied by six equal PV strings. The PV strings are directly connected to each converter, $V_{pv1} = V_{pv2} = V_{pv3} = V_{pv4} = V_{pv5} = V_{pv6} = V_{pv}$, the MPPT must be achieved by the converters, and the DC-link voltages continuously fluctuate. Because the PV strings are supposed to be identical, being created by a single PV string divided into six identical parts, a single MPPT algorithm can be considered. For this reason, the same DC-link voltage reference for the six converters has been considered. The DC-link voltage reference is compared to the sum of the actual six DC-link voltages, and the error is passed through a PI controller to determine the control parameter u_{dc}.

5.3. Modulation Strategy. To obtain an output voltage with low total harmonic distortion (THD), a multicarrier phase-shifted PWM (PS-PWM) switching strategy [24, 25] is implemented to drive each IGBT in the HMC. Optimum harmonic cancellation is accomplished by shifting each carrier

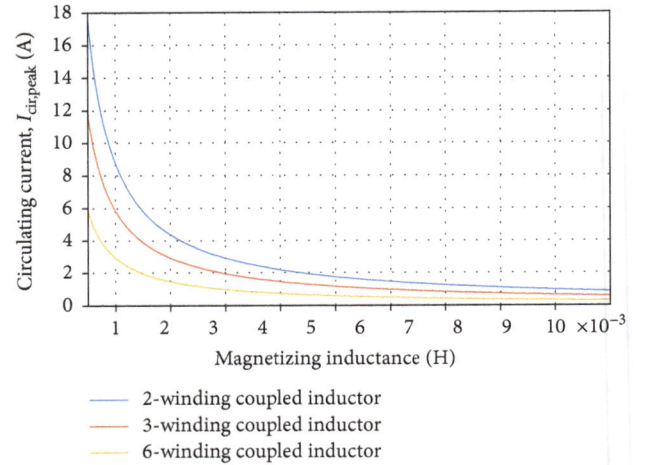

FIGURE 21: Peak circulating current versus magnetizing inductance.

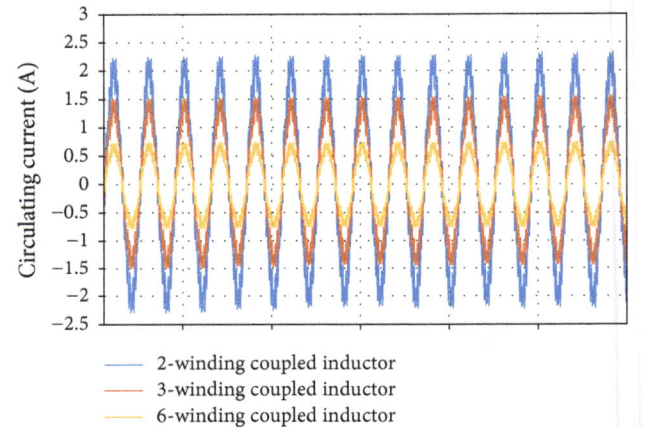

FIGURE 22: Circulating current inside the HMC with multiwinding coupled inductors.

cell by $2\pi T_s/3T$ in sequence, where T_s is the switching time and T is the cycle modulation time. A reference signal of 50 Hz is generated using the control algorithm represented in Figure 15, with the switching frequency fixed at 2500 Hz.

Figure 16 shows the relationship between the modulation waveforms and the three groups of carriers within the HMC. As shown in Figure 16, triangular carriers (i.e., carrier #1, carrier #2, and carrier #3) are phase-shifted 120° to each other and directly compared with the modulation signals to drive the IGBTs within module #1, module #2, and module #3. In order to generate the switching signals used to drive the IGBTs within module #2, module #4, and module #6, the modulation signals are inverted and then compared with the triangular carriers.

6. Simulation Results

In order to demonstrate the performance of the HMC and its control algorithm, the complete grid-connected PV system

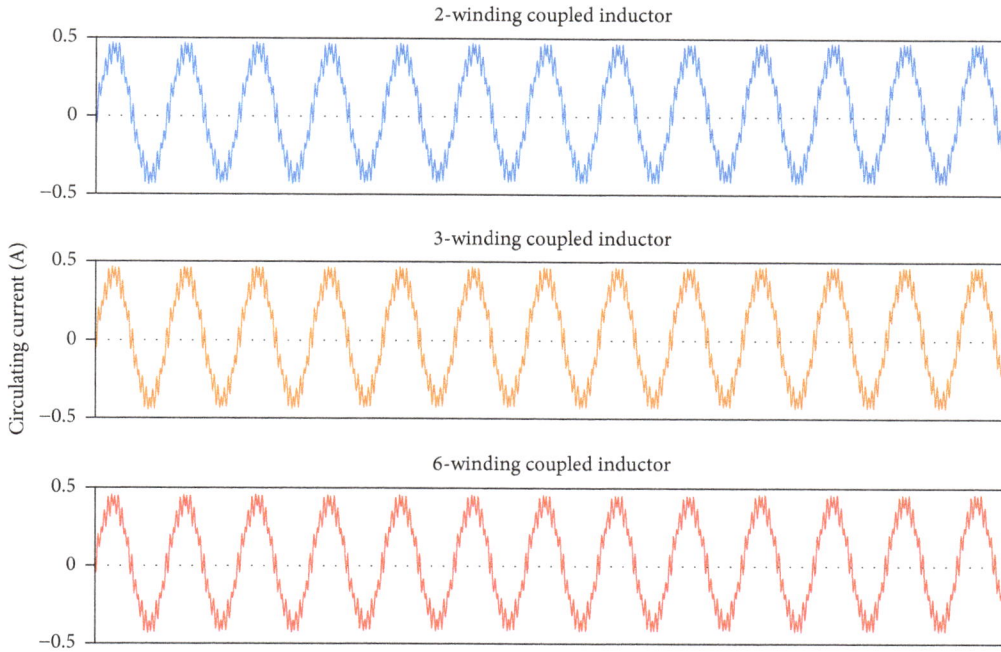

FIGURE 23: Circulating current inside the HMC.

was simulated using the MATLAB/Simulink environment. The technical characteristics of the system parameters and the coupled inductors are provided in Table 2. In order to validate the performance of the HMC, a PS-PWM technique has been implemented, as shown in Figure 16. The output active and reactive power supplies in response to the fluctuations in solar radiation value are shown in the following subsections.

6.1. Performance Analysis.

The seventeen-level phase voltages of the HMC are generated at the steady state and shown in Figure 17. It can be seen that the voltages are balanced, as shown in the voltage phasor diagram. Figure 18 demonstrates the currents of modules 1, 3, and 5, and the currents of modules 2, 4, and 6. Given the configuration of the three-phase HMC, the line currents of every VSC module within the HMC are symmetrical, hence verifying the relationship in (5).

From this figure, it can also be acknowledged that the direction of the currents of modules 2, 4, and 6 is reversed from those of modules 1, 3, and 5. Moreover, the currents inside every VSC module within the HMC are the same as the output currents.

The THD values of the output voltage and current are calculated using the following equation.

$$\%THD_x = 100\sqrt{\sum_{h\neq1}\left(\frac{x_{sh}}{x_{s1}}\right)^2},\qquad(64)$$

where the subscript x indicates the THD in the signal (voltage or current), x_{s1} is the fundamental component, x_{sh} is the component at the h harmonic frequency.

The harmonic spectra of the output voltage and current are shown in Figures 19 and 20, respectively. The THD of

the HMC current is 0.68%, which is fewer than 5% and meets the power quality standard. Using the suggested modulation technique, the highest harmonic family of the phase voltage appears at the band of the 100th harmonic order. Consequently, the effective switching frequency of the phase voltage is two times higher than the switching frequency.

As a comparison, the HMC using two-, three-, and six-winding coupled inductors are simulated under the conditions. The magnetizing inductance of each coupled inductor is 3.5 mH. According to the current from the equations provided in Section 4, if the voltage of the DC links is unbalanced the circulating current will be introduced on the converter output currents. Therefore, to intentionally produce a circulating current, the DC link voltage of Module 3 is decreased from 52.6 to 26.3 V. The created loop voltage is computed using the following equation:

$$V_{\text{loop,rms}} = \frac{\sqrt{3}}{2\sqrt{2}}(52.6 - 26.3)m_a.\qquad(65)$$

Thus, using (42) and (65), the circulating current inside the HMC is expressed as

$$I_{\text{cir,peak}} = \frac{\sqrt{3}}{24\pi}(52.6 - 26.3)\frac{m_a}{fkL_m}.\qquad(66)$$

Investigations with different magnetizing inductance levels have been carried. Figure 21 shows the theoretical values of the peak circulating currents with different magnetizing inductance levels. Figure 22 shows the simulation results with a relatively large magnetizing inductance (3.5 mH). The circulating currents are found by measuring the difference between the output currents of module 1 and module 3. As shown in the waveforms, the circulating

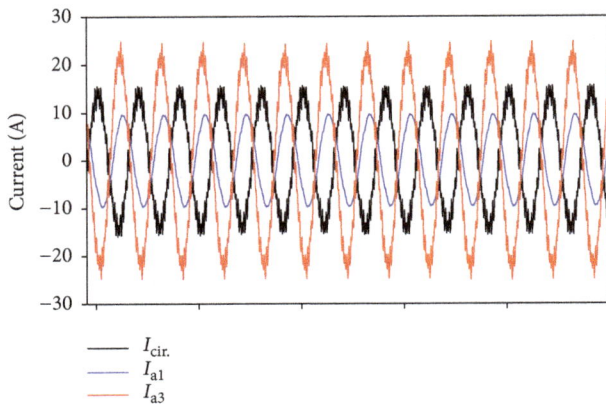

FIGURE 24: Circulating current with two-winding coupled inductors with a magnetizing inductance of 0.5 mH.

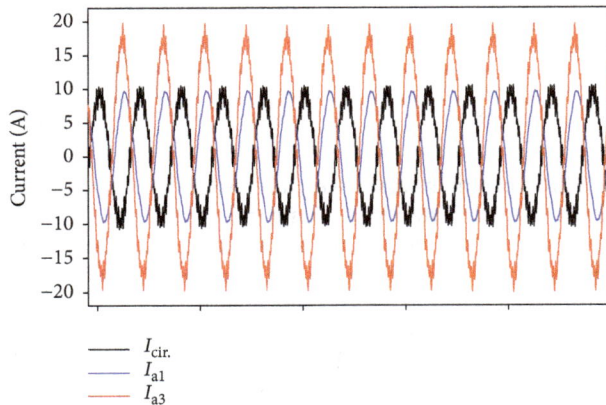

FIGURE 25: Circulating current with three-winding coupled inductors with a magnetizing inductance of 0.5 mH.

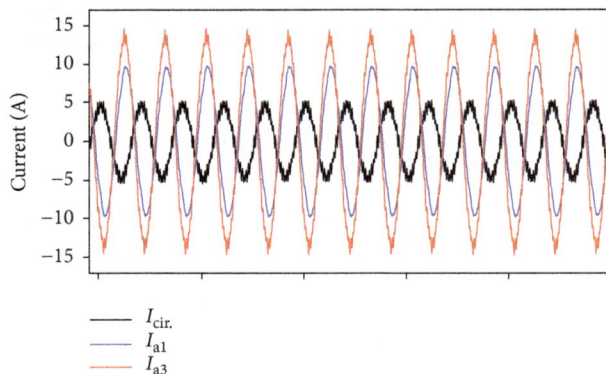

FIGURE 26: Circulating current with six-winding coupled inductors with a magnetizing inductance of 0.5 mH.

TABLE 3: Total harmonic distortion under different solar radiation conditions.

Solar radiation	THD (%)		Fundamental value	
	Voltage	Current	Voltage (V)	Current (A)
1000 (W/m^2)	13.64	0.68	184.0	8.80
600 (W/m^2)	13.65	1.15	181.8	5.20
400 (W/m^2)	12.78	2.80	180.9	3.44
0 (W/m^2)	23.94	0.35	148.3	9.98

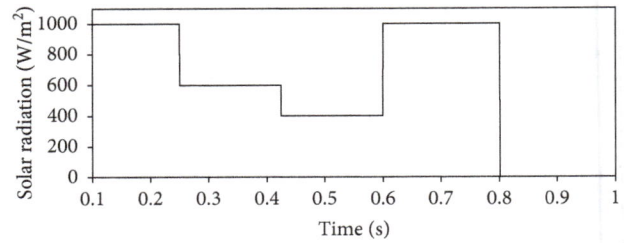

FIGURE 27: Step changes in solar radiation.

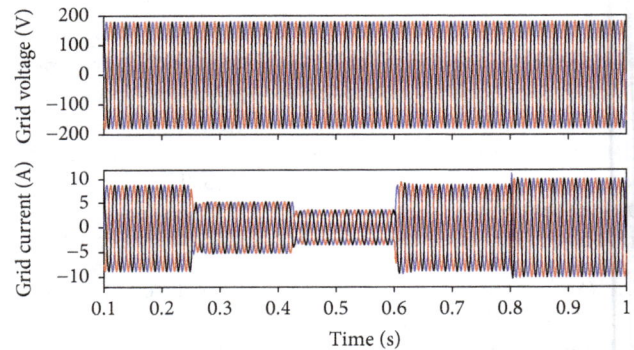

FIGURE 28: Response to changes in solar irradiance in the hexagram converter.

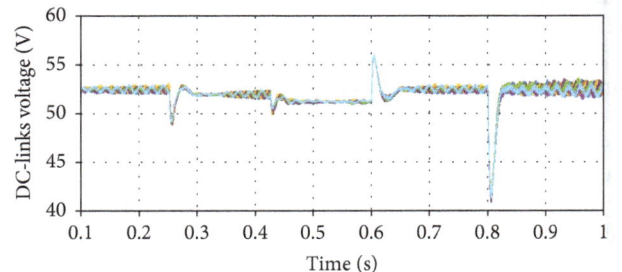

FIGURE 29: Variations in DC-link voltages in response to the changes in solar radiation.

currents inside the HMC with two-, three-, and six-winding coupled inductors are 2.3, 1.5, and 0.76 A peaks as calculated in (66). From the waveforms in Figure 22, using a six-winding coupled inductor, the circulating current is efficiently minimized.

In order to achieve the same circulating currents (e.g., 0.5 A), the magnetizing inductance of the two-winding, three-winding and six-winding coupled inductors should be

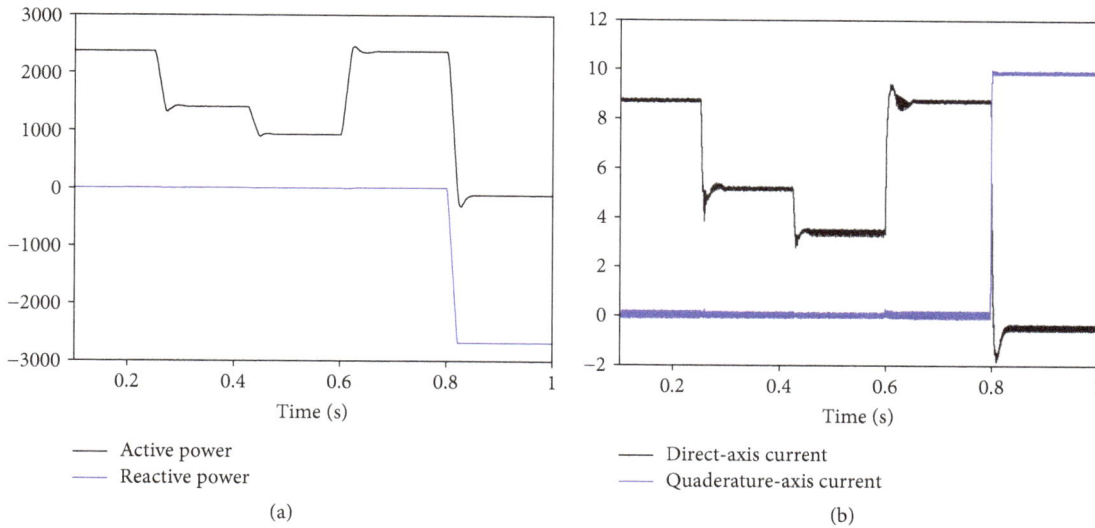

FIGURE 30: Response to changes in solar radiation. (a) Output active and reactive power, and (b) direct and quadrature axis current.

increased to 17.5, 11.6, and 6 mH, respectively. As shown in Figure 23, the waveform matches the theoretical analysis in Figure 21. The output current of module 1, module 3, and the circulating current inside the HMC with two-winding, three-winding, and six-winding coupled inductors under an unbalanced DC link voltage are shown in Figures 24, 25, and 26, respectively. The waveforms show the simulation result with an extremely small magnetizing inductance of 0.5 mH, which is 1/7 of that in Figure 22. The simulations validated the quantitative relationship that the circulating current will be too big to maintain a normal operation if the coupled inductors with low magnetizing inductance are used. As shown, the current ia3 nearly doubles the circulating current, which evidently proves that the HMC is sensitive to an unbalanced DC voltage. The waveforms match the theoretical analysis in Figure 21.

6.2. Active Power Variation. The system is tested for different solar radiation conditions and the results are shown in Table 3. Figure 27 shows the solar radiation variation which is considered in this study. The output current and grid voltage variation in response to the changes in solar radiation is shown in Figure 28. The transient behavior of the total DC-link voltage is presented in Figure 29. The parameters are successfully attuned by the controller to keep the DC-link voltage at the desired level of 52.6 V. Fluctuations are observed in the DC-links, due to the step variations in the solar radiation. Nevertheless, the controller brings the voltage to the reference level within 0.02 s.

The active power supplied by the HMC is directly proportional to the magnitude of direct axis, i_d, as indicated by (38). The solar radiation at all DC links is reduced by 40% at 0.25 s. The fluctuation in solar radiation is imitated through the reduction in the direct-axis current by approximately 40%, in a step. It decreased to 5.216 A from the initial value of 8.823 A, as shown in Figure 30. This reduction in i_d was to keep the DC-link voltage at the reference level by decreasing the output power drawn from the PV. Moreover,

to guarantee maximum utilization of the PV system, the quadrature axis current was kept at zero. At 0.45 s, the solar radiation is further reduced to 400 W/m². The direct-axis current is reduced to 3.267 A from the original value of 5.216 A. Later, the direct-axis current is increased to 8.823 A, because of the increment in the solar radiation.

6.3. Reactive Power Compensation. The described control algorithm permits the HMC to act as DSTATCOM in the absence of solar radiation. The output reactive power can be calculated as

$$Q_{\text{out}} = -\frac{3}{2} I_q V_d. \tag{67}$$

In this condition, the reactive power is increased by 2700 VAR in a step, in the absence of solar radiation. The influence, for the DSTATCOM mode operation, is presented in Figure 31.

The direct-axis current is found to be zero, indicating the fact that no active power is being transferred to the grid in the absence of solar radiation. However, due to the step change in reactive power, the capacitors consume current from the grid. It is found that the direct-axis current takes 0.02 s to stabilize. The nature of the fluctuation in the DC-link voltage is depicted in Figure 29. The DC-link controller effectively keeps the DC-link voltage by regulating the power flow through the capacitor. Hence, it can be mentioned that the HMC effectively operates as DSTATCOM in the absence of solar radiation. Figure 31 shows the source voltage and converter output current in DSTATCOM mode. The waveforms show that the output current increases after the reactive power command comes at 0.8 s. Moreover, Figure 31 approves that the phase difference of the converter output current with grid voltage is 90° in this mode of operation.

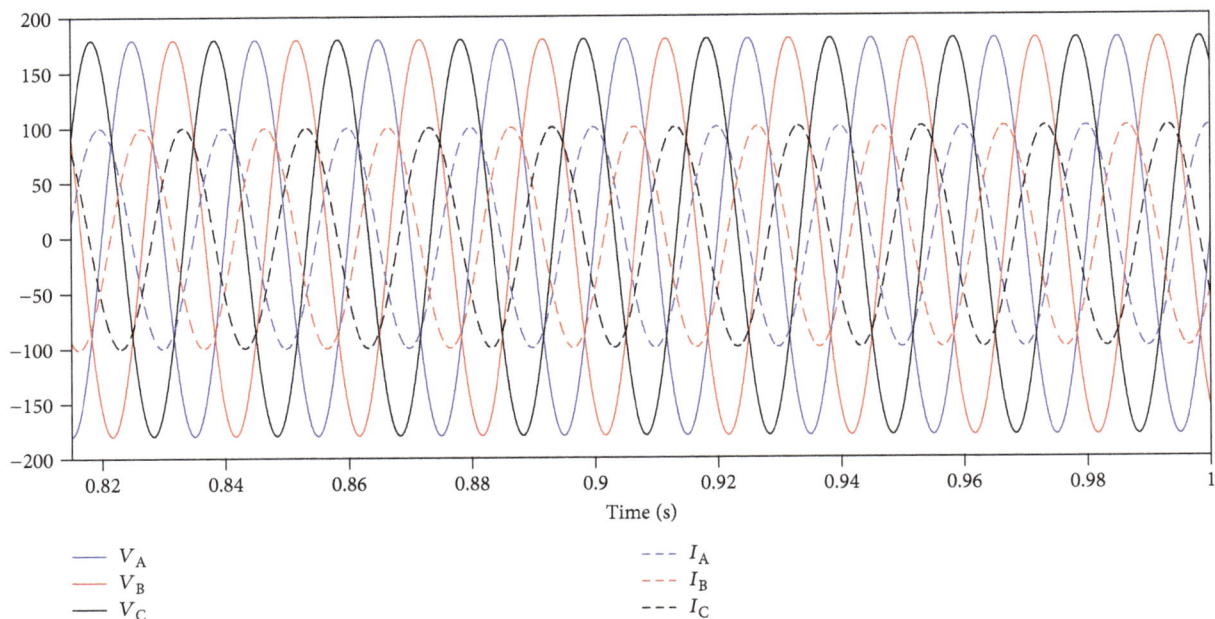

FIGURE 31: System performance with the increase in the reactive power supply in DSTATCOM mode.

7. Conclusion

The HMC for a grid-connected PV system shares many advantages of the CHB but uses fewer switches and reduces the size of the DC-link capacitor. However, the HMC is sensitive to the loop voltage produced by the probable DC-link voltage unbalance. The unbalanced conditions of the DC-link voltages will initiate a line-frequency circulating current. Therefore, to minimize the circulating current as well as to reduce an extra voltage drop on the inductors that affects the terminal voltage, the inductance value should be adequate. The multiwinding coupled inductors are the key to control the circulating current and ensure the proper operation of the HMC. The equivalent circuit model of the HMC configuration is derived to recognize the control scheme. The two-winding coupled inductor imposes twelve times the magnetizing inductance for the circulating current, while the impedance to the circulating current using three-winding coupled inductors is eighteen times as much as the magnetizing inductance. Owing to the six-winding coupled inductor, the impedance on the circulating current path is thirty-six times the magnetizing inductance. The results show the good performance of the control algorithm in both steady state and transient conditions. Moreover, it is interesting to note that in the absence of solar radiation, the controller acts in DSTATCOM mode to supply reactive power to the grid. The performance of the controller under different solar radiation conditions is found to be satisfactory.

Conflicts of Interest

The authors declare that they have no conflicts of interest.

Acknowledgments

The authors would like to express their thanks to King Abdulaziz City for Science and Technology (KACST) for providing financial and technical support to this study.

References

[1] Z. M. Shahrbabak, A. Tabesh, and G. R. Yousefi, "Economical design of utility-scale photovoltaic power plants with optimum availability," *IEEE Transactions on Industrial Electronics*, vol. 61, no. 7, pp. 3399–3406, 2014.

[2] A. A. Al-Shamma'a and K. E. Addoweesh, "Techno-economic optimization of hybrid power system using genetic algorithm," *International Journal of Energy Research*, vol. 38, no. 12, pp. 1608–1623, 2014.

[3] J. H. R. Enslin and P. J. M. Heskes, "Harmonic interaction between a large number of distributed power inverters and the distribution network," *IEEE Transactions on Power Electronics*, vol. 19, no. 6, pp. 1586–1593, 2004.

[4] Y. Riffonneau, S. Bacha, F. Barruel, and S. Ploix, "Optimal power flow management for grid connected PV systems with batteries," *IEEE Transactions on Sustainable Energy*, vol. 2, no. 3, pp. 309–320, 2011.

[5] P. G. Arul, V. K. Ramachandaramurthy, and R. K. Rajkumar, "Control strategies for a hybrid renewable energy system: a review," *Renewable and Sustainable Energy Reviews*, vol. 42, pp. 597–608, 2015.

[6] J. M. Guerrero, P. C. Loh, T.-L. Lee, and M. Chandorkar, "Advanced control architectures for intelligent microgrids—part II: power quality, energy storage, and AC/DC microgrids," *IEEE Transactions on Industrial Electronics*, vol. 60, no. 4, pp. 1263–1270, 2013.

[7] M. A. Elsaharty, H. A. Ashour, E. Rakhshani, E. Pouresmaeil, and J. P. S. Catalao, "A novel DC-bus sensor-less MPPT

technique for single-stage PV grid-connected inverters," *Energies*, vol. 9, no. 4, pp. 1–23, 2016.

[8] J. Rodriguez, J. S. Lai, and F. Z. Peng, "Multilevel inverters: a survey of topologies, controls, and applications," *IEEE Transactions on Industrial Electronics*, vol. 49, no. 4, pp. 724–738, 2002.

[9] J. A. Barrena, L. Marroyo, M. Á. Rodriguez Vidal, and J. R. T. Apraiz, "Individual voltage balancing strategy for PWM cascaded H-bridge converter-based STATCOM," *IEEE Transactions on Industrial Electronics*, vol. 55, no. 1, pp. 21–29, 2008.

[10] K. K. Gupta, A. Ranjan, P. Bhatnagar, L. K. Sahu, and S. Jain, "Multilevel inverter topologies with reduced device count: a review," *IEEE Transactions on Power Electronics*, vol. 31, no. 1, pp. 135–151, 2016.

[11] L. Franquelo, J. Rodriguez, J. Leon, S. Kouro, R. Portillo, and M. Prats, "The age of multilevel converters arrives," *IEEE Industrial Electronics Magazine*, vol. 2, no. 2, pp. 28–39, 2008.

[12] A. Nabae, I. Takahashi, and H. Akagi, "A new neutral-point-clamped PWM inverter," *IEEE Transactions on Industry Applications*, vol. IA-17, no. 5, pp. 518–523, 1981.

[13] M. F. Escalante, J. C. Vannier, and A. Arzandé, "Flying capacitor multilevel inverters and DTC motor drive applications," *IEEE Transactions on Industrial Electronics*, vol. 49, no. 4, pp. 809–815, 2002.

[14] M. A. Pérez, S. Bernet, J. Rodriguez, S. Kouro, and R. Lizana, "Circuit topologies, modeling, control schemes, and applications of modular multilevel converters," *IEEE Transactions on Power Electronics*, vol. 30, no. 1, pp. 4–17, 2015.

[15] Y. Yu, G. Konstantinou, B. Hredzak, and V. G. Agelidis, "Power balance of cascaded H-bridge multilevel converters for large-scale photovoltaic integration," *IEEE Transactions on Power Electronics*, vol. 31, no. 1, pp. 292–303, 2016.

[16] E. Cengelci, P. Enjeti, C. Singh, F. Blaabjerg, and J. K. Pederson, "New medium voltage PWM inverter topologies for adjustable speed AC motor drive systems," in *APEC '98. Conference Proceedings 1998, Thirteenth Annual Applied Power Electronics Conference and Exposition*, pp. 565–571, Anaheim, CA, USA, 1998.

[17] R. Teodorescu, F. Blaabjerg, J. K. Pedersen, E. Cengelci, and P. N. Enjeti, "Multilevel inverter by cascading industrial VSI," *IEEE Transactions on Industrial Electronics*, vol. 49, no. 4, pp. 832–838, 2002.

[18] J. Wen and K. Ma Smedley, "Hexagram inverter for medium-voltage six-phase variable-speed drives," *IEEE Transactions on Industrial Electronics*, vol. 55, no. 6, pp. 2473–2481, 2008.

[19] J. Wen and K. M. Smedley, "Synthesis of multilevel converters based on single- and/or three-phase converter building blocks," *IEEE Transactions on Power Electronics*, vol. 23, no. 3, pp. 1247–1256, 2008.

[20] M. G. Vaillalva, J. R. Gazoli, and E. R. Filho, "Comprehensive approach to modeling and simulation of photovoltaic arrays," *IEEE Transactions on Power Electronics*, vol. 24, no. 5, pp. 1198–1208, 2009.

[21] L. Piegari, R. Rizzo, I. Spina, and P. Tricoli, "Optimized adaptive perturb and observe maximum power point tracking control for photovoltaic generation," *Energies*, vol. 8, no. 5, pp. 3418–3436, 2015.

[22] J. A. Barrena, L. Marroyo, M. A. R. Vidal, and J. R. T. Apraiz, "Individual voltage balancing strategy for PWM cascaded H-bridge converter-based STATCOM," *IEEE Transactions on Industrial Electronics*, vol. 55, no. 1, pp. 21–29, 2008.

[23] S. Debnath, J. Qin, B. Bahrani, M. Saeedifard, and P. Barbosa, "Operation, control, and applications of the modular multilevel converter: a review," *IEEE Transactions on Power Electronics*, vol. 30, no. 1, pp. 37–53, 2015.

[24] M. Moranchel, F. Huerta, I. Sanz, E. Bueno, and F. Rodríguez, "A comparison of modulation techniques for modular multilevel converters," *Energies*, vol. 9, no. 12, pp. 1–20, 2016.

[25] D. Holmes and T. Lipo, *Pulse width modulation for power converters*, IEEE Press, Piscataway, NJ, USA, 2003.

13

Study of Temperature Coefficients for Parameters of Photovoltaic Cells

Daniel Tudor Cotfas⑩, Petru Adrian Cotfas⑩, and Octavian Mihai Machidon ⑩

Electronics and Computers Department, Transilvania University of Brasov, Brasov, Romania

Correspondence should be addressed to Daniel Tudor Cotfas; dtcotfas@unitbv.ro

Academic Editor: Leonardo Sandrolini

The temperature is one of the most important factors which affect the performance of the photovoltaic cells and panels along with the irradiance. The current voltage characteristics, *I-V*, are measured at different temperatures from 25°C to 87°C and at different illumination levels from 400 to 1000 W/m^2, because there are locations where the upper limit of the photovoltaic cells working temperature exceeds 80°C. This study reports the influence of the temperature and the irradiance on the important parameters of four commercial photovoltaic cell types: monocrystalline silicon—mSi, polycrystalline silicon—pSi, amorphous silicon—aSi, and multijunction InGaP/InGaAs/Ge (Emcore). The absolute and normalized temperature coefficients are determined and compared with their values from the related literature. The variation of the absolute temperature coefficient function of the irradiance and its significance to accurately determine the important parameters of the photovoltaic cells are also presented. The analysis is made on different types of photovoltaics cells in order to understand the effects of technology on temperature coefficients. The comparison between the open-circuit voltage and short-circuit current was also performed, calculated using the temperature coefficients, determined, and measured, in various conditions. The measurements are realized using the SolarLab system, and the photovoltaic cell parameters are determined and compared using the LabVIEW software created for SolarLab system.

1. Introduction

An increasing number of countries have introduced renewable energy policies to reduce the greenhouse gas emissions and to avoid an energetic crisis created by the exhaustion of the fossil fuels. Most of them have fixed targets for using different types of renewable energy, and for this, they offer financial support [1]. The ways to improve the renewable energy domain are to develop hybrid renewable energy systems [2, 3], to solve the problems created when the renewable energy is inserted in the electrical power system [3], to achieve a very good integration of the renewable energy in buildings [4], to solve the storage problem, and to increase the efficiency of the existing ones.

The important role that the photovoltaic technology plays in the renewable energy domain is demonstrated by the dynamics, by the photovoltaic capacity installed worldwide (which is over 40 GW each year over the last years),

and by the growth in the number of jobs created, which is over 2.8 million and represents 30% from the total new jobs created in the renewable energy domain [1].

Due to the major interest for photovoltaic technology, the researchers have developed various types of photovoltaic cells, such as multijunction, perovskite, and quantum well [5–9]. Although these types of photovoltaic cells are very promising, the monocrystalline, polycrystalline, and the amorphous silicon photovoltaic cells and panels are still more widely used in terrestrial applications. The multijunction photovoltaic cells are highly efficient, but because of their rather high price, they are generally used in space applications and in concentrated light applications.

The photovoltaic cells and panels can be characterized using their important dc parameters: the photogenerated current, I_{ph}; the short-circuit current, I_{sc}; the open-circuit voltage, V_{oc}; the maximum power, P_{max}; the fill factor, FF; the efficiency, η; the series resistance, R_s; the shunt resistance,

R_{sh}; the ideality factor of diode, m; and the reverse saturation current, I_o [10]. Using the I-V characteristic, the equivalent circuit and one or more of the methods developed by researchers in the last 40 years, [10], the important parameters of the photovoltaic cells can be determined.

All the photovoltaic cell parameters are influenced by the temperature variation. If the temperature of the photovoltaic cells increases, most of them being influenced negatively—they decrease. The others increase with temperature, such as the short-circuit current, which slightly increases, and the reverse saturation current which increases exponentially [11–14].

The temperature of the photovoltaic cells in most of the locations varies from 0°C to 60°C. There are locations where the lower limit of the working temperature can be below −20°C and the upper limit can be over 80°C in semiarid areas [15]. These limits can be exceeded in other applications such as the spatial applications and concentrated light applications or extreme locations [16, 17].

The behavior of the photovoltaic cell parameter function of the temperature is very well described by the temperature coefficients [11–21]. The temperature coefficients, TC, can be absolute and normalized as in the following [13, 18, 21]:

$$\text{TC}_a(p) = \frac{dp}{dT},$$
$$\text{TC}_n(p) = \frac{1}{p}\frac{dp}{dT}\bigg|_{T=25°C}, \qquad (1)$$

where p represents the parameter of the photovoltaic cell and T is the temperature.

The dependence of the photovoltaic cell parameter function of the temperature is approximately linear [21], and thus, the temperature coefficients of the parameters can be determined experimentally using the linear regression method [22]. The mechanisms which influence the performance of the photovoltaic cell can be better studied if the normalized temperature coefficient of the P_{max} is considered as a sum of the normalized temperature coefficients of the I_{sc}, V_{oc}, and FF [12, 21].

Four types of commercial photovoltaic cells are taken into consideration for this study: three from the silicon family—the monocrystalline, polycrystalline, and the amorphous silicon photovoltaic cells—and one from the multijunction family—InGaP/InGaAs/Ge photovoltaic cell. The important parameters of these photovoltaic cells, like I_{sc}, V_{oc}, P_{max}, FF, η, R_s, and m were studied related to the temperature, which was varied from 25°C to 87°C. The temperature coefficients of the photovoltaic cell parameters are determined and compared with the reference ones found in the related literature. The dependence of the temperature coefficients for I_{sc}, V_{oc}, P_{max}, FF, and η upon the irradiance was also studied.

2. Theoretical Considerations

The I-V characteristic and the equivalent circuit with the suitable mathematical model are important tools to study and to determine the parameters of the photovoltaic cells in

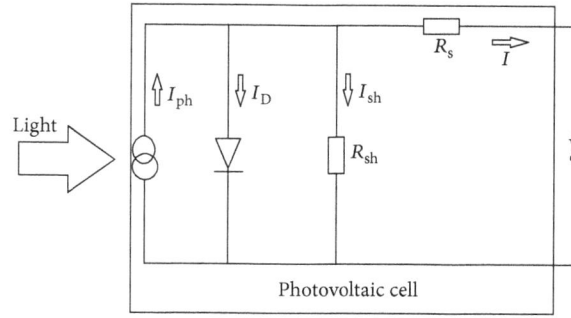

FIGURE 1: The equivalent circuit for one-diode model.

different conditions. There are three models: one-, two-, and three-diode model function of the electric current conduction mechanism from the photovoltaic cell as the diffusion mechanism, the generation-recombination mechanism, and the thermionic mechanism [10]. The generally accepted model is the one-diode model [10, 23]. The equivalent circuit for this model can be seen in Figure 1, and the model is described mathematically by the following equation:

$$I = I_{ph} - I_o\left(e^{(V+IR_s)/V_T m} - 1\right) - \frac{V + IR_s}{R_{sh}}, \qquad (2)$$

where $V_T = kT/q$ is the thermal voltage, T is the temperature, k is the Boltzmann constant, and q represents the elementary charge.

The generally used equivalent circuit and the model for analyzing the multijunction InGaP/InGaAs/Ge photovoltaic cell consist of three one-diode models, one for each junction, connected in series [24].

The open-circuit voltage can be obtained using (2) where $I = 0$ and $I_{ph} \approx I_{sc}$, m is equal to 1, and the shunt resistance is considered very high. In this case, the last term of (2) tends towards zero and can be neglected. Deriving the equation obtained in function of the temperature, the absolute temperature coefficient of the open-circuit voltage is given by the following equation [13]:

$$\frac{dV_{oc}}{dT} = \frac{V_{oc}}{T} + V_T\left(\frac{1}{I_{sc}}\frac{dI_{sc}}{dT} - \frac{1}{I_o}\frac{dI_o}{dT}\right). \qquad (3)$$

The reverse saturation current of the photovoltaic cell is a parameter strongly dependent on the temperature. This dependence is given by the following empirical equation which was simplified by Green [25]:

$$I_0 = Ae^{\left(-(qE_g/kT)\right)}, \qquad (4)$$

where A is a constant equal to $1.5 \times 10^8 \text{ mA cm}^{-2}$ and E_g represents the band gap energy.

The variation of the reverse saturation current function of the temperature [16] is given as follows:

$$\frac{1}{I_o}\frac{dI_o}{dT} = -\frac{1}{V_T}\left(-\frac{E_g}{T} + \frac{dE_g}{dT}\right). \qquad (5)$$

The fill factor temperature coefficient can be calculated using the fill factor formula given by Green [25] or using

FIGURE 2: The SolarLab system used to characterize the photovoltaic cells.

the formula which takes into account the series resistance [12]. Using the Green formula, the absolute temperature coefficient is obtained as follows:

$$\frac{dFF}{dT} = \frac{(dV_{oc}/dT) - (V_{oc}/T)}{V_{oc} + V_T} \left(\frac{(V_{oc}/V_T) - 0.28}{(V_{oc}/V_T) + 0.72} - FF \right). \quad (6)$$

The absolute temperature coefficient of the photovoltaic cell efficiency can be determined by linear fitting of the efficiency dependence on the temperature. The efficiency is calculated as follows:

$$\eta = \frac{P_{max}}{I_t \times A}, \quad (7)$$

where A represents the area of the photovoltaic cell and I_t is the irradiance.

The V_{oc}, I_{sc}, and P_{max} are parameters which can be determined very easily from the I-V characteristic of the photovoltaic cell and also FF using (8). The normalized temperature coefficient of the reverse saturation current, $TC_n(I_o)$, can be calculated, using (3), and the temperature coefficients, $TC_a(V_{oc})$ and $TC_n(I_{sc})$ can be determined experimentally. The advantage of this method is that it does not require the

determination of I_o. Also, the absolute coefficient of the energy band gap, $TC_a(E_g)$, can be determined using (5), $TC_n(I_o)$ and the value of E_g at 25°C from the literature [26] without determination of the E_g.

$$FF = \frac{P_{max}}{I_{sc} \times V_{oc}}. \quad (8)$$

The influence magnitude of the series resistance and the ideality factor of diode can be found by the comparison of the results obtained for $TC_a(FF)$ using (6) and by linear regression of the dependency FF versus T obtained experimentally.

3. Experimental Set-Up

Four types of commercial photovoltaic cells—monocrystalline silicon 3 cm/3 cm, polycrystalline silicon 2.7 cm/1.3 cm, amorphous silicon 3 cm/3 cm, and triple junction InGaP/InGaAs/Ge 1 cm/1 cm—were measured function of temperature and irradiance and also analyzed and compared. The band gap energies of the InGaP/InGaAs/Ge photovoltaic cell junctions are 1.86 eV/1.40 eV/0.67 eV. For each type of photovoltaic cells, several ones were measured in order to choose a representative one.

The I-V characteristics of the photovoltaic cells were measured using the SolarLab system developed by the authors [27] (see Figure 2).

The SolarLab system allows varying the temperature of the photovoltaic cells, and the I-V characteristic is measured at constant temperature. The temperature was maintained constant using a PID (proportional-integral-derivative controller) algorithm which is implemented in the software which serves the system.

The system is covered with a black box to avoid the reflection or other light sources and to minimize the temperature variation of the photovoltaic cell under test.

The I-V characteristics of the four photovoltaic cells are measured at different levels of illuminations from 400 W/m^2 to 1000 W/m^2, and for each level of illumination, the temperature of the photovoltaic cell was varied from 25°C to 87°C.

The temperature of the photovoltaic cell and the irradiance are measured simultaneously with the I-V characteristics. The accuracy of the temperature measurement is ±0.5°C, and the accuracy of the irradiance is ±3 W/m^2.

The SolarLab software is used to control the measurement system, to make the data acquisition, and it has modules implemented to determine the important parameters of the photovoltaic cells using different methods developed by researchers [10, 27]. The software application was created in the graphical programming language LabVIEW, and its interface is presented in Figure 3.

4. Results

The important parameters of the four photovoltaic cells are obtained using the SolarLab system, the I-V characteristics,

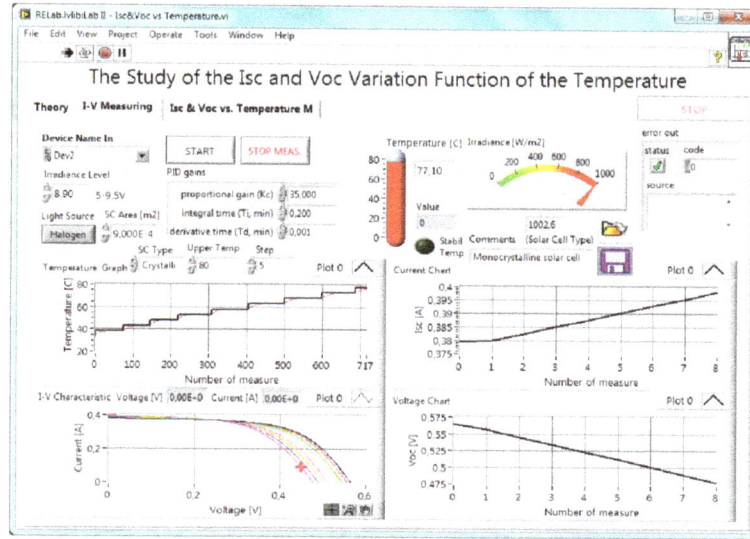

FIGURE 3: The interface of the SolarLab software.

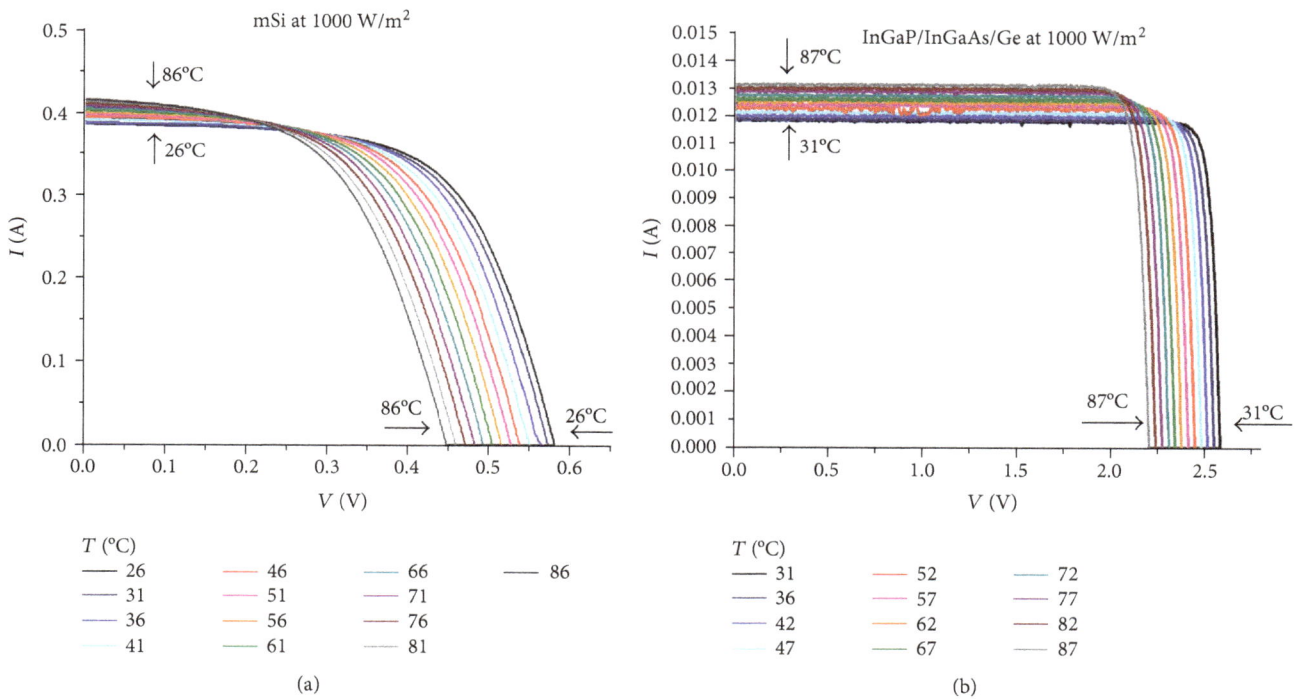

(a)

(b)

FIGURE 4: The I-V characteristics versus temperature at 1000 W/m^2 irradiance for (a) mSi and (b) InGaP/InGaAs/Ge.

the-one diode model, and the analytical five-point method [10, 28] at different temperatures and different irradiance.

The I-V characteristics measured under illumination at 1000 W/m^2 irradiance and at different temperatures, which varies from 26°C to 87°C, are presented in Figure 4(a) for mSi photovoltaic cell and in Figure 4(b) for InGaP/InGaAs/Ge photovoltaic cell.

The open-circuit voltage is strongly affected by the temperature variation. The increase in temperature led to decreasing the value of the band gap energy (see (9)). The intrinsic carrier concentration, n_i, increases because it

depends inversely exponentially on the band gap energy [14]. Due to the proportionality of the I_o with n_i^2, the reverse saturation current increases with temperature. The increase is exponential (see (4)), leading to the decrease of the open-circuit voltage and the fill factor.

$$E_g(T) = E_g(0) - \frac{\alpha T^2}{T + \beta},$$

(9)

where $E_g(0)$, α, and β are material constants [14].

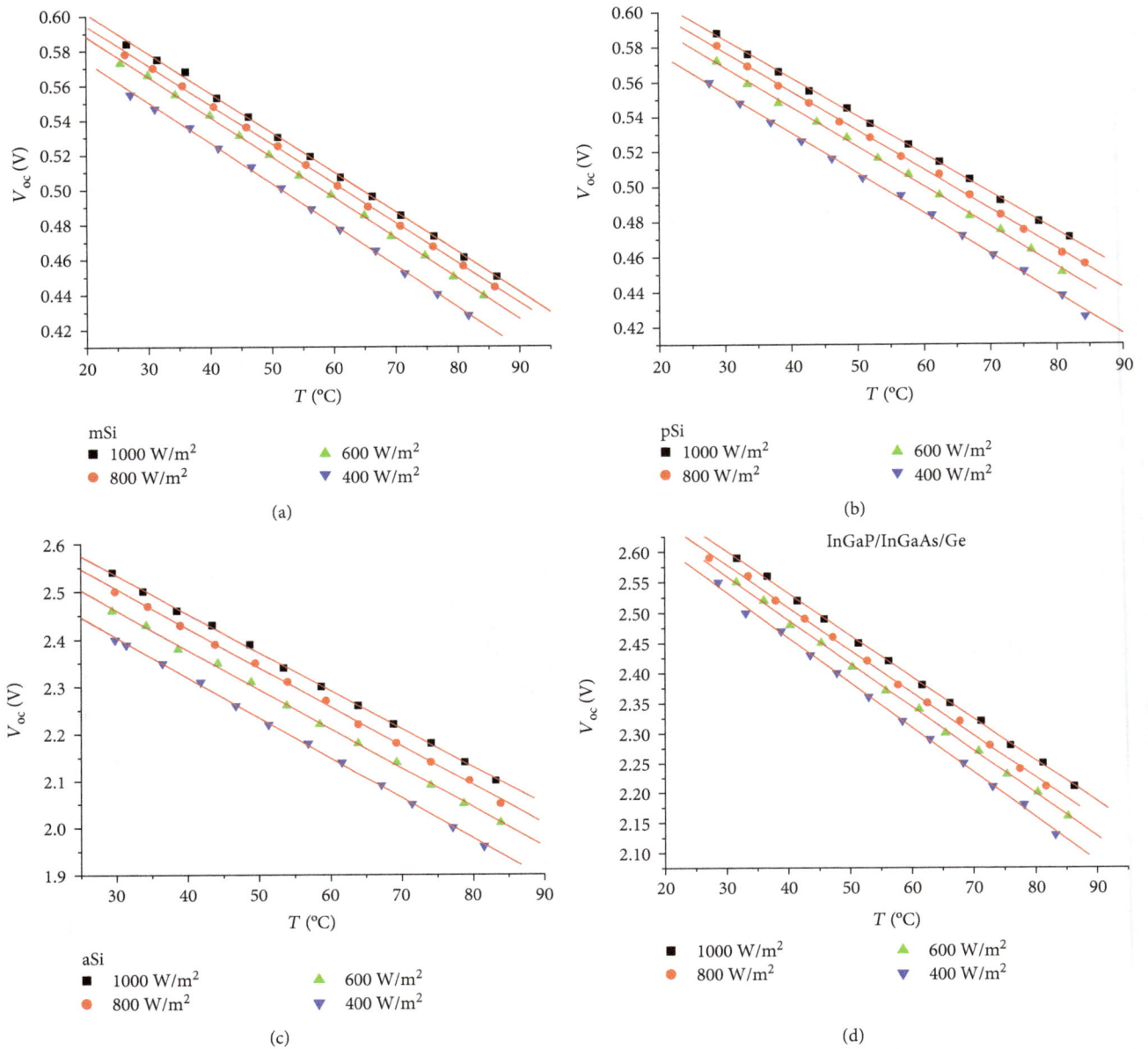

FIGURE 5: The open-circuit voltage versus temperature and irradiance for (a) mSi, (b) pSi, (c) aSi, and (d) InGaP/InGaAs/Ge. The lines represent the curve obtained by linear regression.

The behavior of the open-circuit voltage function of the temperature and irradiance is presented in Figure 5(a) for the monocrystalline photovoltaic cell, Figure 5(b) for the polycrystalline photovoltaic cell, Figure 5(c) for the amorphous photovoltaic cell, and Figure 5(d) for the InGaP/InGaAs/Ge photovoltaic cell.

The absolute temperature coefficients for the important parameters of the photovoltaic cells at different levels of illumination are presented in Table 1.

The short-circuit current slightly increases with temperature, as a consequence of the band gap energy reduction. In this case, the number of photons which have enough energy to create electron-hole pairs is higher. The behavior of the open-circuit voltage temperature coefficient and the short-circuit current temperature coefficient function of the irradiance is reverse.

The maximum power of the photovoltaic cells decreases when the temperature of the photovoltaic cells increases because the increase in the maximum current does not compensate for the decrease in the maximum voltage.

The values of the absolute and normalized temperature coefficients for different types of photovoltaic cells under $1000\,W/m^2$ illumination obtained in this paper (TI) and the ones from literature [13, 18, 20, 29, 30] are presented in Table 2.

Analyzing the values of the temperature coefficients, a good matching can be observed for the $TC_a(V_{oc})$ and $TC_n(V_{oc})$, but there are differences for the $TC_a(I_{sc})$. This

TABLE 1: The absolute temperature coefficients for the important parameters of the photovoltaic cells.

PV	I_T (W/m²)	dV_{oc}/dT (mV °C⁻¹)	dI_{sc}/dT (mA °C⁻¹)	dP_{max}/dT (mW °C⁻¹)	dFF/dT (% °C⁻¹)
mSi	400	−2.36	0.265	−0.251	−0.138
	600	−2.32	0.375	−0.419	−0.141
	800	−2.27	0.465	−0.551	−0.147
	1000	−2.24	0.537	−0.691	−0.162
pSi	400	−2.29	0.108	−0.07	−0.122
	600	−2.27	0.147	−0.103	−0.130
	800	−2.24	0.191	−0.128	−0.139
	1000	−2.19	0.215	−0.163	−0.147
aSi	400	−8.53	0.00727	−0.0101	−0.026
	600	−8.36	0.0106	−0.0158	−0.031
	800	−8.3	0.0151	−0.0208	−0.0411
	1000	−8.09	0.0187	−0.0268	−0.0487
InGaP/InGaAs/Ge	400	−7.46	0.0125	−0.016	−5.49E − 4
	600	−7.21	0.0158	−0.027	−5.6E − 4
	800	−7.05	0.0194	−0.032	−5.76E − 4
	1000	−6.93	0.0253	−0.040	−5.99E − 4

TABLE 2: Comparison between the temperature coefficients at 1000 W/m².

PV cell	Ref.	dV_{oc}/dT (mV °C⁻¹)	$(dV_{oc}/dT)/V_{oc}$ (ppm/°C⁻¹)	dI_{sc}/dT (mA °C⁻¹)	$(dI_{sc}/dT)/I_{sc}$ (ppm/°C⁻¹)	dP_{max}/dT (mW °C⁻¹)	$(dP_{max}/dT)/P_{max}$ (ppm/°C⁻¹)	dFF/dT (% °C⁻¹)	(dFF/dT)/FF (ppm/°C⁻¹)
mSi	TI	−2.24	−3835	0.537	138	−0.691	−4798	−0.162	−2189
	[16]c₁	−2.19	—	0.326	—	—	—	—	—
	[16]c₂	−2.36	—	0.401	—	—	—	—	—
	[18]	—	−3413	—	130	—	−5035	—	−1642
pSi	TI	−2.19	−3725	0.215	870	−0.163	−3713	−0.147	−2046
	[18]	—	−3675	—	675	—	−4690	—	−1732
aSi	TI	−8.09	—	0.0187	—	−0.0268	—	−0.0487	—
	[29] a	−5.8	—	0.007	—	−0.0138	—	—	—
	[29] b	−5.57	—	0.00788	—	−0.0175	—	—	—
InGaP	TI	−6.93	—	0.0253	—	−0.04	—	−5.99E − 4	—
InGa	[30]	−6.2	—	0.005	—	—	—	−6E − 4	—
As/Ge	[20]	−5.6	—	0.0098	—	−0.055	—	−6.3E − 4	—

difference influences the matching for the other temperature coefficients. The differences can be explained by the differences in fabrication process for the same type of photovoltaic cells and the difference between the measurement conditions, and a very important role is played by the illumination source. The differences for $TC_a(I_{sc})$ can be higher than 50% due to the spectral mismatch between the light sources used [18]. A very good matching is observed for the multijunction photovoltaic cell.

Analyzing the open-circuit voltage absolute temperature coefficient variation with the irradiance, for all types of the analyzed photovoltaic cells, a linear dependence is observed (see Figure 6(a)) and also, a linear dependency is obtained for the absolute temperature coefficient of the maximum power (see Figure 6(b)).

The positive slope of the linear dependency of the absolute temperature coefficient of the open-circuit voltage upon the irradiance shows a reduction in voltage drop if the temperature increases when the irradiance increases. The highest slope is obtained for the InGaP/InGaAS/Ge photovoltaic cell. The absolute temperature coefficient of the short-circuit current has the same behavior.

The negative slope of the linear dependency of the absolute temperature coefficient of the maximum power upon the irradiance shows a growth in power drop if the temperature increases when the irradiance increases. The best slope of photovoltaic cells' efficiency is obtained for the InGaP/InGaAS/Ge photovoltaic cell. The amorphous silicon has a very good behavior, but it has a minor impact because the amorphous photovoltaic cell has the smallest maximum

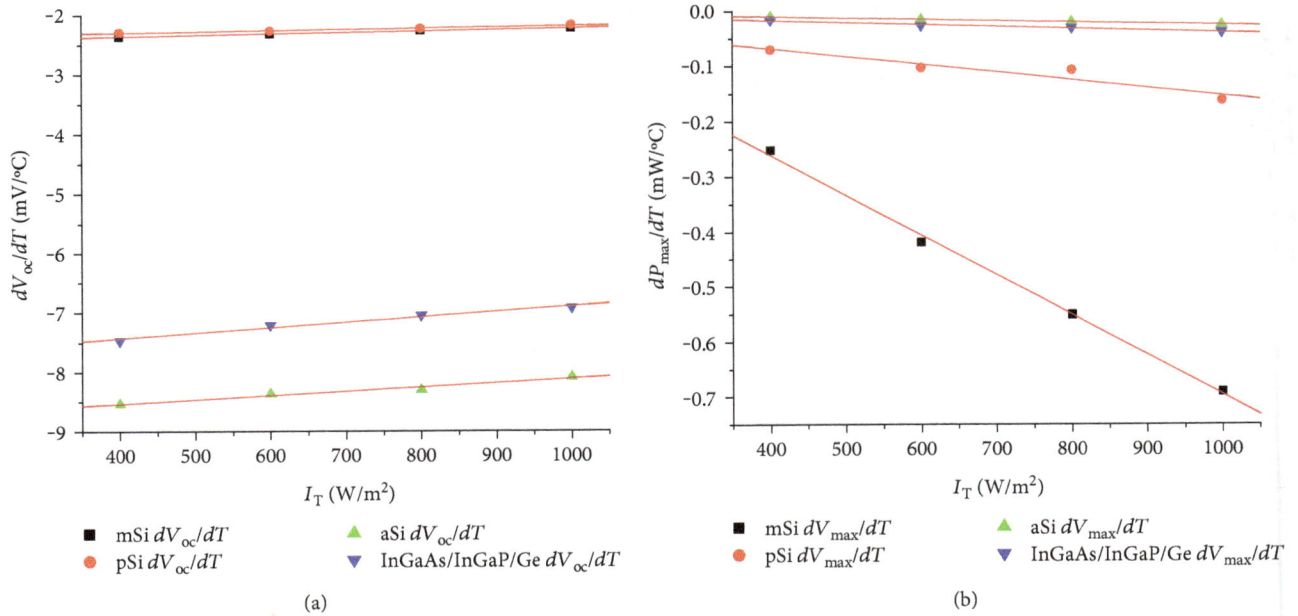

(a)

(b)

FIGURE 6: The dependency of the absolute temperature coefficient versus irradiance for all photovoltaic cells (a) for the open-circuit voltage and (b) for the maximum power.

power from all photovoltaic cells taken into account in this study. The similar results are obtained for the dependence $TC_a(V_{oc})$, $TC_a(I_{sc})$, and $TC_a(FF)$ on the irradiance for two photovoltaic cells polycrystalline silicon and Elkem Solar Silicon by Tayyib et al. [31].

Prediction of the photovoltaic cell and panel output in real work conditions is very important for the final users. There are some models developed which can give the maximum power generated by the photovoltaic panels, the short-circuit current and the open-circuit voltage function of the irradiance and temperature using the values given for the manufacturers in the data sheet, determined at standard test conditions (STC)—global irradiance $1000\,W/m^2$, AM 1.5, and panel temperature 25°C [32–36]. These models take into account the absolute temperature coefficients obtained at $1000\,W/m^2$ irradiance. Using the results obtained by the data analyzed, these models can be improved if the variation function of the irradiance is taken into account for the absolute temperature coefficients. The improved models to predict the maximum power generated by the photovoltaic panel at different irradiances and temperatures are given by (10), for the short-circuit current by (11) and for the open-circuit voltage by (12).

$$P_{max} = P_{max(STC)} \frac{G}{1000}[1 + TC(P_{max}, G)(T_c - 25)], \quad (10)$$

where $P_{max(STC)}$ is the maximum power at STC conditions, G is the irradiance, T_c represents the temperature of the photovoltaic cells, and $TC(P_{max}, G)$ is the normalized temperature coefficient for the maximum power at irradiance G.

$$I_{sc} = I_{sc(STC)} \frac{G}{1000}[1 + TC(I_{SC}, G)(T_c - 25)], \quad (11)$$

where $I_{SC(STC)}$ is the short-circuit current at STC conditions and $TC(I_{sc}, G)$ is the normalized temperature coefficient for the short-circuit current at irradiance G.

$$V_{oc} = V_{oc(STC)}[1 + TC(V_{OC}, G)(T_c - 25)] + TC(G)\ln\frac{G}{1000}, \quad (12)$$

where $V_{OC(STC)}$ is the open-circuit voltage at STC conditions, $TC(V_{oc}, G)$ is the normalized temperature coefficient for the open-circuit voltage at irradiance G and $TC(G)$ represents the temperature coefficient for the irradiance.

The I-V characteristics for all photovoltaic cells under test were performed at temperature room which is 21°C and at $1000\,W/m^2$ irradiance (Figure 7). Using (11) and (12), the short-circuit current and the open-circuit voltage were calculated to compare the obtained results with measured ones. The results obtained for the representative photovoltaic cells are presented in Table 3.

The difference between the measured and calculated value with (11) and (12) of the short-circuit current and the open-circuit voltage for all photovoltaic cells under test is under 0.1%. Better results are obtained for mSi photovoltaic cell.

Two additional experiments were performed to verify the results obtained for the short-circuit current and the open-circuit voltage, one in lab conditions using a mSi photovoltaic cell with sizes $6\,cm \times 8\,cm$ (Figure 8). Both photovoltaic cells, $3\,cm \times 3\,cm$ and $6\,cm \times 8\,cm$, were cut from the mSi by the same lot. The halogen bulbs used to illuminate the photovoltaic cells are the same with halogen bulb of SolarLab. The other experiment was performed in natural sunlight conditions using mSi from the same lot with the cells tested with SolarLab, and the sizes are $12.5\,cm \times 12.5\,cm$ (Figure 9).

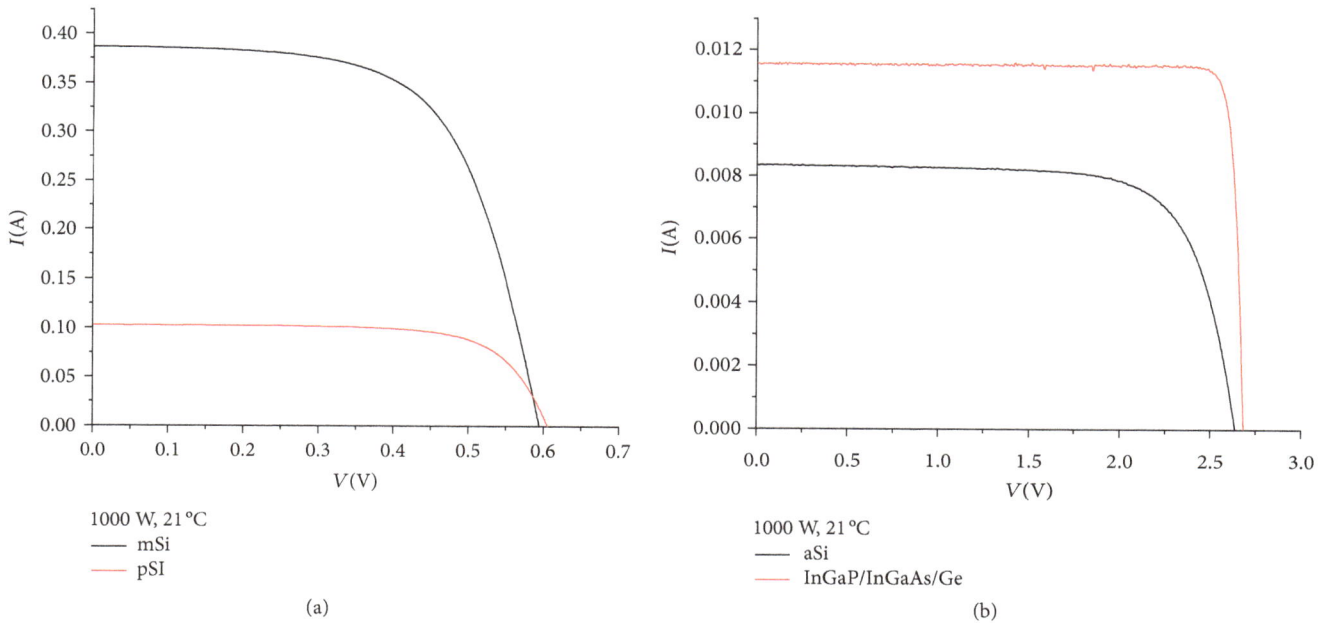

(a)

(b)

FIGURE 7: The *I-V* characteristics at $1000 \, W/m^2$ and room temperature 21°C: (a) mSi and pSi; (b) aSi and InGaP/InGaAs/Ge.

TABLE 3: The calculated and measured values for V_{oc} and I_{sc}.

	mSi		pSi		aSi		InGaP/InGaAs/Ge	
	V_{oc} (V)	I_{sc} (A)	V_{oc} (V)	I_{sc} (A)	V_{oc} (V)	I_{sc} (A)	V_{oc} (V)	I_{sc} (A)
Calculated	0.59494	0.38664	0.60593	0.10273	2.6382	0.008346	2.6863	0.01155
Measured	0.59483	0.38647	0.6058	0.102626	2.6396	0.008351	2.6878	0.01156

The photovoltaic cells, in lab and outside, are cooled using Peltier module and water collector.

The *I-V* characteristics were measured for two levels of irradiance $900 \, W/m^2$ and $1000 \, W/m^2$, and at 30°C temperature of the photovoltaic cells. The results obtained are presented in Table 4.

The difference between V_{oc} calculated and measured is under 1.1 mV which means a 0.5°C temperature difference in laboratory conditions and over 1°C in sunlight conditions. For I_{sc}, this difference increases, almost doubling. The short-circuit current is very sensitive to the level of illumination. The irradiance was measured with three different sensors, one for each case. This can be explained by the increasing difference for I_{sc}. The difference between the results in laboratory and sunlight conditions can be due to the spectral difference between the light sources.

The extrapolation from the monocrystalline photovoltaic cells considered to a 15.6 cm × 15.6 cm one is as follows: the open-circuit voltage temperature coefficient is the same, and the short-circuit current and maximum power temperature coefficients can be obtained by multiplying the determined temperature coefficient with the ratio between the areas of the two cells.

The maximum power generated by the photovoltaic cells is very important from the energy point of view. The decreasing percent in °C for the maximum power of the photovoltaic cells at $1000 \, W/m^2$ is presented in Table 5, where the result

obtained for the photovoltaic panel is also presented. This percent slightly varies with the irradiance variation; for example, it varies from 0.38%/°C at $1000 \, W/m^2$ to 0.44%/°C at $400 \, W/m^2$ for polycrystalline photovoltaic cells. The best behavior is obtained for the multijunction photovoltaic cell.

The fill factor of all the photovoltaic cells considered in this study decreases with the temperature increase, and this dependence for $1000 \, W/m^2$ is being presented in Figure 10. The behavior of the fill factor function of the temperature is determined by the dependency of the open-circuit voltage, short-circuit current, and the maximum power on the temperature. By analyzing the FF dependency function of the temperature, it is observed that the FF temperature coefficient of the amorphous photovoltaic cell is the smallest and the FF temperature coefficient of the monocrystalline photovoltaic cell is the highest. This situation is the same for all illumination levels taken into consideration.

The FF temperature coefficient in absolute value decreases almost by half for the amorphous photovoltaic cell, while for the monocrystalline silicon, polycrystalline silicon and InGaP/InGaAS/Ge photovoltaic cells, the decrease proportion is much smaller (see the values for $TC_a(FF)$ from Table 1).

The series resistance and the ideality factor of diode are determined using the five-parameter method. The parameter can be determined with very good accuracy with the five-parameter method, if the maximum power is determined

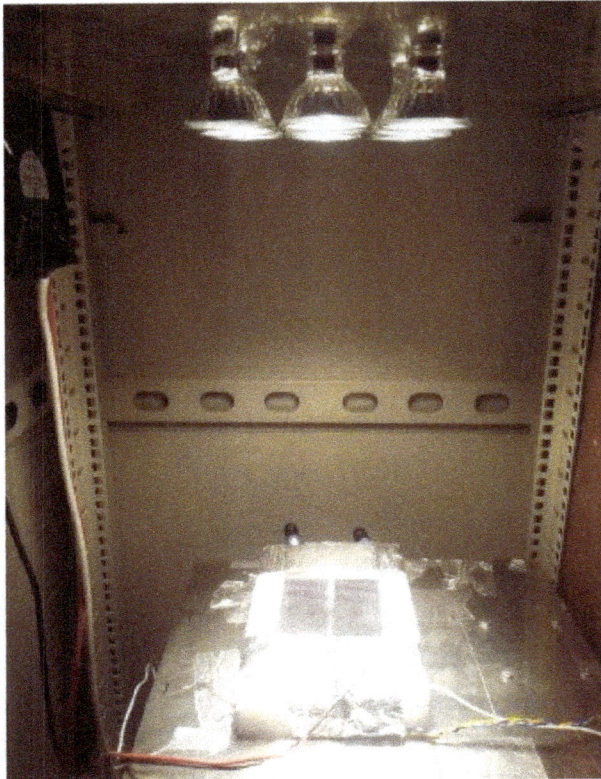

FIGURE 8: The laboratory experiment.

FIGURE 9: The natural sunlight experiment.

TABLE 4: The calculated and measured values for V_{oc} and I_{sc} of mSi at 30°C.

		900 W/m²		1000 W/m²	
		V_{oc} (V)	J_{sc} (mA)	V_{oc} (V)	J_{sc} (mA)
Laboratory	Calculated	0.5727	39.138	0.57482	43.487
	Measured	0.5719	39.08	0.574	43.422
Sunlight	Calculated	0.5729	39.14	0.5748	43.486
	Measured	0.5704	39.04	0.5721	43.378

TABLE 5: The decreasing percent for P_{max} at 1000 W/m².

The photovoltaic cell		mSi	pSi	aSi	InGaP/InGaAs/Ge
Percent/°C	TI	−0.47%	−0.38%	−0.18%	−0.14%
	[37]	−0.45%	−0.39%	−0.23%	—

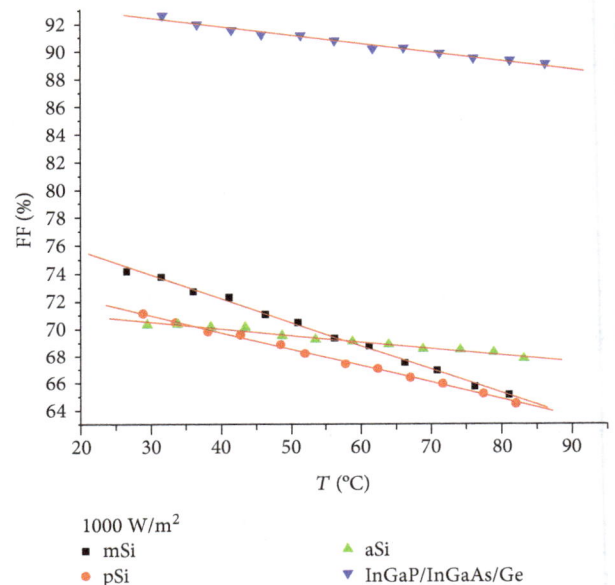

FIGURE 10: The FF versus temperature at 1000 W/m² irradiance.

with very good accuracy [38]. The standard uncertainty of the module used to measure the I-V characteristics is 0.016 mV, and the number of I-V points is higher than 1000, to minimize the errors in determining the maximum power.

The dependence of the series resistance for all photovoltaic cells on temperature at 1000 W/m² irradiance is presented in Figure 11(a). The series resistance decreases linearly with the increase in temperature. The series resistance is determined by the semiconductor resistivity and the resistance of the metal conductors. These two components depend on the temperature, first exponentially and then linearly. Figure 11(b) presents the variations of the ideality factor of all four photovoltaic cells with temperature at

1000 W/m² irradiance. A decrease in the ideality factor is observed with the increase in temperature. The ideality factor of diode m decreases due to the decrease of the active layer resistance of the semiconductors or the temperature influence on surface recombination rate mechanism and Shockley–Read–Hall. The results obtained (Figure 11(c)) also indicate that the shunt resistance of all four photovoltaic cells decreases nearly linearly when the temperature of the cells increases. The R_{sh} decreases because the shunt defects, as recombination centers or traps, and manufacturing defects increase with temperature growth. The decrease rate of the shunt resistance differs function of the type of the photovoltaic cells. The linear temperature dependence obtained for the three parameters of the photovoltaic cells is well consistent with ones found in literature [14, 39].

The values of the $TC_n(I_o)$ were determined using (3), and the temperature coefficients $TC_a(V_{oc})$ and $TC_n(I_{sc})$ are presented in Table 6. This method was verified using the values

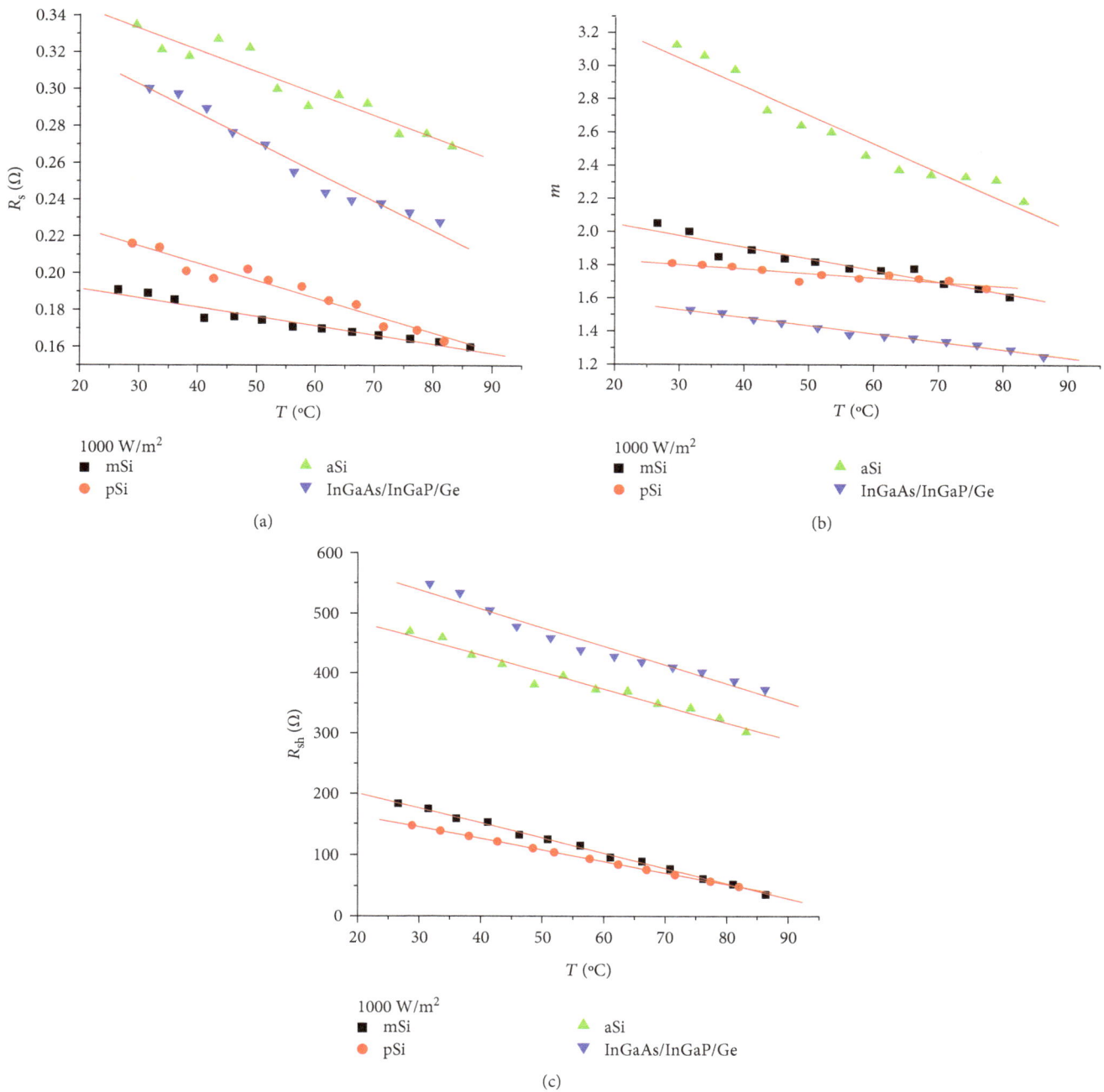

FIGURE 11: (a) The R_s versus temperature at 1000 W/m^2 irradiance (the values of the aSi R_s are divided with 30). (b) The m versus temperature at 1000 W/m^2 irradiance. (c) The R_{sh} versus temperature at 1000 W/m^2 irradiance (the values of the aSi R_{sh} are divided with 30 and for multijunction with 40).

TABLE 6: The normalized temperature coefficient of the I_o at 1000 W/m^2.

The photovoltaic cell		mSi	pSi	aSi	InGaP/InGaAs/Ge
$(dI_o/dT)/I_o$	°C^{-1}	0.164	0.161	0.636	0.591

from reference [13]. The values calculated for the TC$_n$(I_o) for the three cases reported in [13] are 0.166 K^{-1}, 0.166 K^{-1}, and 0.155 K^{-1} and using (3) are 0.170 K^{-1}, 0.170 K^{-1}, and 0.158 K^{-1}. There are some differences due to the approximations made to obtain (3), but it is a very simple and rapid method to find out TC$_n$(I_o). The error in determining the normalized temperature coefficient of the I_o is smaller than 3%.

5. Conclusions

The performance of the four photovoltaic cells, mSi, pSi, aSi, and InGaP/InGaAs/Ge, is analyzed depending upon the temperature and irradiance, by investigating the most important parameters, such as the open-circuit voltage, the short-circuit current, the maximum power, the series resistance, the diode's ideality factor, the reverse saturation current, and the fill factor. The photovoltaic cell temperature was varied from 25°C to 87°C, and the irradiance was varied from 400 W/m^2 to 1000 W/m^2.

The temperature coefficients and their behavior in function of the irradiance of the enumerated parameters were calculated and compared with related literature results, and a good consistency is obtained. The analysis of the results underlines the fact that the temperature is an important factor which influences the performance of the photovoltaic cells. The maximum power decreases with values between 0.14% and 0.47% if the temperature increases with 1°C for the photovoltaic cells analyzed.

The analysis shows that the absolute temperature coefficients of the parameters taken into consideration are dependent on the irradiance level. The irradiance has a small influence on the absolute temperature coefficient of the open-circuit voltage, but for the others, such as the absolute temperature coefficients of the short-circuit current and of the maximum power the influence, the irradiance is much higher. The variation of the last two absolute temperature coefficients with irradiance is almost proportional.

The improved models to predict the maximum power, short-circuit current, and the open-circuit voltage function of the temperature and irradiance are developed. A very simple and rapid way to calculate $TC_n(I_o)$ is also described in this paper. The comparative results obtained for the open-circuit voltage, the short-circuit current measured, and the maximum power in different conditions, laboratory and outside, and the values calculated (estimated) with the temperature coefficients found, validate the improved models presented.

Consequently, in order to maximize the power generated by the photovoltaic cells, it is necessary to choose the optimum technology for the respective location, increasing the irradiance, for example, using the concentrated light and assuring a good cooling.

Conflicts of Interest

The authors declare that they have no conflicts of interest.

Acknowledgments

This work was supported by a grant of the Romanian National Authority for Scientific Research and Innovation, CNCS, UEFISCDI, Project no. PN-II-RU-TE-2014-4-1083 and Contract no. 135/1.10.2015.

References

[1] J. L. Sawin, F. Sverrisson, K. Seyboth et al., *Renewables 2017 Global Status Report*, REN21 Secretariat, Paris, France, 2017.

[2] G. M. Shafiullah, "Hybrid renewable energy integration (HREI) system for subtropical climate in Central Queensland, Australia," *Renewable Energy*, vol. 96, Part A, pp. 1034–1053, 2016.

[3] X. Wang, N. H. El-Farra, and A. Palazoglu, "Optimal scheduling of demand responsive industrial production with hybrid renewable energy systems," *Renewable Energy*, vol. 100, pp. 53–64, 2017.

[4] A. Chel and G. Kaushik, "Renewable energy technologies for sustainable development of energy efficient building," *Alexandria Engineering Journal*, 2017, https://ac.els-cdn.com/S1110016817300911/1-s2.0-S1110016817300911-main.pdf?_tid=2de225ab-cefd-48f5-a087-7480a4cfe9f0&acdnat=1520273225_372f323900114420e46ba92190113366.

[5] P.-P. Zhang, Z.-J. Zhou, D.-X. Kou, and S.-X. Wu, "Perovskite thin film solar cells based on inorganic hole conducting materials," *International Journal of Photoenergy*, vol. 2017, Article ID 6109092, 10 pages, 2017.

[6] A. Aissat, H. Arbouz, S. Nacer, F. Benyettou, and J. P. Vilcot, "Efficiency optimization of the structure pin-InGaN/GaN and quantum well-InGaN for solar cells," *International Journal of Hydrogen Energy*, vol. 41, no. 45, pp. 20867–20873, 2016.

[7] C. Zhang, J. Gwamuri, R. Andrews, and J. M. Pearce, "Design of multijunction photovoltaic cells optimized for varied atmospheric conditions," *International Journal of Photoenergy*, vol. 2014, Article ID 514962, 7 pages, 2014.

[8] A. E.-M. Aly and A. Nasr, "Theoretical study of one-intermediate band quantum dot solar cell," *International Journal of Photoenergy*, vol. 2014, Article ID 904104, 10 pages, 2014.

[9] S. Fara, P. Sterian, L. Fara, M. Iancu, and A. Sterian, "New results in optical modelling of quantum well solar cells," *International Journal of Photoenergy*, vol. 2012, Article ID 810801, 9 pages, 2012.

[10] D. T. Cotfas, P. A. Cotfas, and S. Kaplanis, "Methods to determine the dc parameters of solar cells: a critical review," *Renewable and Sustainable Energy Reviews*, vol. 28, pp. 588–596, 2013.

[11] S. Chander, A. Purohit, A. Sharma, Arvind, S. P. Nehra, and M. S. Dhaka, "A study on photovoltaic parameters of monocrystalline silicon solar cell with cell temperature," *Energy Reports*, vol. 1, pp. 104–109, 2015.

[12] O. Dupre, R. Vaillon, and M. A. Green, "Experimental assessment of temperature coefficient theories for silicon solar cells," *IEEE Journal of Photovoltaics*, vol. 6, no. 1, pp. 56–60, 2016.

[13] P. Singh and N. M. Ravindra, "Temperature dependence of solar cell performance—an analysis," *Solar Energy Materials and Solar Cells*, vol. 101, pp. 36–45, 2012.

[14] A. B. Or and J. Appelbaum, "Dependence of multi-junction solar cells parameters on concentration and temperature," *Solar Energy Materials and Solar Cells*, vol. 130, pp. 234–240, 2014.

[15] M. Benghanem, A. A. Al-Mashraqi, and K. O. Daffallah, "Performance of solar cells using thermoelectric module in hot sites," *Renewable Energy*, vol. 89, pp. 51–59, 2016.

[16] P. Singh, S. Singh, M. Lal, and M. Husain, "Temperature dependence of *I–V* characteristics and performance parameters of silicon solar cell," *Solar Energy Materials and Solar Cells*, vol. 92, no. 12, pp. 1611–1616, 2008.

[17] D. J. Friedman, "Modelling of tandem cell temperature coefficients," in *Conference Record of the Twenty Fifth IEEE Photovoltaic Specialists Conference - 1996*, pp. 39–42, Washington, DC, USA, 1996.

[18] K. Emery, J. Burdick, Y. Caiyem et al., "Temperature dependence of photovoltaic cells, modules and systems," in *Conference Record of the Twenty Fifth IEEE Photovoltaic Specialists Conference - 1996*, pp. 1275–1278, Washington, DC, USA, 1996.

[19] E. Radziemska, "The effect of temperature on the power drop in crystalline silicon solar cells," *Renewable Energy*, vol. 28, no. 1, pp. 1–12, 2003.

[20] M. Cui, N. Chen, X. Yang, and H. Zhang, "Fabrication and temperature dependence of a GaInP/GaAs/Ge tandem solar cell," *Journal of Semiconductors*, vol. 33, no. 2, article 024006, 2012.

[21] O. Dupré, R. Vaillon, and M. A. Green, "Physics of the temperature coefficients of solar cells," *Solar Energy Materials and Solar Cells*, vol. 140, pp. 92–100, 2015.

[22] Y. Riesen, M. Stuckelberger, F.-J. Haug, C. Ballif, and N. Wyrsch, "Temperature dependence of hydrogenated amorphous silicon solar cell performances," *Journal of Applied Physics*, vol. 119, no. 4, article 044505, 2016.

[23] D. T. Cotfas, P. A. Cotfas, D. Ursutiu, and C. Samoila, "The methods to determine the series resistance and the ideality factor of diode for solar cells-review," in *2012 13th International Conference on Optimization of Electrical and Electronic Equipment (OPTIM)*, pp. 966–972, Brasov, Romania, 2012.

[24] G. Segev, G. Mittelman, and A. Kribus, "Equivalent circuit models for triple-junction concentrator solar cells," *Solar Energy Materials and Solar Cells*, vol. 98, pp. 57–65, 2012.

[25] M. A. Green, "Solar cells—operating principles, technology and system applications," *Solar Energy*, vol. 28, no. 5, p. 447, 1982.

[26] M. A. Green, "Intrinsic concentration, effective densities of states, and effective mass in silicon," *Journal of Applied Physics*, vol. 67, no. 6, pp. 2944–2954, 1990.

[27] P. A. Cotfas and D. T. Cotfas, "Design and implementation of RELab system to study the solar and wind energy," *Measurement*, vol. 93, pp. 94–101, 2016.

[28] D. S. H. Chan, J. R. Phillips, and J. C. H. Phang, "A comparative study of extraction methods for solar cell model parameters," *Solid-State Electronics*, vol. 29, no. 3, pp. 329–337, 1986.

[29] C. R. Osterwald, T. Glatfelter, and J. Burdick, "Comparison of the temperature coefficients of the basic I-V parameters for various types of solar cells," in *Conference Record of the Nineteenth IEEE Photovoltaic Specialists Conference-1987*, pp. 188–193, New Orleans, LA, USA, 1987.

[30] K. Nishioka, T. Takamoto, T. Agui, M. Kaneiwa, Y. Uraoka, and T. Fuyuki, "Annual output estimation of concentrator photovoltaic systems using high-efficiency InGaP/InGaAs/Ge triple-junction solar cells based on experimental solar cell's characteristics and field-test meteorological data," *Solar Energy Materials and Solar Cells*, vol. 90, no. 1, pp. 57–67, 2006.

[31] M. Tayyib, J. O. Odden, and T. O. Saetre, "Irradiance dependent temperature coefficients for MC solar cells from Elkem solar grade silicon in comparison with reference polysilicon," *Energy Procedia*, vol. 55, pp. 602–607, 2014.

[32] F. Mavromatakis, F. Vignola, and B. Marion, "Low irradiance losses of photovoltaic modules," *Solar Energy*, vol. 157, pp. 496–506, 2017.

[33] C. Morcillo-Herrera, F. Hernández-Sánchez, and M. Flota-Bañuelos, "Method to calculate the electricity generated by a photovoltaic cell, based on its mathematical model simulations in MATLAB," *International Journal of Photoenergy*, vol. 2015, Article ID 545831, 12 pages, 2015.

[34] M. Fuentes, G. Nofuentes, J. Aguilera, D. L. Talavera, and M. Castro, "Application and validation of algebraic methods to predict the behaviour of crystalline silicon PV modules in Mediterranean climates," *Solar Energy*, vol. 81, no. 11, pp. 1396–1408, 2007.

[35] M. E. A. Slimani, M. Amirat, I. Kurucz, S. Bahria, A. Hamidat, and W. B. Chaouch, "A detailed thermal-electrical model of three photovoltaic/thermal (PV/T) hybrid air collectors and photovoltaic (PV) module: comparative study under Algiers climatic conditions," *Energy Conversion and Management*, vol. 133, pp. 458–476, 2017.

[36] E. Skoplaki and J. A. Palyvos, "On the temperature dependence of photovoltaic module electrical performance: a review of efficiency/power correlations," *Solar Energy*, vol. 83, no. 5, pp. 614–624, 2009.

[37] P. K. Dash and N. C. Gupta, "Effect of temperature on power output from different commercially available photovoltaic modules," *International Journal of Engineering Research and Applications*, vol. 5, no. 1, pp. 148–151, 2015.

[38] G. M. Tina and C. Ventura, "Evaluation and validation of an electrical model of photovoltaic module based on manufacturer measurement," in *Sustainability in Energy and Buildings*, pp. 15–24, Springer, Berlin, Heidelberg, 2012.

[39] E. Cuce, P. M. Cuce, and T. Bali, "An experimental analysis of illumination intensity and temperature dependency of photovoltaic cell parameters," *Applied Energy*, vol. 111, pp. 374–382, 2013.

14

A Solution-Processed Tetra-Alkoxylated Zinc Phthalocyanine as Hole Transporting Material for Emerging Photovoltaic Technologies

Gloria Zanotti (ID),[1] **Giuseppe Mattioli,**[1] **Anna Maria Paoletti,**[1] **Giovanna Pennesi,**[1] **Daniela Caschera** (ID),[2] **Nitzan Maman,**[3,4] **Iris Visoly-Fisher,**[3,4] **Ravi K. Misra,**[5] **Lioz Etgar,**[5] and **Eugene A. Katz**[3,4]

[1]*Istituto di Struttura della Materia (ISM), CNR, Via Salaria km 29.300, 00015 Monterotondo, Rm, Italy*
[2]*Istituto per lo Studio dei Materiali Nanostrutturati (ISMN), CNR, Via Salaria km 29.300, 00015 Monterotondo, Rm, Italy*
[3]*Department of Solar Energy and Environmental Physics, Swiss Institute for Dryland Environmental and Energy Research, Jacob Blaustein Institutes for Desert Research, Ben-Gurion University of the Negev, Midreshet Ben-Gurion 8499000, Israel*
[4]*Ilse Katz Institute for Nanoscale Science and Technology, Ben-Gurion University of the Negev, 84105 Be'er Sheva, Israel*
[5]*Center for Applied Chemistry, the Institute of Chemistry, The Hebrew University of Jerusalem, Jerusalem, Israel*

Correspondence should be addressed to Gloria Zanotti; gloria.zanotti@ism.cnr.it

Academic Editor: Yanfa Yan

A tetra-n-butoxy zinc phthalocyanine (n-BuO)$_4$ZnPc has been synthesized in a single step, starting from commercial precursors, and easily purified. The molecule can be solution processed to form an effective and inexpensive hole transport layer for organic and perovskite solar cells. These appealing features are suggested by the results of a series of chemical, optical, and voltammetric characterizations of the molecule, supported by the results of ab initio simulations. Preliminary measurements of (n-BuO)$_4$ZnPc-methylammonium lead triiodide perovskite-based devices confirm such suggestion and indicate that the interface between the photoactive layer and the hole transporting layer is characterized by hole-extracting and electron-blocking properties, potentially competitive with those of other standards de facto in the field of organic hole transport materials, like the expensive Spiro-OMeTAD.

Hole transporting layers (HTLs) have demonstrated to be fundamental to achieve high efficiency and long-term durability in new-generation hybrid organic photovoltaic devices. They must provide a good chemical stability, a high hole mobility, and an ideal matching between their HOMO levels and the valence band maximum of the active layer, promoting an efficient collection of photogenerated holes at the electrode. In particular, they have been extensively employed in perovskite solar cells (PSCs) [1–3], contributing to boost their impressive evolution in the last ten years from an initial 3.8% [4] to a 22.1% [5] certified efficiency. More specifically, the presence of a HTL increases the extraction rate of photogenerated holes from the perovskite film and improves

their transport to the back contact metal electrode. This, in turn, minimizes recombination losses and charge-induced decomposition of the active layer, thus increasing the overall PSC performance. Several small molecules, polymers, and inorganic materials have been studied and tested for this purpose [6]. 2,2′,7,7′-Tetrakis(N,N-di-p-methoxyphenilamine)-9,9′-spirobifluorene (Spiro-OMeTAD) and poly(triarylamine) (PTAA) are among the most effective organic hole transporting materials (HTMs), yielding highly efficient PSCs [7, 8]. However, their synthesis requires complex multistep procedures resulting in a significant increase of the device price. Phthalocyanines (Pcs) and similar macrocyclic compounds are well-known p-type organic semiconductors and have found applications as hole transporters in organic photovoltaics (OPV) [9, 10] and in organic field effect transistors (OFETs) [11] due to their high chemical stability, large conjugated π-system, and tunable optical/electrochemical properties through chemical functionalization. Such well-known features promoted their use as HTLs in PSCs, and several examples have been reported for vacuum-evaporated and spin-coated layers, depending on their molecular structure [12–15]. Recently, an impressive 17.5% efficiency scored by tetra-5-hexyl-2,2′-bisthiophene zinc phthalocyanine in a mixed-ion perovskite solar cell has been reported by Cho et al. [16], due to the effect of a favorable packing of the molecule on its transport properties.

Aimed at pushing further down the fabrication cost, a key feature of OPV and PSC technologies, we developed a solution-processable HTL with the target molecule prepared in a one-step synthesis from easily available, cheap precursors. The molecule is a symmetric tetra-n-butoxy-substituted zinc phthalocyanine, hereinafter named $(n\text{-BuO})_4\text{ZnPc}$.

Its electron-donating alkoxy peripheral substituents have been chosen to accomplish two main tasks: (i) tuning the $(n\text{-BuO})_4\text{ZnPc}$ electronic properties, thus optimizing the alignment of HOMO-LUMO levels to the perovskite band edges, and (ii) assuring the $(n\text{-BuO})_4\text{ZnPc}$ processability within a device by formation of spin-coated homogeneous films with favorable molecular packing.

$(n\text{-BuO})_4\text{ZnPc}$ has been synthesized with a 53% yield in a one-step reaction, using commercially available 4-n-butoxyphthalonitrile and ZnCl_2 as templating salt, as shown in Scheme 1.

The crude has been purified by filtration on silica gel using dichloromethane and THF as eluents, and the resulting purple crystalline solid has been characterized by different techniques. UV-Vis spectra have been measured in transmission mode both in solution (solid black line) and as spin-coated film (solid red line), as shown in Figure 1(a).

The solution spectrum in acetonitrile is dominated by the typical phthalocyanine Q-band absorption, with its maximum peak at 675 nm and a shoulder peak at 610 nm, which was also observed in gas-phase measurements of unsubstituted ZnPc [17]. The latter should therefore be considered as part of the vibrational envelope of the $S_0 \rightarrow S_1^*$ transition rather than a fingerprint of H-type (cofacial) aggregation of the molecules. As expected, the same Q-band is much broader in the case of the $(n\text{-BuO})_4\text{ZnPc}$ film and composed by two redshifted bands centered at 692 and 737 nm. This behavior is compatible with an "edge-to-edge" macrocycle arrangement typical of J-type (coplanar) aggregation [18, 19]. The steady-state emission spectrum of a $(n\text{-BuO})_4\text{ZnPc}$ solution in acetonitrile, recorded at $\lambda_{ex} = 600$ nm, shows a maximum centered at 686 nm, with a small Stokes shift of 11 nm. The related time-correlated single-photon counting measurements, shown in Figure 1(b), were performed at 686 nm, and the resulting decay curve was monoexponentially fitted, providing a $\tau_1 = 2.4$ ns. An optical band gap of 1.82 eV was estimated from the interception of the normalized absorption and emission spectra. This result, combined with cyclovoltammetric measurements, allowed to estimate the reduction potentials of the $(n\text{-BuO})_4\text{ZnPc}/(n\text{-}$

$\text{BuO})_4\text{ZnPc}^+$ and $(n\text{-BuO})_4\text{ZnPc}^-/(n\text{-BuO})_4\text{ZnPc}$ couples indicated as E_{HOMO} and E_{LUMO}, respectively. The optical and electrochemical characterizations of the compound are summarized in Table 1.

To evaluate the processability of $(n\text{-BuO})_4\text{ZnPc}$ in terms of quality, homogeneity, and roughness of its films, atomic force microscopy topographic measurements have been performed. Molecular films have been prepared by spin coating, using 20 mg/ml of toluene solutions on nonconductive glass plates. Large-area scans of $5\,\mu m \times 5\,\mu m$ surfaces, an example is reported in Figure 2, indicated a flat and uniform film in the explored regions, suggesting that the molecule can be successfully layered in a planar device. The root mean square roughness of the reported film was found to be 4.94 nm, with an average height of 4.76 nm and a surface skewness of 0.24 nm. The presence of rare overlying crystallites up to 288 nm (143 nm in Figure 2) has sometimes been detected.

Optical and electrochemical measurements are in agreement with ab initio results based on density functional theory (DFT) [20] and are compared to simulations of Spiro-OMeTAD and tetra-tert-butyl zinc phthalocyanine (hereinafter referred to as $(t\text{-Bu})_4\text{ZnPc}$) properties, chosen as internal standards. The theoretical results are summarized in Table 2 and support the idea that the n-butoxy substituents have a combined structural and electronic effect on the properties of $(n\text{-BuO})_4\text{ZnPc}$ molecular films.

The peripheral n-butoxy substituents induce an extension of the spatial distribution of the $(n\text{-BuO})_4\text{ZnPc}$ HOMO orbital responsible for the transport of holes to the back electrode, as opposed to the t-Bu substituents of the similar $(t\text{-Bu})_4\text{ZnPc}$ molecule (see Figure 3). Such a resonant effect is accompanied with the strong electron-donating behavior of O-R groups that stabilizes positive-charge carriers and induces a 0.25 eV rising of the HOMO with respect to the weaker alkyl substituents. This indication is confirmed by an analysis of the charge displacement induced by a hole injected in $(n\text{-BuO})_4\text{ZnPc}$ (Figure 3), which underlines the role of the peripheral n-butoxy substituents. Theoretical estimates of reduction potentials in acetonitrile [21] are displayed in Figure 4, left side, and offer further comparison between the properties of $(n\text{-BuO})_4\text{ZnPc}$ and Spiro-OMeTAD. A calculated E_{HOMO} value of −5.23 eV (vs vacuum) in the case of $(n\text{-BuO})_4\text{ZnPc}$ approaches the −4.96 eV value estimated in the case of Spiro-OMeTAD, whereas a corresponding reduction potential of −3.31 eV for the E_{LUMO} value is suitable for blocking free electrons that diffuse in the active perovskite layer. These results confirm that $(n\text{-BuO})_4\text{ZnPc}$ is a suitable hole transport material for organic and perovskite solar cells. In particular, as shown in Figure 4, right side, in a methylammonium lead triiodide-based solar cell, the −5.18 eV oxidation potential of the phthalocyanine allows the hole transport from the photoactive layer from the perovskite/HTL interface to the gold electrode, while the reduction potential, higher than that of the perovskite, prevents the electrons to diffuse back and recombine on it.

As a proof of concept, several $\text{CH}_3\text{NH}_3\text{PbI}_3$ PSCs with an active area of $0.04\,\text{cm}^2$ have been prepared using a two-step deposition protocol: a widely employed device architecture

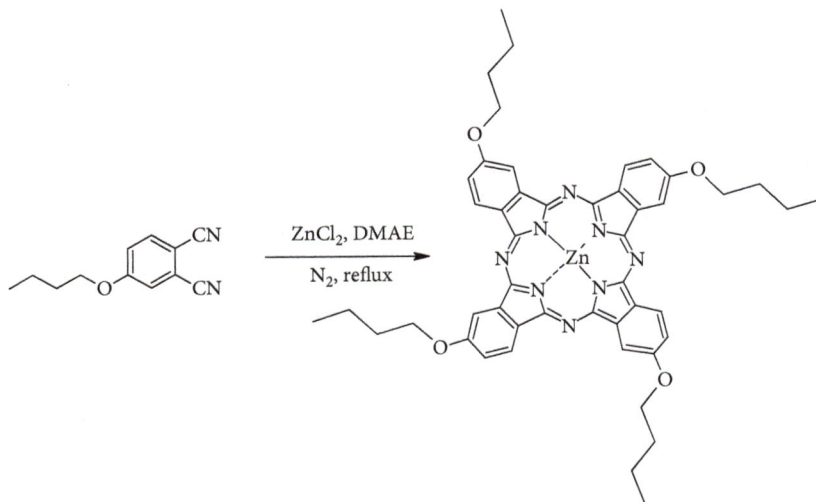

SCHEME 1: Synthesis of (n-BuO)$_4$ZnPc in dimethylaminoethanol (DMAE).

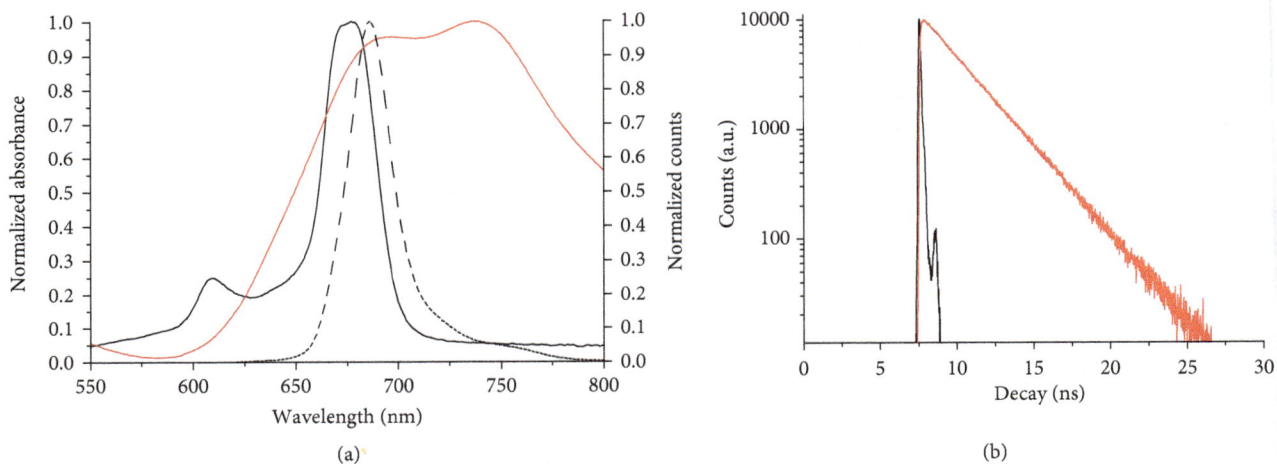

(a)

(b)

FIGURE 1: (a) Black solid line: normalized (n-BuO)$_4$ZnPc UV-Vis absorption in acetonitrile. Black dashed line: emission spectrum in acetonitrile. Red solid line: Q-band absorption spectrum of a spin-coated film on nonconductive glass. (b) Red solid line: (n-BuO)$_4$ZnPc fluorescence decay in acetonitrile. Black solid line: prompt.

TABLE 1: Optical and electrochemical properties of (n-BuO)$_4$ZnPc.

	λ_{max} (nm)[a]	E_{0-0} (eV)[b]	$E(1)_{1/2ox}$ (V)[c]	E_{HOMO} (eV)[d]	E_{LUMO} (eV)[e]
(n-BuO)$_4$ZnPc	675	1.82	0.741	−5.18	−3.36

[a]Absorption maximum in acetonitrile. [b]0-0 transition energy. [c]Solvent: acetonitrile vs NHE. [d]$E_{HOMO} = E(1)_{1/2ox} + 4.44$ (eV) [16]. [e]$E_{LUMO} = E_{HOMO} + E_{0-0}$.

with a mesoporous TiO$_2$ electron transporting layer (ETL) and a tert-butylpyridine- (TBP-), lithium bis(trifluoromethanesulfonyl)imide- (LiTFSI-), and tris(2-(1H-pyrazol-1-yl)-4-tert-butylpyridine)-cobalt(III) tris(bis(trifluoromethylsulfonyl)imide))-doped HTL has been used. Spiro-OMeTAD control cells have also been prepared and measured as benchmarks. The obtained data have been subjected to the Peirce's criterion [22] to discard anomalous results. The phthalocyanine scored average efficiencies of 6.06% ± 0.74 (forward bias) and 9.00% ± 0.21 (reverse bias), while 6.75% ± 0.19 (forward bias) and 8.61% ± 0.60 (reverse bias) are the average

values obtained using Spiro-OMeTAD. These values, along with the averaged V$_{oc}$, J$_{sc}$, and FF, are reported in Table 3.

Open-circuit voltages average around 0.91 V for both the HTMs, with acceptable standard deviations. Values around 1.0 V are reported in several articles for CH$_3$NH$_3$PbI$_3$-based solar devices [23, 24]; our 10% discrepancy may be due to several factors such as a large reverse saturation current occurring because of a high recombination rate at the TiO$_2$/perovskite and HTL/perovskite interfaces and to the presence of pinholes in the charge transport layers that create small short circuits with the metal contacts. J$_{sc}$ and the fill factor

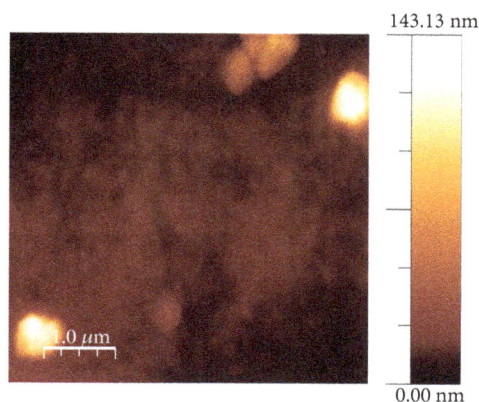

FIGURE 2: $5\,\mu m \times 5\,\mu m$ AFM image of a $(n\text{-}BuO)_4ZnPc$ film on nonconductive glass. The color scale ranges from 0 nm (black) to 143 nm (white).

TABLE 2: HOMO (LUMO) values of isolated molecules calculated at the B3LYP level of theory and E_{HOMO} (E_{LUMO}) voltammetric potentials of the same molecules in CH_3CN calculated at the M06-2X level of theory in the case of $(n\text{-}BuO)_4ZnPc$, $(t\text{-}Bu)_4ZnPc$, and Spiro-OMeTAD. Oxidation and reduction potentials have been calculated as ΔG values between neutral and charged species, as detailed in [21].

	HOMO (eV)	LUMO (eV)	E_{HOMO} (eV)	E_{LUMO} (eV)
$(n\text{-}BuO)_4ZnPc$	−4.95	−2.89	−5.23	−3.31
$(t\text{-}Bu)_4ZnPc$	−5.10	−2.96	−5.31	−3.72
Spiro-OMeTAD	−4.61	−1.14	−4.96	−1.63

have lower values than the state-of-the-art and high standard deviations, indicating that the examined batches are subjected to some degree of structural inhomogeneity likely ascribable to the quality of the perovskite layer in terms of crystallinity and evenness.

The current density-voltage $(J - V)$ curves and the principal photovoltaic parameters of the best $n\text{-}(BuO)_4ZnPc\text{-}$ and Spiro-OMeTAD-based cells are reported in Figures 5(a) and 5(b), respectively.

Both devices show a remarkable hysteresis with consistent performance discrepancies depending on the scan direction, mostly ascribable to differences in J_{sc} and fill factors. No investigations have yet been performed to rationalize this phenomenon in this specific case; but according to the literature, it can be caused by a number of factors such as scan direction and rate, cell preconditioning, and device architecture [25, 26]. As frequently reported, these cells yielded a higher PCE in reverse bias scans rather than in forward bias scans, likely due to the uprising of dynamic trapping and detrapping processes of charge carriers at the perovskite/HTL and perovskite/ETL interfaces and to ion migration that alters the interfacial energy barriers and consequently modifies their height for the carrier collection [27]. The external quantum efficiency (EQE) spectra, shown in Figure 6, have been recorded between 300 and 850 nm.

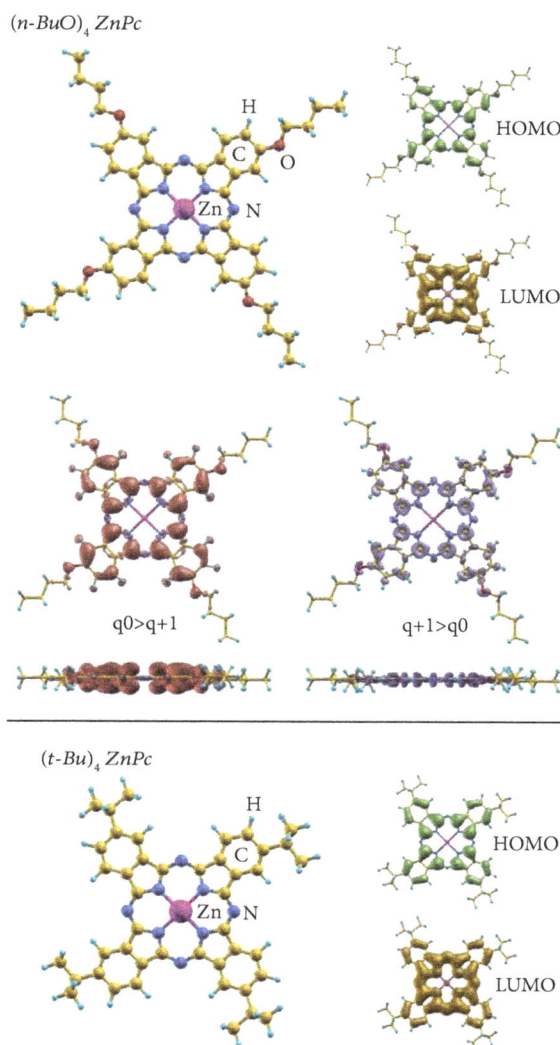

FIGURE 3: (a) Optimized geometry and $|\Psi|^2$ plots of frontier orbitals of $(n\text{-}BuO)_4ZnPc$. Difference density maps (positive and negative values, top and side views) of the charge density of $(n\text{-}BuO)_4ZnPc$ and $[(n\text{-}BuO)_4ZnPc]^+$ sampled at 0.0005 electrons/$Å^3$. Charge is depleted from red regions, basically corresponding to the $(n\text{-}BuO)_4ZnPc$ HOMO, when the molecule is oxidized. Part of such depleted charge is accumulated in blue regions, mainly localized in the molecular σ plane. Such accumulation of charge around the nuclei is connected with the lower availability of electrons in positive molecular ions for further ionization processes. (b) Optimized geometry and $|\Psi|^2$ plots of frontier orbitals of $(t\text{-}Bu)_4ZnPc$.

The photocurrent onset is around 760 nm in both cases, which is in agreement with the 1.55 eV band gap of $CH_3NH_3PbI_3$. In the Spiro-based device, we observe a maximum efficient charge extraction at 570 nm (EQE = 63%), while the phthalocyanine-based cell peaks at 540 nm (EQE = 58%).

For both Spiro-OMeTAD and $(n\text{-}BuO)_4ZnPc$, the integrated photocurrent is lower than the averaged one measured in the JV curves. This effect can occur in many different photovoltaic devices and may be due to several factors [28]. In

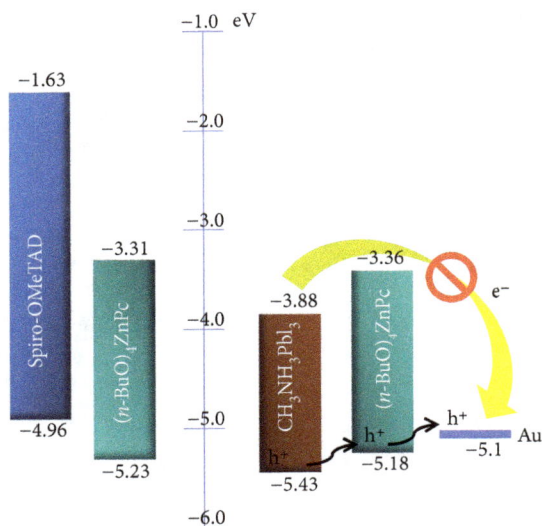

FIGURE 4: Left side: theoretically estimated reduction potentials of $(n\text{-BuO})_4\text{ZnPc}$ and Spiro-OMeTAD. Right side: energetic positioning of $(n\text{-BuO})_4\text{ZnPc}$ within a PSC.

particular, a significant role may be played by a difference in intensity of the incident light in EQE and JV measurement setups (in orders of magnitude). This points to the sublinear dependence of photocurrent in both cells [29] and need for further optimization of current collection and in particular the cell contact system (including contact/conducting layer interfaces). These preliminary evidences suggest that there is wide room for performance improvements focusing the attention on the optimization of the cell construction, which was beyond our initial goal. Nevertheless, in a nonoptimized device, our phthalocyanine-based HTL is capable of forming an interface with the photoactive layer characterized by hole-extracting and electron-blocking properties. These promising results can be further improved with a careful study of the layer deposition parameters and of the dopant ratio to enhance its hole transport properties.

Since we have stressed the importance of lowering the fabrication costs of novel hole transport materials to foresee their future commercialization, we propose a cost-per-gram analysis as a function of the synthetic complexity of $(n\text{-BuO})_4\text{ZnPc}$ and compare the obtained results with the current price of Spiro-OMeTAD. We point out that moving from lab to fab requires a complete cost analysis that must include parameters like energy, facility maintenance and personnel costs, taxes, and other charges that we have not taken into account. The total material cost (in euros) for our hole transport material has been estimated according to a paper published in 2013 by Osedach et al. [30]. For the sake of simplicity, Sigma-Aldrich has been chosen as the only supplier for reagents, solvents, and purification materials with the exception of 4-n-butoxyphthalonitrile that has been quoted from TCI chemicals. In all cases, the bulkiest batches available on the online catalogs have been considered—the list of which is reported in table I in the supplementary materials. The costs of our synthesis have been estimated assuming linearity when scaling the procedure from grams to kilograms, calculating the required amounts of chemicals to

produce 1 kg of $(n\text{-BuO})_4\text{ZnPc}$ in a 53% yield (see the synthesis flowchart in Figure S3 in the supplementary materials) and multiplying them by the respective prices. The resulting costs for reagents, solvents, and purification materials are reported in Table 4, given as euro per gram of the synthesized product.

Given that the experimental protocol can be further optimized and that solvents and reagents can be partially recovered and recycled, two total costs are provided. The "worst-case scenario" (wcs) accounts that any unreacted reagent and all the purification materials are lost as wastes. In the "best-case scenario" (bcs), the purification costs are not included because the chromatographic filtration is replaced by a vacuum treatment (10–3 mmHg at 250°C) that completely avoids the use of silica gel and solvents. Even if we did not include them in our analysis, it is worth to mention that diminishing the amount of wastes is beneficial for reducing their disposal costs.

Spiro-OMeTAD is commercially available and has been quoted from three suppliers as reported in table II in the supplementary materials. With respect to the less-expensive batch, supplied by Sigma-Aldrich at 242.40 €/g, our molecule is 2.8 times cheaper in the worst-case scenario and 3.3 times cheaper in the best-case scenario. Since a large-scale production requires kilograms of input reagents and 4-n-butoxyphthalonitrile is sold in 5 g batches at the most, we requested a bulk quotation for a batch of 1 kg, reducing its price from 42.40 €/g to 17.50 €/g. The recalculated synthesis costs, given as euro per gram of synthesized product, are summarized in Table 5.

In this way, $(n\text{-BuO})_4\text{ZnPc}$ results to be 7.9 times cheaper than Spiro in the best-case scenario, and given the limited number of suppliers that we have taken into consideration and the possibility to request bulk quotations of all the other materials, we are confident that its fabrication costs can be further lowered.

In conclusion, we have synthesized and characterized an easily processable, symmetrically tetrasubstituted zinc phthalocyanine, which is a cheap and promising hole transporting material for new-generation photovoltaics. We have also provided a first proof of concept assessment of its implementation in perovskite solar cells as proof of concept, obtaining promising results even in nonoptimized devices. We suggest therefore that improvements in the cell construction as well as the optimization of the molecular layer can promote $(n\text{-BuO})_4\text{ZnPc}$ as a possible alternative to commercially available organic p-type semiconductors thanks to its straightforward, potentially scalable, and less-expensive synthetic procedure.

1. Experimental and Computational Details

1.1. $(n\text{-BuO})_4\text{ZnPc}$ Synthesis and Characterization. ZnCl_2 and all the solvents (RPE grade) were purchased from Carlo Erba except for N,N-dimethylethanolamine (99.5%) which was purchased from Aldrich. Acetonitrile was distilled over calcium hydride. 4-n-Butoxyphthalonitrile (95%) was purchased from Tokyo Chemical Industry Co. (TCI). 1,5-Diazabicyclo[5.4.0]undec-5-ene (DBU) (≥99%) was purchased from Fluka. Reactions were monitored by thin-layer

TABLE 3: Average values of the main photovoltaic characteristics of $(n\text{-BuO})_4$ZnPc and Spiro-OMeTAD.

HTM		V_{oc} (V)	J_{sc} (mA/cm^2)	FF (%)	PCE (%)
$(n\text{-BuO})_4$ZnPc	Forward bias	0.896 ± 0.009	12.1 ± 1.2	57.0 ± 3.1	$6.06\% \pm 0.74$
	Reverse bias	0.928 ± 0.04	14.5 ± 1.9	61.8 ± 2.6	$9.00\% \pm 0.21$
Spiro-OMeTAD	Forward bias	0.903 ± 0.02	13.6 ± 0.9	55.2 ± 4.0	$6.75\% \pm 0.19$
	Reverse bias	0.928 ± 0.04	14.5 ± 1.9	61.8 ± 2.6	$8.61\% \pm 0.60$

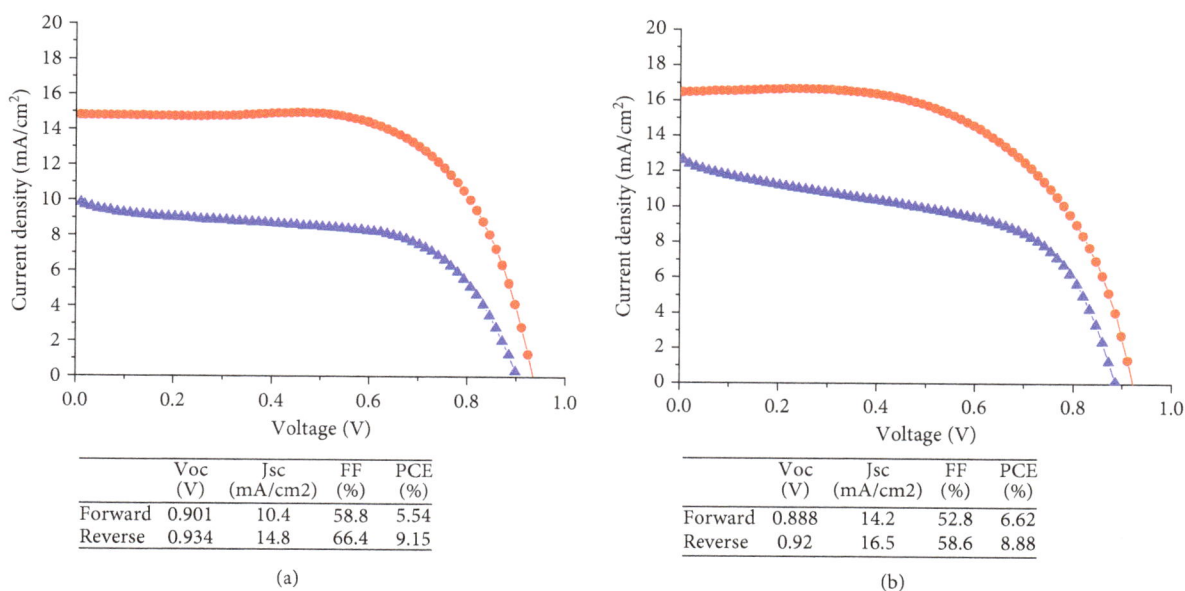

	Voc (V)	Jsc (mA/cm2)	FF (%)	PCE (%)
Forward	0.901	10.4	58.8	5.54
Reverse	0.934	14.8	66.4	9.15

(a)

	Voc (V)	Jsc (mA/cm2)	FF (%)	PCE (%)
Forward	0.888	14.2	52.8	6.62
Reverse	0.92	16.5	58.6	8.88

(b)

FIGURE 5: J-V characteristics of (a) $(n\text{-BuO})_4$ZnPc and (b) Spiro-OMeTAD-based PSCs.

chromatography (TLC) employing a polyester layer coated with 250 mm F254 silica gel. The crude purification was performed by filtration using silica gel Carlo Erba Reactifs SDS 60A C.C. 35–70 mm and dichloromethane and THF as eluents. To perform the fluorescence measurements, a further purification has been performed by size exclusion chromatography using Bio-Beads S-X Resin (Bio-Rad) as the stationary phase and THF as the mobile phase. ^{1}H NMR spectra were recorded in THF d8 on a Bruker AVANCE 600 NMR spectrometer operating at a proton frequency of 600.13 MHz; chemical shifts (δ) are given in ppm relative to the residual solvent peaks of the deuterated solvent (3.58, 1.73 ppm). Infrared spectra were recorded on a Shimadzu FT-IR prestige-21 spectrometer using an attenuated total reflectance (ATR) unit. All the UV-Vis measurements were performed in direct transmission mode. Absorption spectra of 10^{-5} M solution in acetonitrile were recorded on a PerkinElmer LAMBDA 950 UV-Vis/NIR spectrophotometer, while transparent sensitized films have been analyzed using a Cary 5000 UV-Vis-NIR spectrophotometer (Agilent Technologies). MALDI-TOF spectra have been recorded at Toscana Life Sciences facility in a MALDI-TOF/TOF Ultraflex III (Bruker) using α-cyano-4-hydroxycinnamic acid (HCCA) as matrix. Isotopic pattern simulations have been performed by Bruker Daltonics program. Steady-state fluorescence spectra were recorded on a 10^{-6} M solution in

acetonitrile with a Jobin Yvon Fluorolog3 spectrofluorometer, in the range 620–850 nm, using 5 nm grids for excitation and 5 nm for emission ($\lambda_{ex} = 600$ nm). The corresponding excitation spectrum has been collected in the range 350–700 nm, under an excitation wavelength of 750 nm, with 10 nm grids. No filters have been used. Time-resolved fluorescence measurements were carried out by a time-correlated single-photon counting (TCSPC) system (HORIBA Jobin Yvon), using a 405 nm pulsed laser diode and collecting the emission decay at the corresponding maximum emission wavelength (686 nm). The fluorescence decay profile was analyzed through decay analysis software (DAS6a HORIBA Scientific). The curve has been fitted with a monoexponential decay, with a final time decay of 2.4 ns. The quality of the fit was checked by examining the residual distribution having a χ^2 value of 1.4. Voltammograms were recorded at 25°C with a potentiostat-galvanostat Metrohm PGStat 204 in a conventional three-electrode cell; a platinum disk (~1 mm) was used as working electrode together with a platinum wire as auxiliary electrode. The reference electrode was Ag/AgNO$_3$ (0.01 M) in distilled acetonitrile (ACN) (Aldrich) (E_0 vs NHE = 0.548 V), and the Fc$^+$/Fc (ferrocenium/ferrocene) couple was used as external standard. The sample solutions were ~10^{-4} M in distilled anhydrous solvents, and dry tetra(n-butyl)ammonium tetrafluoroborate (TBATFB) (Aldrich) was used as the supporting electrolyte

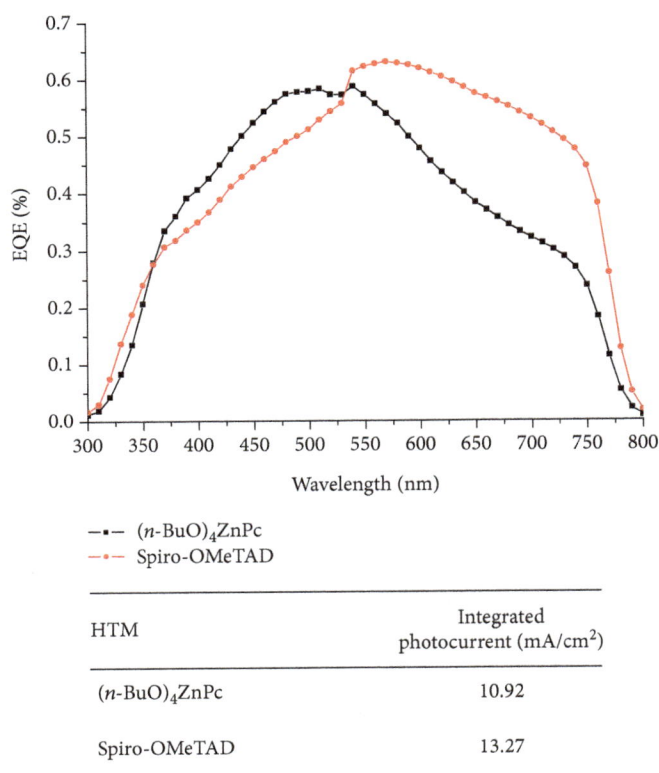

—■— $(n\text{-BuO})_4\text{ZnPc}$
—●— Spiro-OMeTAD

HTM	Integrated photocurrent (mA/cm^2)
$(n\text{-BuO})_4\text{ZnPc}$	10.92
Spiro-OMeTAD	13.27

FIGURE 6: External quantum efficiency graphs of $(n\text{-BuO})_4\text{ZnPc}$ (black line) and Spiro-OMeTAD-based PSCs (red line).

TABLE 4: Calculated synthesis cost for 1 g of $(n\text{-BuO})_4\text{ZnPc}$.

Synthetic steps	Reagents	Solvent	Purification	Total (wcs)	Total (bcs)
1	74.20	0.03	11.26	85.49	74.23

TABLE 5: Calculated synthesis cost for 1 g of $(n\text{-BuO})_4\text{ZnPc}$ with a bulk quotation of 1 kg of 4-n-butoxyphthalonitrile.

Synthetic steps	Reagents	Solvent	Purification	Total (wcs)	Total (bcs)
1	30.67	0.03	11.26	41.96	30.70

at 0.1 M. The solutions were previously purged for 10 minutes with nitrogen, and all measurements were performed under nitrogen. Cyclic voltammetry (CV) investigations were carried out at scan rates typically ranging from 0.05 to 2 Vs^{-1}. AFM images have been acquired in contact mode using a Keysight (formerly Agilent/MI) 5500 AFM with AppNano HYDRA6R-200N probes in air at room temperature and analyzed with the freely downloadable software WSxM 5.0 develop 8.4 December2016 [31].

Synthesis of $(n\text{-BuO})_4\text{ZnPc}$ (Mixture of Regioisomers). 200 mg of 4-n-butoxyphthalonitrile (1.0 mmol, 4 eq), 34 mg of zinc chloride (0.250 mmol, 1 eq), and a catalytic amount of 1,5-diazabiciclo[5.4.0]undec-5-ene (DBU) (1%mol, 1.7 μl)

have been dissolved in 1 ml of N,N-dimethylethanolamine in a round bottom flask. The mixture has been refluxed for 15 hours and then cooled down, diluted with water, and filtered to obtain a solid which has been washed with water until the pH is neutral and the washings are colorless. The desired molecule has been purified by filtration on silica gel using dichloromethane and THF as eluents, affording 114 mg (yield 53%) of the target molecule as a shiny purple crystalline solid. ^{1}H NMR (THF-d8, ppm), δ: 1.21–1.25 (m, 12H; CH$_3$), 1.80–1.88 (m, 8H; CH$_2$), 2.08–2.16 (m, 8H; CH$_2$), 4.50 (m, 8H; OCH$_2$), 7.45–7.55 (m, 4H; macrocycle), 8.39–8.49 (m, 4H; macrocycle), 8.79–8.90 (m, 4H; macrocycle); IR (cm^{-1}): 2954–2870 (cluster), 1604, 1236, 1089, 1047, 742, UV-Vis (acetonitrile, nm,): 675, 354, MALDI-TOF (m/z) [M$^+$H$^+$] 866.3, elemental analysis: calcd = C (66.54%), N (12.93%), H (5.58%), found: C (66.07), N (13.05%), H (5.70%).

1.2. Device Fabrication. All the chemicals used for the device fabrication were purchased from Aldrich except methylamine (40% in methanol) which was purchased from Tokyo Chemical Industry Co. (TCI). TiO$_2$ paste was purchased from Dyesol. Methylammonium iodide (MAI) was synthesized by stirring 55.6 ml of methylamine (40% in methanol) with 60 ml of hydroiodic acid (57 wt% in water) in a 250 ml round bottom flask at 0°C for 2 h. The solvent was then evaporated at 50°C for 1 h using a rotary evaporator; the resulting solid was collected and stirred with ethanol for 30 minutes. It was then filtered and repeatedly washed with diethyl ether to obtain a white solid that has been dried at 60°C in a vacuum oven for 24 h.

The fluorine-doped tin oxide-coated SnO$_2$:F (FTO) conducting glass (15, Pilkington) substrates have been cut in 2.5 × 1.5 cm size. The substrates were then etched with Zn/ HCl and cleaned properly. A hole-blocking layer was spin coated on FTO using a solution of titanium diisopropoxide bis(acetylacetonate) (TiDIP) (75% in 2-propanol) in ethanol and then annealed at 450°C for 30 minutes. A mesoporous titania layer was then deposited by spin coating using TiO$_2$ paste (DSL-90T TiO$_2$ NPs paste) diluted in ethanol (0.1 g of paste in 0.32 g of absolute ethanol) and calcinated at 500°C for 30 minutes. The TiO$_2$-deposited substrates were then treated with TiCl$_4$ for 30 minutes at 70°C, then rinsed, and annealed at 500°C for 30 minutes. The perovskite has been deposited using a two-step method. In the first step, 1 M solution of PbI$_2$ in a DMF:DMSO mixture (85:15 ratio) was spin coated on the substrates and annealed at 70°C for 45 min. The PbI$_2$-coated films were then immersed in a 10 mg/ml solution of MAI in 2-propanol for 30 seconds, resulting in the formation of dense black-brown MAPbI$_3$. The cells were further annealed at 90°C for 30 min. 40 μl of chlorobenzene solutions of $(n\text{-BuO})_4\text{ZnPc}$ (40 mg/ml) and Spiro-OMeTAD (72.3 mg/ml) was deposited by spin coating. 26.3 μl of Bis(trifluoromethane)sulfonimide lithium salt in acetonitrile (520 mg/ml), 29 μl of tris(2-(1H-pyrazol-1-yl)-4-tert-butylpyridine)-cobalt(III) tris(bis(trifluoromethylsulfonyl)imide)) in acetonitrile (300 mg/ml), and 19.2 μl of 4-tert-butilpyridine were added as dopants. Finally, 70 nm-thick gold-back contacts have been evaporated under pressure of 5×10^{-6} Torr. The active area was $0.09 \, cm^2$, and a $0.04 \, cm^2$ screen mask has been used during the measurements.

1.3. Photovoltaic Characterization.

Photovoltaic measurements were made on a Newport system, composed of an Oriel I–V test station with an Oriel Sol3A simulator. The solar simulator is class AAA for spectral performance, uniformity of irradiance, and temporal stability. The solar simulator is equipped with a 450 W xenon lamp. The output power is adjusted to match AM1.5 global sunlight (100 mW·cm²). The spectral match classifications are IEC60904-9 2007, JIC C 8912, and ASTM E927-05. I–V curves were obtained by applying an external bias to the cell and measuring the generated photocurrent with a Keithley model 2400 digital source meter. A 10 mV voltage step and a 40 ms photocurrent delay time were used. External quantum efficiency was calculated by measuring the I_{sc} of both the tested cell and a reference silicon cell at every wavelength, according to the following equation:

$$EQE = \left[\frac{I(\lambda)}{I_{ref}(\lambda)} \right] EQE_{ref}. \qquad (1)$$

1.4. Computational Details.

The structural, electronic, redox, and optical properties of (n-BuO)$_4$ZnPc, (t-Bu)$_4$ZnPc, and Spiro-OMeTAD have been investigated using ab initio simulations based on density functional theory (DFT). In detail, the calculations have been performed by using the ORCA suite of programs [20] in a localized-basis-set framework. Kohn-Sham orbitals have been expanded on a def2-TZVPP Gaussian-type basis set [32]. Fully decontracted def2-TZVPP/J has been also used as an auxiliary basis set for Coulomb fitting in a resolution-of-identity/chain-of-spheres (RIJCOSX) framework [33]. Molecular geometries have been fully optimized at the B3LYP level of theory [34, 35], including dispersion forces calculated by using the DFT-D3 approach [36]. Redox potentials of the molecules have been calculated by using the M06-2X functional [37], as discussed in detail elsewhere [21]. The molecules and their positive and negative ions have been embedded in an implicit CH_3CN solvent using the conductor-like polarizable continuum model (CPCM) [38] and fully optimized. The corresponding redox potentials have been calculated as ΔG values between neutral and charged species.

Data Availability

The phthalocyanine photoelectrochemical characterization data and the photovoltaic characteristics of the best-performing phthalocyanine-based and Spiro-based perovskite solar cells, used to support the findings of this study, are included within the article. The complete IV measurements can be furnished upon request. Furthermore, the experimental protocols to reproduce the synthesis and prepare the photovoltaic devices, the ^1H NMR chemical shifts, the IR main peaks, the UV-Vis absorption maxima, the MALDI-TOF molecular ion mass to charge ratio, the elemental analysis values of our molecule, and the computational details for the ab initio simulations are made available in the experimental section. The ^1H NMR full spectrum, MALDI spectrum, and anodic voltammogram of (n-BuO)$_4$ZnPc, used to confirm the molecular structure of our compound and support the findings of this study, are included in the supplementary materials.

Conflicts of Interest

The authors declare no competing financial interests.

Authors' Contributions

Gloria Zanotti and Giuseppe Mattioli contributed equally to the work.

Acknowledgments

G.Z. is thankful to the Ente Nazionale Energia e Ambiente (ENEA) and to the Italian Ministry of Foreign Affairs for a visitor post-doc fellowship to Ben Gurion University of the Negev. The authors are thankful to Sara Notarantonio for her precious assistance in the experimental work and in the electrochemical measurements. The research was funded in part by Israel's Ministry of National Infrastructures, Water and Energy Resources (grant no. 0399202/215-11-037) and by the Adelis Foundation.

References

[1] Z. Yu and L. Sun, "Recent progress on hole-transporting materials for emerging organometal halide perovskite solar cells," *Advanced Energy Materials*, vol. 5, no. 12, article 1500213, 2015.

[2] S. Ameen, M. A. Rub, S. A. Kosa et al., "Perovskite solar cells: influence of hole transporting materials on power conversion efficiency," *ChemSusChem*, vol. 9, no. 1, pp. 10–27, 2016.

[3] J. Yang, B. D. Siempelkamp, D. Liu, and T. L. Kelly, "Investigation of $CH_3NH_3PbI_3$ degradation rates and mechanisms in controlled humidity environments using in situ techniques," *ACS Nano*, vol. 9, no. 2, pp. 1955–1963, 2015.

[4] A. Kojima, K. Teshima, Y. Shirai, and T. Miyasaka, "Organometal halide perovskites as visible-light sensitizers for photovoltaic cells," *Journal of the American Chemical Society*, vol. 131, no. 17, pp. 6050-6051, 2009.

[5] W. S. Yang, B.-W. Park, E. H. Jung et al., "Iodide management in formamidinium-lead-halide–based perovskite layers for efficient solar cells," *Science*, vol. 356, no. 6345, pp. 1376–1379, 2017.

[6] L. Calió, S. Kazim, M. Grätzel, and S. Ahmad, "Hole-transport materials for perovskite solar cells," *Angewandte Chemie International Edition*, vol. 55, no. 47, pp. 14522–14545, 2016.

[7] M. Saliba, T. Matsui, J.-Y. Seo et al., "Cesium-containing triple cation perovskite solar cells: improved stability, reproducibility

and high efficiency," *Energy & Environmental Science*, vol. 9, no. 6, pp. 1989–1997, 2016.

[8] S. F. Völker, S. Collavini, and J. L. Delgado, "Organic charge carriers for perovskite solar cells," *ChemSusChem*, vol. 8, no. 18, pp. 3012–3028, 2015.

[9] P. Peumans, S. Uchida, and S. R. Forrest, "Efficient bulk heterojunction photovoltaic cells using small-molecular-weight organic thin films," *Nature*, vol. 425, no. 6954, pp. 158–162, 2003.

[10] K. Cnops, B. P. Rand, D. Cheyns, B. Verreet, M. A. Empl, and P. Heremans, "8.4% efficient fullerene-free organic solar cells exploiting long-range exciton energy transfer," *Nature Communications*, vol. 5, no. 1, p. 3406, 2014.

[11] O. A. Melville, B. H. Lessard, and T. P. Bender, "Phthalocyanine-based organic thin-film transistors: a review of recent advances," *ACS Applied Materials & Interfaces*, vol. 7, no. 24, pp. 13105–13118, 2015.

[12] G. Sfyri, C. V. Kumar, G. Sabapathi et al., "Subphthalocyanine as hole transporting material for perovskite solar cells," *RSC Advances*, vol. 5, no. 85, pp. 69813–69818, 2015.

[13] F. J. Ramos, M. Ince, M. Urbani et al., "Non-aggregated Zn(II)octa(2,6-diphenylphenoxy) phthalocyanine as a hole transporting material for efficient perovskite solar cells," *Dalton Transactions*, vol. 44, no. 23, pp. 10847–10851, 2015.

[14] J.-J. Guo, X.-F. Meng, J. Niu et al., "A novel asymmetric Phthalocyanine-based hole transporting material for perovskite solar cells with an open-circuit voltage above 1.0 V," *Synthetic Metals*, vol. 220, pp. 462–468, 2016.

[15] K. T. Cho, K. Rakstys, M. Cavazzini, S. Orlandi, G. Pozzi, and M. K. Nazeeruddin, "Perovskite solar cells employing molecularly engineered Zn(II) phthalocyanines as hole-transporting materials," *Nano Energy*, vol. 30, pp. 853–857, 2016.

[16] K. T. Cho, O. Trukhina, C. Roldán-Carmona et al., "Molecularly engineered phthalocyanines as hole-transporting materials in perovskite solar cells reaching power conversion efficiency of 17.5%," *Advanced Energy Materials*, vol. 7, no. 7, article 1601733, 2017.

[17] L. Edwards and M. Gouterman, "Porphyrins: XV. Vapor absorption spectra and stability: phthalocyanines," *Journal of Molecular Spectroscopy*, vol. 33, no. 2, pp. 292–310, 1970.

[18] F. Würthner, T. E. Kaiser, and C. R. Saha-Möller, "J-aggregates: from serendipitous discovery to supramolecular engineering of functional dye materials," *Angewandte Chemie International Edition*, vol. 50, no. 15, pp. 3376–3410, 2011.

[19] G. Mattioli, S. B. Dkhil, M. I. Saba et al., "Interfacial engineering of P3HT/ZnO hybrid solar cells using phthalocyanines: a joint theoretical and experimental investigation," *Advanced Energy Materials*, vol. 4, no. 12, article 1301694, 2014.

[20] F. Neese, "The ORCA program system," *Wiley Interdisciplinary Reviews: Computational Molecular Science*, vol. 2, no. 1, pp. 73–78, 2012.

[21] M.-H. Baik and R. A. Friesner, "Computing redox potentials in solution: density functional theory as a tool for rational design of redox agents," *The Journal of Physical Chemistry A*, vol. 106, no. 32, pp. 7407–7412, 2002.

[22] S. M. Ross, "Peirce's criterion for the elimination of suspect experimental data," *Journal of Engineering Technology*, vol. 20, pp. 38–41, 2003.

[23] J. Burschka, N. Pellet, S.-J. Moon et al., "Sequential deposition as a route to high-performance perovskite-sensitized solar cells," *Nature*, vol. 499, no. 7458, pp. 316–319, 2013.

[24] C. Roldán-Carmona, P. Gratia, I. Zimmermann et al., "High efficiency methylammonium lead triiodide perovskite solar cells: the relevance of non-stoichiometric precursors," *Energy & Environmental Science*, vol. 8, no. 12, pp. 3550–3556, 2015.

[25] H.-S. Kim and N.-G. Park, "Parameters affecting *I–V* hysteresis of $CH_3NH_3PbI_3$ perovskite solar cells: effects of perovskite crystal size and mesoporous TiO_2 layer," *Journal of Physical Chemistry Letters*, vol. 5, no. 17, pp. 2927–2934, 2014.

[26] H. J. Snaith, A. Abate, J. M. Ball et al., "Anomalous hysteresis in perovskite solar cells," *Journal of Physical Chemistry Letters*, vol. 5, no. 9, pp. 1511–1515, 2014.

[27] B. Chen, M. Yang, S. Priya, and K. Zhu, "Origin of *J–V* hysteresis in perovskite solar cells," *Journal of Physical Chemistry Letters*, vol. 7, no. 5, pp. 905–917, 2016.

[28] R. Scheer and H.-W. Schock, *Chalcogenide Photovoltaics: Physics, Technologies, and Thin Film Devices*, Wiley, 2011.

[29] E. A. Katz, A. Mescheloff, I. Visoly-Fisher, and Y. Galagan, "Light intensity dependence of external quantum efficiency of fresh and degraded organic photovoltaics," *Solar Energy Materials and Solar Cells*, vol. 144, pp. 273–280, 2016.

[30] T. P. Osedach, T. L. Andrew, and V. Bulović, "Effect of synthetic accessibility on the commercial viability of organic photovoltaics," *Energy & Environmental Science*, vol. 6, no. 3, pp. 711–718, 2013.

[31] I. Horcas, R. Fernández, J. M. Gómez-Rodríguez, J. Colchero, J. Gómez-Herrero, and A. M. Baro, "WSXM: a software for scanning probe microscopy and a tool for nanotechnology," *Review of Scientific Instruments*, vol. 78, no. 1, article 013705, 2007.

[32] F. Weigend and R. Ahlrichs, "Balanced basis sets of split valence, triple zeta valence and quadruple zeta valence quality for H to Rn: design and assessment of accuracy," *Physical Chemistry Chemical Physics*, vol. 7, no. 18, pp. 3297–3305, 2005.

[33] F. Weigend, "Accurate coulomb-fitting basis sets for H to Rn," *Physical Chemistry Chemical Physics*, vol. 8, no. 9, pp. 1057–1065, 2006.

[34] A. D. Becke, "Density-functional thermochemistry. III. The role of exact exchange," *The Journal of Chemical Physics*, vol. 98, no. 7, pp. 5648–5652, 1993.

[35] C. Lee, W. Yang, and R. G. Parr, "Development of the Colle-Salvetti correlation-energy formula into a functional of the electron density," *Physical Review B*, vol. 37, no. 2, pp. 785–789, 1988.

[36] S. Grimme, J. Antony, S. Ehrlich, and H. Krieg, "A consistent and accurate ab initio parametrization of density functional dispersion correction (DFT-D) for the 94 elements H-Pu," *The Journal of Chemical Physics*, vol. 132, no. 15, article 154104, 2010.

[37] M. Isegawa, F. Neese, and D. A. Pantazis, "Ionization energies and aqueous redox potentials of organic molecules: comparison of DFT, correlated ab initio theory and pair natural orbital approaches," *Journal of Chemical Theory and Computation*, vol. 12, no. 5, pp. 2272–2284, 2016.

[38] V. Barone and M. Cossi, "Quantum calculation of molecular energies and energy gradients in solution by a conductor solvent model," *The Journal of Physical Chemistry A*, vol. 102, no. 11, pp. 1995–2001, 1998.

15

Photovoltaic Energy-Assisted Electrocoagulation of a Synthetic Textile Effluent

Thelma Beatriz Pavón-Silva ⓘ,[1] Hipólito Romero-Tehuitzil,[2] Gonzálo Munguia del Río,[2] and Jorge Huacuz-Villamar[2]

[1]*Unidad Académica Profesional-Acolman, Universidad Autónoma del Estado de México, Camino de Caleros No. 11, Ejidos de Santa Catarina, Acolman, MEX, Mexico*
[2]*Instituto Nacional de Electricidad y Energías Limpias, Av. Reforma 113, Col. Palmira, 62490 Cuernavaca, MOR, Mexico*

Correspondence should be addressed to Thelma Beatriz Pavón-Silva; th.pavon@gmail.com

Academic Editor: Carlos A. Martínez-Huitle

The feasibility of using photovoltaic modules to power a continuous 14 L electrochemical reactor applied to remove an azo dye with an efficiency of 70% is reported. The photovoltaic modules were directly connected, and the system efficiency was observed properly maintained when currents were applied in the range of 2.5 to 7.9 A. This value depends on solar radiation. Likewise, it was found that the efficiency depends mainly on the current density and the flow rate prevailing in the reactor.

1. Introduction

The problem of industrial wastewater is a topic that deserves special attention, particularly for industries that use dyes and thus generate large volumes of polluted wastewater. Due to their high molecular weights, complex structures, and especially high solubility in water, dyes persist once discharged into a natural environment, for example, textile [1, 2], pharmaceutical [3], cosmetics and food industry [4], and industrial wastewater [5]. Specifically for the textile industry, the chemical structure of these compounds is complex, generally of the type azo [6–8]. This group consists of colored substances with a complex chemical structure (many functional groups) and a high molecular weight. Thus, their removal from industrial effluents is also a subject of major importance from the environmental point of view. The removal of dissolved organic matter by coagulation is widely reported in literature [9], where the primary mechanism consists of two methods [10]. The first is binding the metal species to anionic sites of the organic molecules, thereby neutralizing their charge and resulting in reduced solubility. The second is the absorption of organic substances on amorphous metal hydroxide precipitates.

Chemical coagulation/flocculation is the most widely used technique for textile wastewater treatment; they have some disadvantages, such as needing pH adjustment before and after treatment, producing large amounts of sludge, and adding undesirable inorganic chemicals like aluminum, iron, sulfate, and chloride to the environment [11].

The results for the optimization of the effect of the coagulation–flocculation showed the rate of dye removal increased from 11.25% to 13.20 for the methylene blue and from 27.5% to 29.25 for the indigo carmine when the concentration of coagulant and flocculant were varied from 40 to 120 mg·L^{-1}. These results confirm the poor applicability of this process for the elimination of such dyes. Assadia et al. [12] show that ferric chloride and alum at optimum concentration were capable of removing dye and COD by 79.63% and 84.83% and 53% and 55%, respectively.

The electrocoagulation process is deemed an economical and environmental choice to minimize the drawbacks of conventional wastewater treatment technologies [13]; it provides a number of benefits, such as low cost, compatibility, and safety [14]. In addition, electrocoagulation has been proven to eliminate complex contaminants in wastewater that require a combination of physicochemical and biological methods [15–17]. Its combination with other processes, like ozonation [18], has also been demonstrated to be highly effective for complex matrix wastewater remediation. The application of electrocoagulation, however, has been restricted at some extent by the cost that implies electrical energy usage. Moreover, climate change motivates research and development of new forms of renewable energy [19, 20] such as photovoltaic (PV) power [8], whose advantages include its free use and abundant availability, is a renewable resource, decentralized, with long life span and low maintenance costs, and does not result in contamination.

In this research, a PV system was evaluated in a pilot electrochemical wastewater reactor using a synthetic wastewater model containing azo dye, in this case remazol yellow dye 3GL (RYD), which is mainly used in textile industry [4, 8, 21]. This was first accomplished with batch tests and then with continuous tests in a pilot reactor, both of which are based on previously reported studies [8].

2. Experimental

The experimental setup is shown in Figure 1 and consists of the following: (1) the sun as a renewable energy source; (2) photovoltaic modules to convert sunlight into electricity; (3) an acrylic cylindrical reactor with a 0.15 m diameter and one meter length (14 L of water was treated); (4) direct connection of the sun to the electrochemical reactor, two aluminum electrodes formed by 39 circular blades (0.12 m diameter) with total surface area of 0.44 m^2, separated from each other by one centimeter. (5) Registration data and continuous monitoring was performed by a data acquisition system: current, voltage, and solar radiation. (6) The pH, conductivity, and temperature were measured with an electrical conductivity meter.

2.1. Reagents. The working solution was prepared by dissolving remazol yellow dye (RYD) 3GL (DyStar SA™) in potable water and was used as wastewater model containing azo dye (RYD) as pollutant. Table 1 summarizes its characteristics.

The conductivity was set at 366–380 μS/cm, and this in concordance with Can et al. [22], who stated this parameter to be lower than 500 μS/cm. They investigated the conductivity between 250 and 4000 μS/cm using NaCl as the support electrolyte and noticed that above 500, the remazol red dye removal efficiency decreases. In our case, due to the size of the reactor (14 L pilot reactor), it was decided to work with potable water that already has a conductivity of 302–315 μS/cm, so an electrolyte support was not added. Thus, as reported by Can et al., the decline in the fading efficiency with increasing conductivity can be attributed to a change in ionic strength due to the change in conductivity of the aqueous medium. Ionic strength affects the kinetics and equilibrium of reactions between charged species during electrocoagulation.

2.2. Determination of Remazol Yellow Dye (RYD). RYD concentration was determined by UV-Vis spectrophotometry in a Perkin Elmer Lambda 2 spectrophotometer. The concentration of the molecule was determined by its

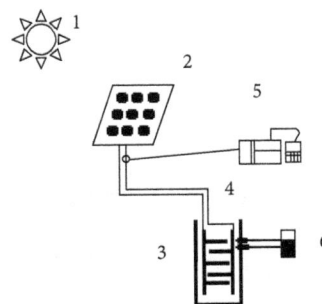

FIGURE 1: Scheme of the experimental system: 1, solar energy; 2, photovoltaic module; 3, electrochemical reactor; 4, direct connection of the power supply; 5, parameter control: current, voltage, and solar radiation; 6, pH, conductivity, and temperature control.

TABLE 1: Characterization of the remazol yellow dye.

Parameter	Valor
Color index	Remazol yellow
Chromophore	Azo
Molar mass (g·mol^{-1})	362.27
Percentage of pure dye	68%
pKa	3.77
Water solubility at 293 K (g·L^{-1})	80
Acute oral toxicity LD50 (mg·kg^{-1})	>2000
pH value (at 10 g·L^{-1} water)	6.1
Conductivity (mS·cm^{-1})	302–308

absorption at 269.2 nm. The calibration curve was carried out between 0 and 100 mg·L^{-1} of RYD obtaining the following model: A = 0.0301(C)–0.0003, where A = absorption and C = concentration of RYD. The determination coefficient was $r^2 = 0.9994$. Color determination was conducted in a Hach DR/3000 spectrophotometer.

2.3. Removal Efficiency of RYD. The removal efficiency of RYD was calculated as

$$E(\%) = \frac{C_1 - C_f}{C_1} * 100, \tag{1}$$

where C_1 is the initial dye concentration and C_f is the final dye concentration, both in mg·L^{-1}; this expression was also used in case of color determination where C_1 and C_f are color intensities in Pt/Co units [23].

2.4. Chemical Oxygen Demand. Chemical oxygen demand was determined according to standard techniques, APHA/AWWA/WPCE [24].

The removal efficiency or percentage of COD removal (%RE) was then calculated as follows [25, 26]:

$$E(\%) = \frac{COD_0 - COD_f}{COD_0} * 100, \tag{2}$$

where COD_0 is the initial chemical oxygen demand and COD_f is the final chemical oxygen demand, both in $mg \cdot L^{-1}$.

2.5. Sampling. The determination of RYD and the removal efficiency of RYD and COD correspond to the samples taken in the upper part of the reactor and later filtered with Whatman paper number 1. The samples were withdrawn from the reactor every 5 min for analysis.

2.6. Applied Current Intensity. The experimental tests were divided into two stages. First, a series of discontinuous tests with current intensities of 4, 6, 8, and 10 A controlled by a power supply during 60 minutes (system DC power supply, Agilent Technologies N5700 series) were carried out and then with current intensities of 2, 3, 4, and 5. The second phase was carried out in a continuous flow, during approximately 6 hours; the current intensity was applied according to the results obtained in the discontinuous tests.

The objective in this stage was to evaluate the intensity of the current that is variable, depending on the following flow rates: 300, 500, 700, and 1000 $mL \cdot min^{-1}$, in order to check the contact time between the dye and the electrode to evaluate the percentage of removal of RYD through photovoltaic live connection.

The electrodes were treated by rinsing them with a 1 M HCl solution and distilled water at the end of each test.

2.7. Photovoltaic Array. The photovoltaic array consisted of two Siemens Solar photovoltaic (PV) modules Solar 75 Wp, SP-75. The solar module characteristics were verified using the I-V Checker MP140 Portable PV Device Evaluation Instrument, for which the PV module was first determined, and afterwards, with two PV modules connected in parallel [8], the short circuit current increases with the number of modules connected in parallel. The experiments were carried out at the Institute of Electrical Research (latitude 18°52' 40.99'' N, longitude 99°13 '6.89'' O, inclination 19°, and south oriented).

Table 2 shows the PV module characterization for one and two PV modules connected in parallel. If one solar module is considered, the acquired current is 3.8 A and is sufficient to obtain similar removal results as accomplished by the power source. In the case of a cloudy day where there is low solar radiation, the application of two modules would be required. Parallel-connected solar PV modules would offer a required minimum current intensity in the reaction performance with reasonable efficiency. The calibration of both modules set in parallel array produces a current intensity of 7.9 A.

3. Results and Discussion

3.1. Batch Tests with Artificial Power Source. Table 3 shows that COD removal is time dependent while applying different current intensities. At lower current intensity, the removal efficiency of COD is higher; this variation can be attributed to an interference during COD determination caused by the presence of aluminum ions when applying more current. The electrochemical reaction will produce a higher number

TABLE 2: Characterization of solar PV modules.

Parameter	1 PV module	2 PV module
Solar irradiance (W/m^2)	1007.3	1046.5
Ambient temperature (°C)	25	25
Cell temperature (°C)	58.4	57
Short current (A)	4.4	9.0
Open voltage (V)	18.5	18.8
Max power (W)	51.5	107.3
Max power current (A)	3.8	7.9
Max power voltage (V)	13.5	13.6
Fill factor (FF)	0.63	0.63

TABLE 3: Removal efficiency of COD $mg \cdot L^{-1}$ by different current intensities.

| Time (min) | Current (A) | | | |
	4.0	6.0	8.0	10.0
15	52.9	48.0	42.2	37.8
30	70.6	48.0	39.0	76.5
45	77.5	41.2	61.8	64.3
60	43.1	40.2	49.0	57.7

of aluminum ions, and this explains why the removal efficiency decreases during the treatment time.

It is known that the current intensity applied to the system determines the amount of released ions and therefore the amount of the resulting coagulant. Thus, the higher the amount of dissolved Al^{3+} ions in a solution, the greater the rate of $Al(OH)_3$ formation, and consequently a higher COD removal efficiency is expected to be achieved. In addition, the increase of current density promotes the generation of H_2 bubbles and decreases its size, which should lead to a higher removal of pollutants by flotation. However, higher current values may promote a higher turbulence in the system, and consequently, the particles responsible for coagulation do not have enough time to agglomerate themselves and remove the pollutants [27] as stated by Fajardo et al. (2015).

The ions generated in the electrode are $Al(aq)^{3+}$ and OH^-; the combination of these ions is expected to form various monomeric species such as $Al(OH)^{2+}$, $Al(OH)_2^+$, $Al_2(OH)_2^{4+}$, and $Al(OH)_4^-$ and polymeric species such as $Al_6(OH)_{15}^{3+}$, $Al_7(OH)_{17}^{4+}$, $Al_8(OH)_{20}^{4+}$, $Al_{13}(OH)_{34}^{5+}$, and $Al_{13}(OH)_{37}^{2+}$ [23, 24]. By relating these species with the report about super-faradaic efficiencies [28, 29], they are more significant at low current densities, that is, $1.75 \, mA \cdot cm^{-2}$. In the case of aluminum, experimental results are significantly over the expected values for a 100%-efficiency process according to Faraday's law, (3). This super-faradaic efficiency is explained in terms of a chemical dissolution process, which corresponds to the oxidation of the aluminum sheets with the simultaneous reduction of water to form hydrogen. It has also been mentioned that the amount of generated aluminum seems to depend on the pH, and this has been explained in terms of chemical dissolution of the aluminum electrode. In this case, the pH was initially between 5.6 and 6.3. However, it was observed that as current increases, pH rapidly decreases. For example, at $I = 4, 6, 8$, and $10\,A$, the pH was 4.8, 2.3, 2.2, and 1, respectively. Gu et al. suggest that (4) and (5) for the oxidation of water will decrease solution pH [30] and this was observed in this work, so, the super-faradaic efficiencies are greater for aluminum than those for iron [29].

The pKa of the remazol yellow dye is 3.77. Therefore, above this pH, the species of ionic form will be found, and its chromophore groups will be negatively charged. When the removal efficiency decreases, it is suggested that the precipitated hydroxide metal may have tended to dissolve in liquid thus adversely affecting the performance of the process. This fact may take place as the medium turns extremely acidic or alkaline as previously reported [31–33]. The cationic metal species are responsible for the destabilization of the particles (charge neutralization), leading to the formation of flocculating particles, which will have the power to sediment contaminants [33]. In situ generation of coagulants has the advantage of reducing the amount of chemical reagents introduced into the system; however, this may lead to a change in pH. In this context, the pH of the batch tests has been reported acidic and in the continuous tests, the pH fluctuated between 7.8 and 8.9, so the predominant reaction will be represented by (3) [29].

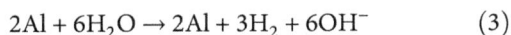

$$2Al + 6H_2O \rightarrow 2Al + 3H_2 + 6OH^- \qquad (3)$$

According to the diagram of Pourbaix [34], at pH less than 4, Al^{3+} ions are expected, and between 4 and 8, there may be passivation by Al_2O_3. For this study, however, it is observed that the pH and therefore the predominant reactions depend on the flow regime, that is, in batch tests, an acidic pH is observed and (4) and (5) predominate, and in tests under continuous flow, the pH is between neutral and slightly alkaline, and without considering the effect of passivation, (3) is the predominant one as already mentioned.

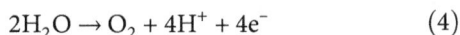

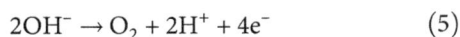

$$2H_2O \rightarrow O_2 + 4H^+ + 4e^- \qquad (4)$$

$$2OH^- \rightarrow O_2 + 2H^+ + 4e^- \qquad (5)$$

Another phenomenon is the production of aluminum, so in Table 3, there is an increase of COD, although the color tends to decrease. As it is known, this production is a function of the applied current. Thus, considering the electrode area ($0.44\,m^2$) and applying Faraday's law (6), the following values of Al^{3+} are calculated, 0.32, 0.48, 0.64, and 0.81 g, for currents of 4, 6, 8, and 10 A, accordingly. It is possible to have an excess of ions and for this reason, the COD value does not fall [5].

$$n = \frac{MIt}{zF}, \qquad (6)$$

where n is the metal (g), M molecular mass of electrode, I current intensity, t operating time, z number of electron transferred, and F Faraday's constant (96,500 C/mol).

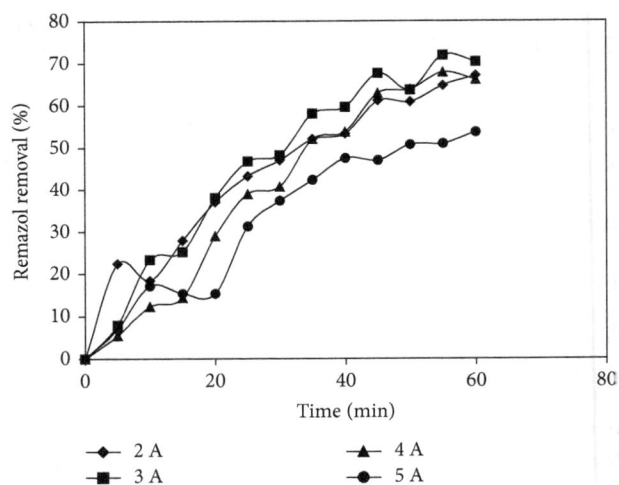

FIGURE 2: Removal of RYD at different current intensities: 2 A, 3 A, 4 A, and 5 A.

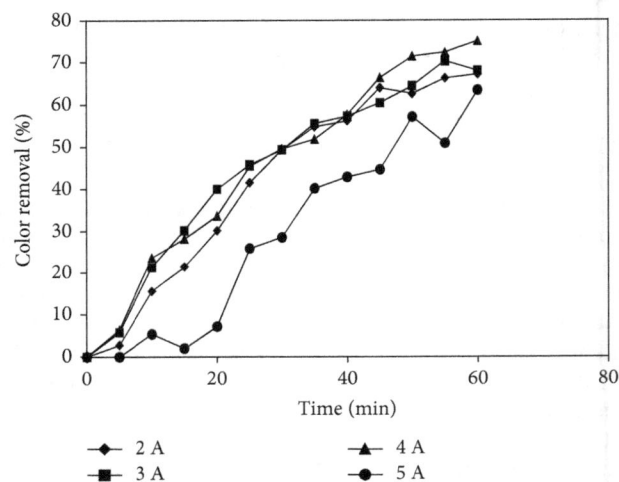

FIGURE 3: Color removal at different current intensities: 2 A, 3 A, 4 A, and 5 A.

In the case of aluminum, experimental results are significantly over the expected values for a 100%-efficiency process according to Faraday's law. This super-faradaic efficiency is explained in terms of a chemical dissolution process, which corresponds to the oxidation of the aluminum sheets with the simultaneous reduction of water to form hydrogen, according to (3) [29].

3.2. Optimization of the Applied Current with the Power Source. According to the results of Table 3, the applied current must be less than 6 A; so for these tests, currents to be evaluated were selected from 2, 3, 4, and 5 A. Figure 2 shows the dye removal percentage at these current intensities where a significant difference does not appear when applying 2, 3, or 4 A, as compared to 5 A ANOVA and Duncan test ($p > 0.5$) [35].

Figure 3 shows the dye color removal percentage at the same currents; these current intensities have the same behavior as in the removal of dye concentration.

In Figures 2 and 3, a similar behavior is observed regarding the elimination of the dye elaborated with the UV-Vis spectrophotometer and the elimination of the color; they have very similar efficiencies. When performing the ANOVA, it is similar to the dye efficiency tests that with a current of 2, 3, and 4 A, there is no significant difference in the results and with respect to 5 A, if there is a significant difference ($p > 0.5$).

3.3. Batch Examinations with PV Modules.

Table 1 shows that one PV module is enough to generate a current of 4 A under good solar radiation conditions, and two solar PV modules in parallel array are capable to generate 7.9 A. This means that in cloudy weather conditions, the current generated by two solar PV modules must be in the necessary range to carry out the RYD removal tests. Then, under this assumption, Figure 4 shows the comparison for RYD removal with the power source in 3 A and 4 A and one solar PV module. During this testing period, the solar PV module had an average solar radiation of 747 W/m^2 and a current of 3.1 A, with an average voltage of 2.37 V; the pH dropped slightly from 8 to 6.5 units, while the temperature in the electrocoagulation reactor increased from 18.5 to 20°C. As shown in Figure 4, there are no differences in the RYD results from the power source and solar PV modules. Observing that the removal efficiency is lower for 5 A, it is established that the appropriate conditions for the work of the system with the solar panel will be sufficient if it reaches between 2 and 4 A of current. This will depend on the solar irradiation of the moment; since the solar modules are connected directly, the solar irradiation impacts directly to the electrochemical reactor, generating greater or less quantity of ions to participate in the electrocoagulation. If there is a sunny and cloudless day, adequate solar radiation is guaranteed, which is why 1 or 2 photovoltaic panels are used. This is a complement of Figures 2 and 3 where we observe the similarity in the color and concentration removals of the RYD, and if we contrast this with Figure 4 which is already an experimental process, it is possible to use one or two photovoltaic modules, considering the use of only one module for the conditions treated in these experiments.

3.4. Examination in Continuous Flow Rate.

Figure 5 shows how the removal of the dye and color removal (65–70%) are kept constant in the test from 1.5 hours and onwards with a flow rate of 300 mL·min^{-1} and the PV current supplied by either one or two PV modules. Contrasting this graph with Figure 6, the radiation in either case, with one and two photovoltaic modules, can be observed. It is worth noting that there is a significant variation with the use of two photovoltaic modules. This can be ascribed to the day being cloudy and although it worked with the two modules reaching currents from 3.08 to 7.93 and 0.78 to 3.92 with one module, a variation is reflected although in an appropriate range for the minimum current needed for the electrochemical process. Thus, the use of two modules allows to attain currents

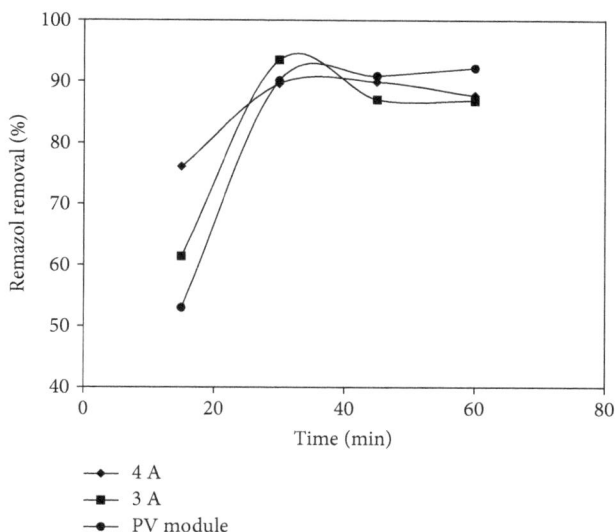

FIGURE 4: Comparison between current intensity controlled with the power source and one module solar: 4 A, 3 A, and PV module.

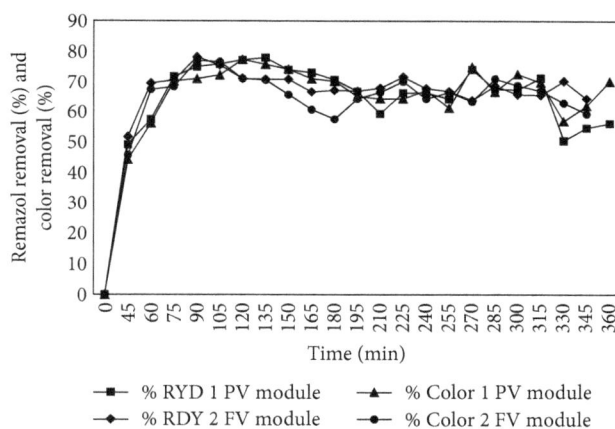

FIGURE 5: Remazol removal and color removal under continuous flow with the application of photovoltaic modules: liquid flow rate: 300 mL·min^{-1}; initial concentration of RYD: 100 mg·L^{-1}. % RYD one PV module, % RYD one PV module, color one PV module, and color two PV module.

higher than 3 A at all times. According to Figure 4, this value is sufficient for the process to occur. When only one module is used, the current will depend on the daytime and the variation may be rather rapid. For example, at 9:00 in the morning, a 0.78 A current was obtained, while 50 minutes later, the measured current was already 2 A. This trend usually prevails until 15:h. On the other hand, when two solar modules are used, the minimum attained current is 3.08 A, and between 10:00 and 15:30, the current is always higher than 3 A. Therefore, it can be concluded that two modules will always provide the adequate current regardless of the solar radiation intensity during daytime.

The tests by different flow rates are shown in Figure 7. As it can be observed, the differences of the dye removal efficiencies are not significant for flow rates of 500, 700,

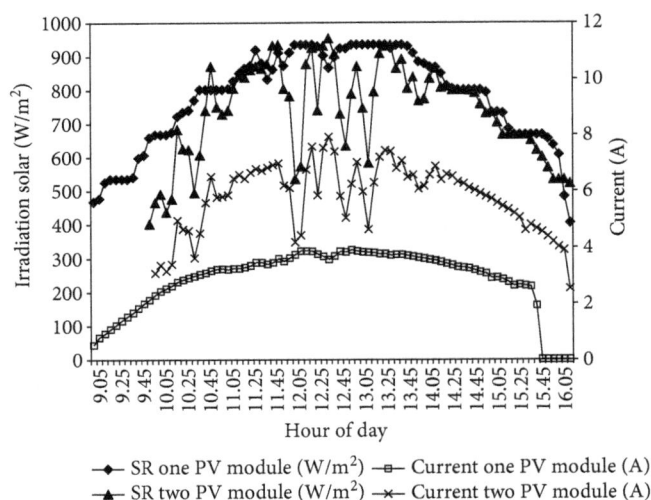

FIGURE 6: Solar irradiation profiles as function of time: SR one PV module, SR two PV module, current one PV module, and current two PV module.

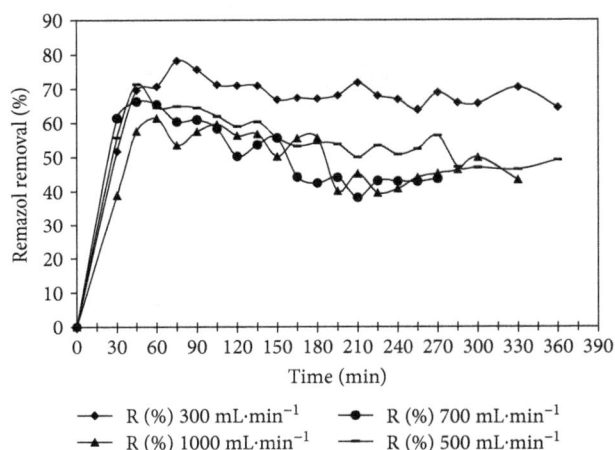

FIGURE 7: Effect of removal percentage rate on RYD initial concentration remazol 100 mg·L^{-1}: % RYD 300 mL·min^{-1}, % RYD 500 mL·min^{-1}, % RYD 700 mL·min^{-1}, and % RYD 1000 mL·min^{-1}.

and 1000 mL·min^{-1}; however, at 300 mL·min^{-1} flow rate, the difference is around 20% [36, 37]. This effect was expected since lower flow rates also imply higher residence time. It can be concluded that RYD removal efficiency is favored when the retention time, electric current, and flow rate are low [38, 39].

Table 4 shows the average, maximum, and minimum values of solar radiation, current intensity, and voltage for the accomplished tests by different flow rates comparing each one of 300 mL·min^{-1} with the remaining other. It is possible to observe that the current intensity is lower at 300 and 500 mL·min^{-1}; however, the removal efficiency is better in the low flow rate. This proves that current above 5 A does not increase the removal efficiency.

As shown, there is variation in the current intensity for each test, because it is not possible to compare them,

TABLE 4: Evaluated parameters during different RYD flow rate.

Flow rate (mL·min^{-1})	Solar radiation (W/m^2)	Voltage (V)	Current intensity (A)
300	741.4 (437.5–954)	3.9 (1.92–4.9)	5.7 (2.53–7.93)
500	624.4 (461–777)	4.3 (3.9–4.9)	5.5 (3.94–6.97)
700	861.7 (688–957)	6.6 (4.5–7.56)	7.4 (3.52–8.41)
1000	857.4 (602–977)	6.2 (3.34–8.23)	7.4 (3.45–8.45)

Average value (minimum–maximum).

however, since they are connected live; it depends on the solar radiation received on the day of the test; it can only be confirmed that it will be necessary to modify the water flow rate according to the received radiation so that it is proportional to what is reported in the literature as suitable to produce the amount of aluminum to react with the dye; nevertheless, it should not be forgotten that the phenomenon of super-faradaic production of the ion that helps the dye removal is also present. Under continuous flow, there is a synergy during treatment since after 1 hour of treatment, a stationary state is reached and a dye removal efficiency of 70% is attained. Worth noticing is that this value is kept constant until the treatment is complete (Figure 4). It is considered that it is necessary to establish a chain of batteries that are charged with photovoltaic energy to be able to carry out the elimination of pollutants continuously throughout the day, without being contingent to the solar radiation. That is, to store the energy and use it at the moment that it is required.

3.5. Energy Consumption.

Figure 8 shows the energy consumption to remove 1 g of RYD by means of direct current and photovoltaic current at a flow rate of 300 mL·min^{-1}. It can be seen that under controlled current supply with a DC power system, the energy consumption to remove RYD was constant and linear, 5.38 Wh·g^{-1}.

For RYD removal by means of PV current, this was dependent on solar radiation, reaching a maximum value of solar radiation around 1:00 p.m. and descended to the lowest value by the sunset.

Valero et al. [39, 40] demonstrated that the use of photovoltaic energy (i) reduces the cost of investment by avoiding the use of batteries, solar inverters, and power supplies and (ii) reduces the cost of maintenance, since there is no waste of batteries to properly dispose.

However, the disadvantage of this process is that in Mexico, it could only be applied from 9:30 am to 4:00 p.m., since it is the time in which there is adequate solar radiation in a sunny day. Sunny days are not constant through the year though. In order to overcome this, the use of hybrid energies combining solar photovoltaic with wind could be considered. This approach, however, should be evaluated taking into account that the increase of air is in the evening, so the working hours of an electrochemical reactor can be increased [41], considering the proposal of [39].

According to Figure 8, energy consumption to remove RYD color was dependent on the generated current by one or two PV modules. However, at the highest energy

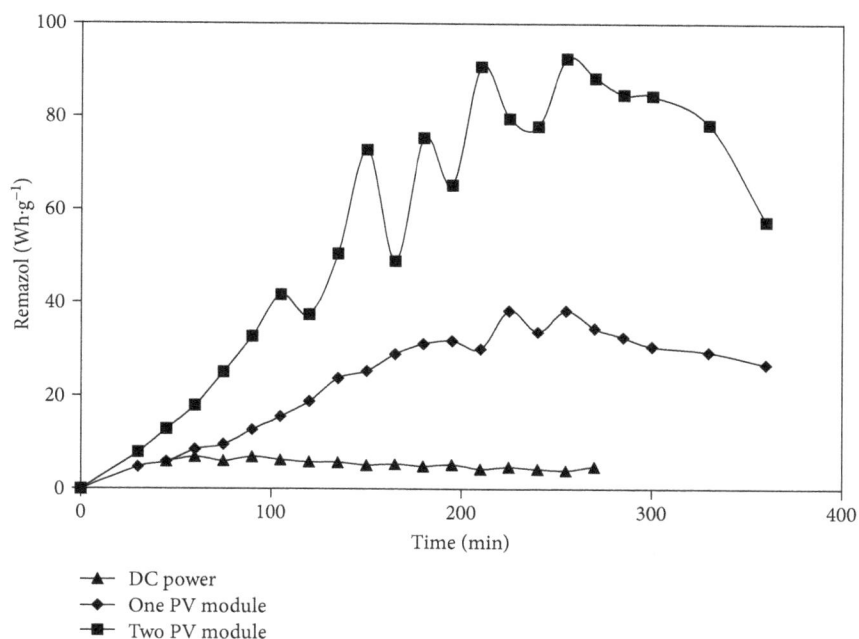

FIGURE 8: Comparison between CD and PV module.

consumption, RYD color removal shows a decay behavior due probably to a passivation phenomenon of the aluminum electrodes (see Section 3.1) and the consequent rising of electrical resistance thus causing the energy consumption increase.

Results reported at industrial scale show a parallel and series arrangement of the photovoltaic modules in order to obtain the electric current for the wastewater pollutant removal. However, the conditions of the photovoltaic arrangement will depend on the climatic conditions [40]. The setup for the pilot reactor of this work was two-module photovoltaic in parallel for a cloudy day and one-module photovoltaic for a sunny day.

Finally, regarding the PV energy costs, it will be possible to get a PV system at a low cost for the wastewater treatment. "Production costs for industry-leading Chinese crystalline-silicon (c-Si) PV module manufacturers – such as Jinko Solar, Renesola, Trina Solar and Yingli Green Energy – will fall from 50 cents per watt in the fourth quarter of 2012 to 36 cents per watt by the end of 2017," according to a new report from GTM Research. The report, *PV Technology and Cost Outlook, 2013-2017* [42], predicts that the majority of these cost declines will derive from technology innovations such as diamond wire sawing for PV wafers, advanced metallization solutions, and increased automation in place of manual labor.

4. Conclusions

It is possible to use the photovoltaic modules to conduct wastewater treatment by electrocoagulation. In this work, the highest attained removal efficiency of RYD and color were 70% for both under continuous flow. The achieved current from one PV module was enough to generate

electrocoagulation and remove RYD and color in a sunny day or two PV modules in parallel on a cloudy day. In order to obtain the electrocoagulation reaction, the current intensity should be between 3 and 4 A. The recommended operation flow rate in the system is 300 mL·min^{-1}; if the flow rate is increased, the removal efficiency decreases.

It is important to take into account the costs and the innovation in photovoltaics, since its implementation for scaling will depend on these, as well as on being able to install a complete system for storing solar energy and control the power to the system, always looking for the better pollutant removal efficiencies.

Conflicts of Interest

The authors declare that they have no conflicts of interest.

References

[1] C. Chienhung, N. Ervin, J. Yaju, and H. Chihpin, "Electrochemical decolorization of dye wastewater by surface-activated boron-doped nanocrystalline diamond electrode," *Journal of Environmental Sciences*, vol. 45, pp. 100–107, 2016.

[2] C. Davide, C. Giacomo, and M. Panizza, "Electrochemical oxidation of a synthetic dye using a BDD anode with a solid polymer electrolyte," *Electrochemistry Communications*, vol. 75, pp. 21–24, 2017.

[3] J. Rajeev, N. Sharma, and K. Radhapyari, "Electrochemical treatment of pharmaceutical azo dye amaranth from waste water," *Journal of Applied Electrochemistry*, vol. 39, pp. 577–582, 2009.

[4] V. David, G.-G. Vicente, E. Eduardo, A. Antonio, and M. Vicente, "Electrochemical treatment of wastewater from almond industry using DSA-type anodes: direct connection

to a PV generator," *Separation and Purification Technology*, vol. 123, pp. 15–22, 2014.

[5] M. Kobya, E. Gengec, and E. Demirbas, "Operating parameters and costs assessments of a real dyehouse wastewater effluent treated by a continuous electrocoagulation process," *Chemical Engineering and Processing: Process Intensification*, vol. 101, pp. 87–100, 2016.

[6] A. Amour, B. Merzouk, L. Jean-Pierre, and F. Lapicque, "Removal of reactive textile dye from aqueous solutions by electrocoagulation in a continuous cell," *Desalination and Water Treatment*, vol. 57, no. 48-49, pp. 22764–22773, 2015.

[7] B. K. Korbahti and K. M. Turan, "Evaluation of energy consumption in electrochemical oxidation of Acid Violet 7 textile dye using Pt/Ir electrodes," *Journal of Turkish Chemical Society, Section A: Chemistry*, vol. 3, no. 3, pp. 75–92, 2016.

[8] D. Valero, J. M. Ortiz, E. Expósito, V. Montiel, and A. Aldaz, "Electrocoagulation of a synthetic textile effluent powered by photovoltaic energy without batteries: direct connection behaviour," *Solar Energy Materials and Solar Cells*, vol. 92, no. 3, pp. 291–297, 2008.

[9] A. Masion, A. Vilgé-Ritter, J. Rose et al., "Coagulation-flocculation of natural organic matter with al salts: speciation and structure of the aggregates," *Environmental Science & Technology*, vol. 34, no. 15, pp. 3242–3246, 2000.

[10] J. Gregory, "Laminar dispersion and the monitoring of flocculation processes," *Journal of Colloid and Interface Science*, vol. 118, pp. 397–409, 1987.

[11] A. Dalvand, E. Gholibegloo, M. R. Ganjali et al., "Comparison of *Moringa stenopetala* seed extract as a clean coagulant with Alum and *Moringa stenopetala*-Alum hybrid coagulant to remove direct dye from textile wastewater," *Environmental Science and Pollution Research*, vol. 23, no. 16, pp. 16396–16405, 2016.

[12] A. Assadia, A. Soudavaria, and M. M. Fazlia, "Comparison of electrocoagulation and chemical coagulation processes in removing reactive red 196 from aqueous solution," *Journal of Human Environment and Health Promotion*, vol. 1, no. 3, pp. 172–182, 2016.

[13] C. Ricordel, A. Darchen, and D. Hadjiev, "Electrocoagulation-electroflotation as a surface water treatment for industrial uses," *Separation and Purification Technology*, vol. 74, no. 3, pp. 342–347, 2010.

[14] A. Othmani, A. Kesraoui, and M. Seffen, "The alternating and direct current effect on the elimination of cationic and anionic dye from aqueous solutions by electrocoagulation and coagulation flocculation," *Euro-Mediterranean Journal for Environmental Integration*, vol. 2, no. 1, 2017.

[15] S. L. Pérez, M. O. Rodriguez, S. Reyna et al., "Oil refinery wastewater treatment using coupled electrocoagulation and fixed film biological processes," *Physics and Chemistry of the Earth, Parts A/B/C*, vol. 91, pp. 53–60, 2016.

[16] H. Ahmad, W. K. Lafi, K. Abushgair, and J. M. Assbeihat, "Comparison of coagulation, electrocoagulation and biological techniques for the municipal wastewater treatment," *International Journal of Applied Engineering Research*, vol. 11, no. 22, pp. 11014–11024, 2016.

[17] C. Barrera-Díaz, G. Roa-Morales, L. Ávila-Córdoba, T. Pavón-Silva, and B. Bilyeu, "Electrochemical treatment applied to food-processing industrial wastewater," *Industrial & Engineering Chemistry Research*, vol. 45, no. 1, pp. 34–38, 2006.

[18] C. Barrera-Díaz, L. A. Bernal-Martínez, R. Natividad, and J. M. Peralta-Hernández, "Synergy of electrochemical/O_3 process with aluminum electrodes in industrial wastewater treatment," *Industrial and Engineering Chemistry Research*, vol. 51, no. 27, pp. 9335–9342, 2012.

[19] ISES, "Transición hacia un futuro basado en las fuentes renovables de energía," *Asociación Internacional de Energía Solar*, 2003, https://mba.americaeconomia.com/sites/mba.americaeconomia.com/files/paper_ises_dieter_holm.pdf.

[20] S. M. Lucas, R. Mosteo, M. I. Maldonado, S. Malato, and J. A. Peres, "Solar photochemical treatment of winery wastewater in a CPC reactor," *Journal of Agricultural and Food Chemistry*, vol. 57, no. 23, pp. 11242–11248, 2009.

[21] A. M. S. Solano, C. K. Costa, D. Araujo et al., "Decontamination of real textile industrial effluent by strong oxidant species electrogenerated on diamond electrode: viability and disadvantages of this electrochemical technology," *Applied Catalysis B: Environmental*, vol. 130-131, pp. 112–120, 2013.

[22] O. T. Can, M. Bayramoglu, and M. Kobya, "Decolorization of reactive dye solutions by electrocoagulation using aluminum electrodes," *Industrial & Engieneering Chemistry Research*, vol. 42, no. 14, pp. 3391–3396, 2003.

[23] P. K. Holt, *Electrocoagulation: Unravelling and Synthesising the Mechanisms behind a Water Treatment Process, [Ph.D. thesis]*, Chemical Engineering, University of Sydney, Australia, 2002.

[24] American Public Health Association, American Water Works Association, and Water Environment Federation, *Standard Methods for the Examination of Water and Wastewater*, L. S. Clesceri, A. E. Greenberg and A. D. Eaton, Eds., Washington, DC, USA, 1995, https://ses.library.usyd.edu.au/handle/2123/624.

[25] P. Aravind, V. Subramanyan, S. Ferro, and R. Gopalakrishnan, "Eco-friendly and facile integrated biological-cum-photo assisted electrooxidation process for degradation of textile wastewater," *Water Research*, vol. 93, pp. 230–241, 2016.

[26] T. Harif and A. Adin, "Characteristics of aggregates formed by electroflocculation of a colloidal suspension," *Water Research*, vol. 41, no. 13, pp. 2951–2961, 2007.

[27] A. S. Fajardo, R. F. Rodrigues, R. C. Martins, L. M. Castro, and R. M. Quinta-Ferreira, "Phenolic wastewaters treatment by electrocoagulation process using Zn anode," *Chemical Engineering Journal*, vol. 275, pp. 331–341, 2015.

[28] P. Cañizares, C. Jiménez, F. Martínez, C. Sáez, and M. A. Rodrigo, "Study of the electrocoagulation process using aluminum and iron electrodes," *Industrial & Engineering Chemistry Research*, vol. 46, no. 19, pp. 6189–6195, 2007.

[29] E. Lacasa, P. Cañizares, and M. A. Rodrigo, "Production of coagulant reagents for electro-coagulation processes at low current densities," *Desalination and Water Treatment*, vol. 45, no. 1–3, pp. 256–262, 2012.

[30] Z. Gu, Z. Liao, M. Schulz, R. J. Davis, C. J. Baygents, and J. Farrell, "Estimating dosing rates and energy consumption for electrocoagulation using iron and aluminum electrodes," *Industrial & Engineering Chemistry Research*, vol. 48, no. 6, pp. 3112–3117, 2009.

[31] H. Inan, A. Dimoglo, H. Şimşek, and M. Karpuzcu, "Olive oil mill wastewater treatment by means of electro-coagulation," *Separation and Purification Technology*, vol. 36, no. 1, pp. 23–31, 2004.

[32] Ü. T. Ün, S. Uğur, A. S. Koparal, and Ü. B. ÖğÜtveren, "Electrocoagulation of olive mill wastewaters," *Separation and Purification Technology*, vol. 52, pp. 136–141, 2006.

[33] A. S. Fajardo, R. C. Martins, and R. M. Quinta-Ferreira, "Treatment of a synthetic phenolic mixture by electrocoagulation using Al, Cu, Fe, Pb, and Zn as anode materials," *Industrial & Engineering Chemistry Research*, vol. 53, no. 47, pp. 18339–18345, 2014.

[34] M. Pourbaix, *Atlas of Electrochemical Equilibria in Aqueous Solutions*, Pergamon Press, Oxford, 1st edition, 1966.

[35] O. Nurgul and S. Y. Adife, "Factorial experimental design for Remazol Yellow dye sorption using apple pulp/apple pulp carbone–titanium dioxide co-sorbent," *Journal of Cleaner Production*, vol. 100, pp. 333–343, 2015.

[36] M. A. Sandoval, J. L. Nava, O. Coreno, G. Carreno, L. A. Arias, and D. Mendez, "Sulfate ions removal from an aqueous solution modeled on an abandoned mine by electrocoagulation process with recirculation," *International Journal of Electrochemical Science*, vol. 12, pp. 1318–1330, 2017.

[37] S. Liu, X. Ye, H. Kuang, Y. Chen, and H. Yongyou, "Simultaneous removal of Ni(II) and fluoride from a real flue gas desulfurization wastewater by electrocoagulation using Fe/C/Al electrode," *Journal of Water Reuse and Desalination*, vol. 7, no. 3, pp. 288–297, 2017.

[38] K. Wang Lawrence, H. Yung-Tse, and K. Shammas Nazih, Eds., "Advanced physicochemical treatment technologies," in *Handbook of Environmental Engineering*, vol. 5, p. 305, Humana Press, 2007.

[39] D. Valero, J. M. Ortiz, V. García, E. Expósito, V. Montiel, and A. Aldaz, "Electrocoagulation of wastewater from almond industry," *Chemosphere*, vol. 84, no. 9, pp. 1290–1295, 2011.

[40] D. Valero, V. García-García, E. Expósito, A. Aldaz, and V. Montiel, "Electrochemical treatment of wastewater from almond industry using DSA-type anodes: direct connection to a PV generator," *Separation and Purification Technology*, vol. 123, pp. 15–22, 2014.

[41] M. M. Quintero, S. T. Pavón, M. G. Roa, and R. Ruiz Meza, "Remoción de un colorante en medio acuoso utilizando un proceso electroquímico y energía eólica," in *Tercer Congreso Internacional de Ciencias Ambientales. Sustentabilidad y Cambio Climático*, Universidad Autónoma del Estado de México, Toluca, Estado de México, 2016.

[42] S. Lacey, Ed., *Producing Solar Below 70 Cents a Watt*, 2010, http://www.renewableenergyworld.com/rea/news/podcast/2010/09/producing-solar-at-70-cents-a-watt.

16

Theoretical Study and Experimental Validation of Energetic Performances of Photovoltaic/Thermal Air Collector

Khaled Touafek⬡,[1] Abdelkrim Khelifa,[1] Lyes Boutina,[1] Ismail Tabet,[2] and Salim Haddad[2]

[1]*Unité de Recherche Appliquée en Energies Renouvelables (URAER), Centre de Développement des Energies Renouvelables (CDER), 47133 Ghardaïa, Algeria*
[2]*Université de Skikda, Algeria*

Correspondence should be addressed to Khaled Touafek; khaledtouafek@yahoo.fr

Academic Editor: Giulia Grancini

This work undertakes both simulation and experimental studies of a new design of a photovoltaic thermal solar air collector (PV/T). In order to improve the thermal and electrical performances for a specific application, the analytical expressions for thermal parameters and efficiency are derived by developing an energy balance equation for each component of the PV/T air collector. This type of hybrid collector can be applied in the facades of buildings. The electricity and heat produced will satisfy the energy needs of the buildings, while ensuring an aesthetic view of its facades. A typical prototype was designed, constructed, and implemented in the applied research unit on renewable energies in Ghardaia, situated in the south of Algeria. This region has semiarid characteristics. Results obtained by an experimental test are presented and compared to those predicted through simulation. Results include the temperature of each component of the PV/T collector and air temperature at the inlet and outlet of the channel. It has been found that the theoretical results predicted by the developed mathematical model, for instance, outlet temperature, agree with those found through experimental work.

1. Introduction

Hybrid photovoltaic thermal (PV/T) collectors convert solar energy both into electrical and thermal energy. This conversion allows firstly the cooling of the solar cells and secondly the exploitation of the resulting heat energy to heat water or space. Applications of hybrid collectors in habitations are beneficial from a space-saving perspective. In fact, instead of separately using photovoltaic modules for electricity and solar thermal energy for heat, hybrid collectors are used for the same surface. In the literature, there are several studies on hybrid sensors applied in the habitat [1–5]. The most important thing in air hybrid sensors is to have the highest air outlet temperature compared to the input air temperature. To have that, several techniques are possible. They can be classified globally in two techniques. The first concerns the use of new coolant materials or the use of phase change materials [6–8]. The second technique is the design of the absorber [9–11]. In this article we are interested in the second technique of optimization of the absorber.

Work on hybrid collectors has been carried out in recent years [12–15].

Farshchimonfared et al. [16] studied a hybrid PV/T air collector related to an air distribution system. The aim was to optimize the channel depth. He found that increases in optimum depth were related to an increase of the collector area. The integration of photovoltaic thermal solar collectors in a building is increasingly becoming an option of first choice. A photovoltaic/thermal integrated collector in a building is able to generate higher energy output per collector unit area compared to conventional solar panels. A hybrid PV/T offers the same advantages as a photovoltaic PV collector, but in addition, it offers a look that is more aesthetic than side-by-side photovoltaic modules and solar thermal collectors and generally produces more energy for the same surface area [17]. A PV/T air collector integrated to buildings can be an outer layer that creates a building envelope which is double layered. Due to the heat source-generated sound, their thermal characteristic is different from that of the usual wall [18].

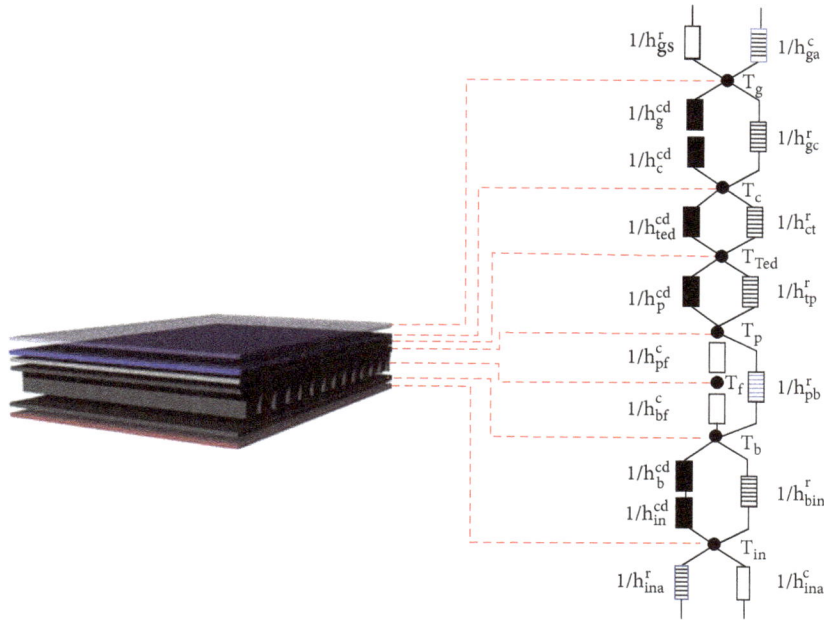

FIGURE 1: Photovoltaic thermal air collector.

In this paper, numerical simulation and experimental validation of a photovoltaic thermal PV/T solar air collector are conducted. The application of this sensor is conducted in a semiarid climate in the south of Algeria where the very high temperature negatively affects the efficiency of solar cells. The evacuation of the solar cells' heat causes their cooling which allows the photovoltaic sensor to perform well. The modeling of the heat transfer in the PV/T air collector is performed to 1D for a transitional regime according to a node approach. A numerical code was developed and used to analyze the thermal behavior as well as the thermal, electrical, and global electrical efficiency. Then, an experimental prototype of the PV/T air collector was set up. The PV/T air collector temperatures, as well as air inlet and outlet temperatures, were measured over a period of a day during the month of June and November and were compared with PV/T air collector model predictions.

2. Theoretical Study

The designed photovoltaic/thermal air collector studied in this work is shown in Figure 1. The PV/T air collector is composed of a photovoltaic module which consists of arrays of monocrystalline circular silicon cells covered with highly transmitting glass. These cell arrays are interposed on an adhesive layer made of Tedlar having good thermal conductivity and poor electrical conductivity. The photovoltaic module is to be mounted on a metal plate painted in black to increase the absorption of the incident solar radiation falling upon an absorbing plate equipped with rectangular fins which serve as channels for air circulation. The circulation of the air in the fins collects the generated heat, and a useful amount of thermal energy is extracted by removing hot air from one end of the duct leaving the cold fluid to enter at

the other end. The bottom of the sheath is covered with good insulation to minimize heat loss to the ambient temperature.

The thermal energy balance equations for the different nodes of the system is shown in Figure 1.

The heat balance which describes the thermal behavior of the PV/T air collector can be written in the form of first order ordinary differential equations and are given for the following essential elements of the hybrid PV/T air collector:

(1) For glass

$$M_g c_g \frac{dT_g}{dt} = q_g(t) - h_{gs}^r A_g \left(T_g - T_s \right)$$
$$- h_{ga}^c A_g \left(T_g - T_a \right)$$
$$- \left(U_1 + h_{gc}^r \right) A_g \left(T_g - T_c \right), \tag{1}$$

where

$$q_g(t) = \alpha_g I_g(t) A_g,$$
$$U_1 = \left(\frac{1}{h_g^{cd}} + \frac{1}{h_c^{cd}} \right)^{-1}. \tag{2}$$

(2) For solar cells

$$M_c c p_c \frac{dT_c}{dt} = q_c(t) - \left(h_{ted}^{cd} + h_{ct}^r \right) A_c \left(T_c - T_{ted} \right)$$
$$+ \left(U_1 + h_{gc}^r \right) A_c \left(T_g - T_c \right), \tag{3}$$

where

$$q_c(t) = \tau_g \alpha_c \left(1 - \eta_{\text{pv}}\right) I_g(t) A_c. \tag{4}$$

The solar cell efficiency depends on the cell temperature and is given by [19]

$$\eta_{\text{pv}} = \eta_{\text{ref}} [1 - 0.0045(T_c - 25)], \tag{5}$$

where η_{ref} is the reference efficiency of a solar cell at a solar irradiance of $1000\ \text{Wm}^{-2}$ and temperature $T_{\text{ref}} = 25°\text{C}$.

In this work, $\eta_{\text{ref}} = 10\%$.

(3) For Tedlar layer

$$M_{\text{ted}} cp_{\text{ted}} \frac{dT_{\text{ted}}}{dt} = \left(h_{\text{ted}}^{cd} + h_{ct}^{r}\right) A_t (T_c - T_{\text{ted}}) \\ - \left(h_{\text{tp}}^{r} + h_{p}^{cd}\right) R_{\text{tp}} A_t \left(T_{\text{ted}} - T_p\right), \tag{6}$$

where

$$R_{\text{tp}} = \frac{A_{\text{tp}}}{A_c}. \tag{7}$$

(4) For upper plate with fins

$$M_p cp_p \frac{dT_p}{dt} = -\left(h_p^{cd} + h_{\text{rtp}}^{r}\right) A_p R_{\text{tp}} (T_p - T_{\text{ted}}) \\ - h_{pf}^{c} A_p \eta_p (T_p - T_f) - h_{pb}^{r} A_p R_{\text{bp}} (T_p - T_b), \tag{8}$$

where η_p is the total efficiency of the absorbing upper plate end and η_f is the fin efficiency [1, 20].

$$\begin{aligned} \eta_f &= \frac{\tanh\left(mh_f\right)}{\lambda_f \delta_f}, \\ m &= \left(\frac{2h_f}{\lambda_f \delta_f}\right)^{1/2}, \\ \eta_p &= \frac{A_c + A_{\text{fin}} \eta_f}{A_{\text{bp}}}, \\ R_{\text{bp}} &= \frac{A_{\text{bp}}}{A_c}. \end{aligned} \tag{9}$$

(5) For fluid in the fins

$$M_f cp_f \frac{dTf}{dt} = -cp_f \dot{m} \frac{dT_f}{w dx} - h_{pf}^{c} A_f R_{\text{bp}} \eta_p (T_f - T_p) \\ - h_{bf}^{c} A_f (T_f - T_b). \tag{10}$$

(6) For lower plate

$$M_b cp_b \frac{dT_b}{dt} = h_{bf}^{c} A_b \left(T_f - T_b\right) - h_{pb}^{r} A_b R_{\text{bp}} \left(T_b - T_p\right) \\ - (h_{\text{bin}}^{r} + U_2) A_b (T_b - T_{\text{in}}), \tag{11}$$

where

$$U_2 = \left(\frac{1}{h_{\text{in}}^{cd}} + \frac{1}{h_b^{cd}}\right)^{-1}. \tag{12}$$

(7) For insulation

$$M_{\text{in}} cp_{\text{in}} \frac{dT_{\text{in}}}{dt} = -h_{\text{ina}}^{r} A_{\text{in}} (T_{\text{in}} - T_s) - h_{\text{ina}}^{c} A_{\text{in}} (T_{\text{in}} - T_a) \\ - (h_{\text{bin}}^{r} + U_4) A_{\text{in}} (T_{\text{in}} - T_b) \\ + A_{\text{in}} (T_b - T_{\text{in}}). \tag{13}$$

The temporal variation of the components of the collector of heat is low, thus $(m\, cpdT)/dt$ can be ignored. We can safely assume a quasistationary operation of the collector. This simplification of the equation suggests that only solar radiation has a significant effect.

By eliminating T_p and T_b, (11) becomes

$$\frac{dT_f}{dx} + C_1 T_f(x) = C_2. \tag{14}$$

If C_1 and C_2 are constants obtained from the performed algebraic calculations, the solution of the equation is

$$T_f(x) = \frac{C_2}{A_1} + \left(T_{fi} - \frac{C_2}{C_1}\right) e^{-C_1 x}, \tag{15}$$

and the outlet temperature of the air is

$$T_{f\text{out}} = \frac{C_2}{C_1} + \left(T_{fi} - \frac{C_2}{C_1}\right) e^{-C_1 l}. \tag{16}$$

In the above equations, radiation and convective heat transfer coefficients are calculated using the relationships as reported in [21–23].

The wind convection heat-transfer coefficient h_{i-a}^{c} for the air flowing over the outside surface depends primarily on the wind velocity v and is defined as

$$h_{i-a}^{c} = 5.7 + 3.8 v. \tag{17}$$

The radiation heat transfer coefficient between the glass cover and the sky is given by the following relationship:

$$h_{g-a}^{r} = \varepsilon_g \sigma \left(T_g^2 + T_s^2\right) \left(T_g + T_s\right). \tag{18}$$

The equivalent sky temperature is calculated by the following simple correlation equation:

$$T_s = T_a - 6. \tag{19}$$

The radiation heat transfer coefficient between the glass cover and the PV panel is given by the following relationship:

$$h^r_{g-c} = \sigma \left(T_g^2 + T_c^2\right)\left(T_g + T_c\right)\left(\left(\frac{1}{\varepsilon_g}\right) + \left(\frac{1}{\varepsilon_c}\right) - 1\right). \tag{20}$$

The radiation heat transfer coefficient between the PV panel and the Tedlar layer is given by the following relationship:

$$h^r_{c-\text{ted}} = \sigma \left(T_c^2 + T_\text{ted}^2\right)\left(T_c + T_\text{ted}\right)\left(\left(\frac{1}{\varepsilon_c}\right) + \left(\frac{1}{\varepsilon_\text{ted}}\right) - 1\right). \tag{21}$$

The radiation heat transfer coefficient between the upper plate and the Tedlar layer is given by the following relationship:

$$h^r_{\text{ted}-p} = \sigma \left(T_\text{ted}^2 + T_p^2\right)\left(T_\text{ted} + T_p\right)\left(\left(\frac{1}{\varepsilon_p}\right) + \left(\frac{1}{\varepsilon_\text{ted}}\right) - 1\right). \tag{22}$$

The radiation heat transfer coefficient between the upper plate and the lower plate is given by the following relationship:

$$h^r_{b-p} = \sigma \left(T_b^2 + T_p^2\right)\left(T_b + T_p\right)\left(\left(\frac{1}{\varepsilon_p}\right) + \left(\frac{1}{\varepsilon_b}\right) - 1\right). \tag{23}$$

The radiation heat transfer coefficient between the lower plate and the insulation is given by the following relationship:

$$h^r_{b-\text{in}} = \sigma \left(T_b^2 + T_\text{in}^2\right)\left(T_b + T_\text{in}\right)\left(\left(\frac{1}{\varepsilon_\text{in}}\right) + \left(\frac{1}{\varepsilon_b}\right) - 1\right). \tag{24}$$

The conduction heat transfer coefficient is given by the following relationship:

$$h^{cd} = \frac{\lambda}{e}. \tag{25}$$

The convection heat transfer coefficient of the fluid is given by the following relationship:

$$h^c = \left(\frac{\lambda}{D_h}\right)\left[0.0158 \, \text{Re}^{0.8} + (0.00181 \, \text{Re} + 2.92)\exp\left(-\frac{0.0379x}{D_h}\right)\right], \tag{26}$$

where Re is the Reynolds number

$$\text{Re} = \frac{\rho D_h v}{\mu}, \tag{27}$$

where D_h is the hydraulic diameter.
For the rectangular channel

$$D_h = \frac{(2 \times h \times w)}{(h + w)}. \tag{28}$$

The thermal efficiency of photovoltaic solar collector is calculated by the following relationship [24, 25]:

$$\eta_\text{th} = \frac{Q_u}{A_c I_t} = \frac{\dot{m} c p_f \int (T_o - T_i)dt}{A_c \int I_t dt}. \tag{29}$$

The global efficiency of the photovoltaic thermal collector is defined as the sum of all thermal efficiency and electrical efficiency:

$$\eta_t = \eta_\text{th} + \eta_\text{pv}. \tag{30}$$

3. Numerical Simulation

It is obvious from the theoretical model given above that a numerical solution can be calculated for the temperatures T_g, T_c, T_p, T_b, T_in, and T_f as most of the heat transfer coefficients are functions of these temperatures. Therefore, an iterative numerical method based on Runge-Kutta method is used. Indeed, a program was written in Fortran 90 in order to find the values of the temperatures of the PV/T air collector, as well as the values of the thermal and electrical efficiency. This program is summarized by the flowchart presented in Figure 2.

Furthermore, the solar PV/T air collector dimensions and properties of working fluids, as well as the operating conditions, are also needed. System properties and operating conditions which are employed in this study are tabulated in Table 1. The following physical properties of air are assumed to vary linearly with temperature owing to the low temperature range that can be encountered [20, 23]:

(1) Dynamic viscosity

$$\mu_\text{air} = (0.0146T + 1.8343)10^{-5}. \tag{31}$$

(2) Density

$$\rho_\text{air} = 1.1774 - 0.00359 \times (T - 27). \tag{32}$$

(3) Thermal conductivity

$$\lambda_\text{air} = 0.02624 + 0.0000758 \times (T - 27). \tag{33}$$

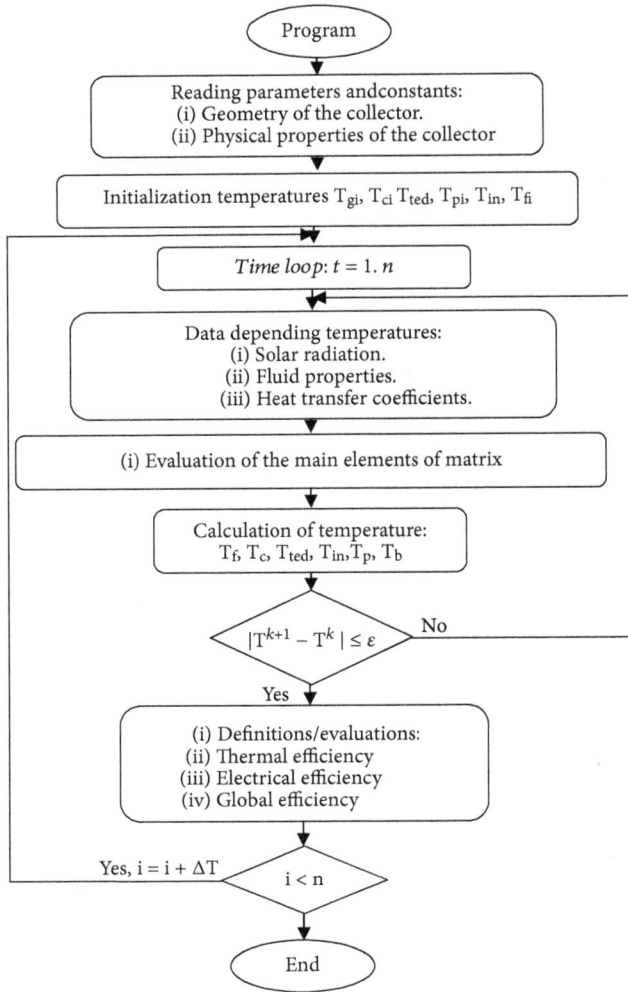

FIGURE 2: Flowchart for the computer program.

TABLE 1: Input parameters for numerical calculations.

Parameter	Values	Parameter	Values
A_i	0.5×1	cp_c	700
h	0.1	cp_{ted}	560
$\delta_f \times h_f \times l_f$	$0.001 \times 0.05 \times 1$	cp_p	465
τ_g	0.90	cp_{in}	880
α_g	0.05	ρ_c	2700
α_c	0.95	ρ_c	2330
λ_g	1	ρ_p	7833
λ_c	148	ρ_{in}	15
λ_{ted}	0.033	ρ_{ted}	1200
λ_{in}	0.041	ε_g	0.88
λ_p	54	ε_c	0.35
σ	$5.675e^{-8}$	ε_{ted}	0.35
v_w	2	ε_p	0.95
cp_g	840	ε_{in}	0.05

(4) Specific heat

$$cp_{air} = [1.0057 + 0.00066 \times (T - 27)] \times 1009. \quad (34)$$

Figure 3 presents the hourly temperature variation of essential elements of the hybrid PV/T air collector (glass cover, solar cell, layer in Tedlar, upper absorber plate, lower absorber plate, and insulation). It can be seen from this figure that all temperatures take maximum values in an interval of time ranging between 10 and 14 hours when the solar intensity is also maximal.

Equation (17) was used to calculate the outlet temperature of the hybrid PV/T air collector. The adopted mass flow rate values are equal to 0.01 kg/s, 0.02 kg/s, and 0.04 kg/s. Figure 4 displays the hourly variation of the fluid temperature at the outlet for different mass flow rates. It can be observed from this figure that maximal temperatures occurred at 43°C, 38.5°C, and 34°C for the respective mass flow rates of 0.01 kg/s, 0.02 kg/s, and 0.04 kg/s. It can be inferred that these temperatures vary with solar radiation intensity.

Figure 5 displays the influence of the channel length on the outlet air temperature at constant channel width and mass flow rate being equal to 0.04 kg/s. It is observed that the outlet air temperature increases with an increase in the channel length. The increased outlet air temperature is due to the increase in the heat exchange area which results in a decrease in air velocity; hence, a low heat transfer coefficient is favorable.

The effect of channel length on thermal, electrical, and global efficiency in the case of a fluid with a mass flow rate equal to 0.04 kg/s and a channel depth equal to 0.1 m is graphed in Figure 6.

It is observed that the thermal efficiency increases with an increase in channel length, while the electrical output decreases with an increase in channel length. The decrease in electrical efficiency is due to the increased PV temperature.

4. Experimental Validation

In Figure 7, a prototype of the hybrid photovoltaic thermal air collector is shown. It has a south orientation and is tilted at an angle equal to that of the site where it is installed, in this case 32.49° in the south of Algeria (Ghardaia) to maximize the intensity of the solar radiation that falls on it. The PV/T air collector was made with a monocrystalline PV module of 40 W, with a 0.5 m surface area and an air layer of 0.1 m for collecting hot air. Instrumentation devices are installed to measure the data relating to the thermal and electrical performances of the PV/T air collector.

A type k thermoelectric device for measuring temperature is used to measure the temperature of each component of the PV/T air collector. Each thermocouple is placed on the surface of each element of the PV/T collector and photovoltaic module. A current and voltage sensor has been installed for measuring the electrical performance. A data-

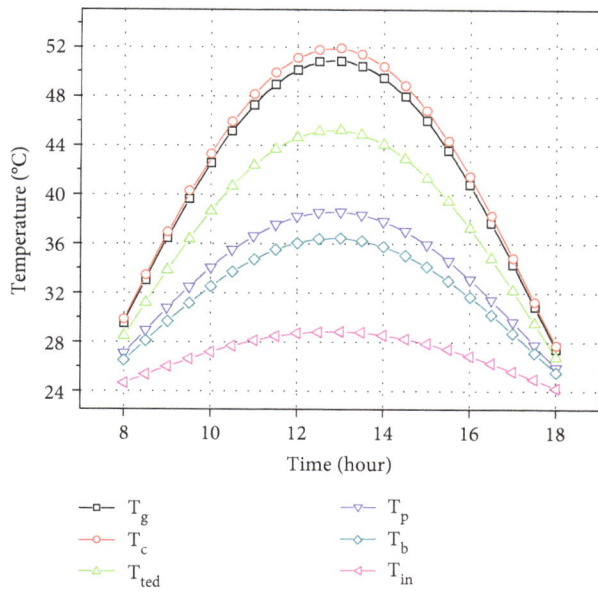

FIGURE 3: Hourly variation of the PV/T air collector element temperature.

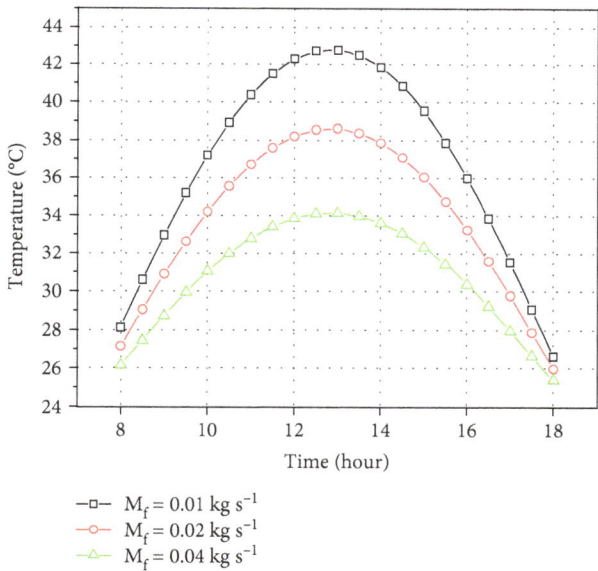

FIGURE 4: Hourly variation of the temperature of the fluid.

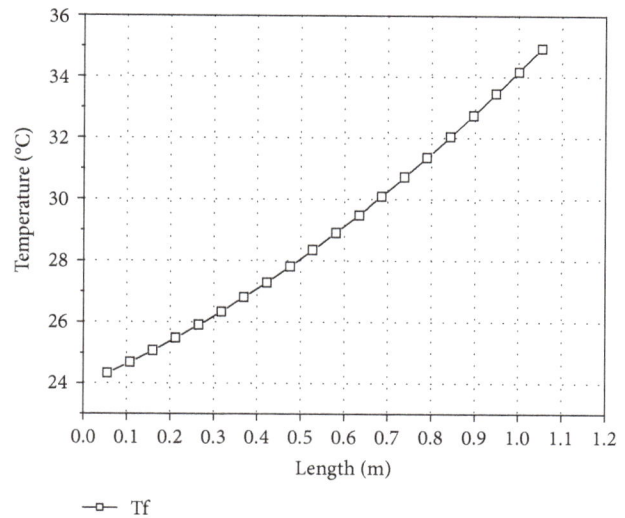

FIGURE 5: Variation of the fluid temperature in the PV/T air collector.

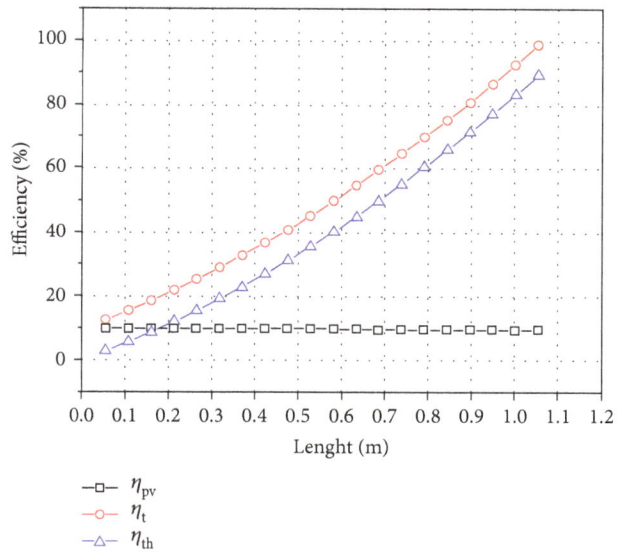

FIGURE 6: Variation of the efficiency of the PV/T air collector.

acquisition instrument was also connected to record all data on the performance of the PV/T collector.

Test standards such as that of ASHRE are used mainly for the determination of instantaneous thermal efficiency.

Figure 8 displays the hourly variation of the inlet and outlet air temperatures for the day 15 June 2014. Figure 9 displays the variation of temperature elements of the PV/T air collector.

It is clearly noticed that the temperature of each constituent element of the PV/T collector reaches it maximum value at noon of the day. This is obviously due to the considerable amount of solar radiation incident on the PV/T at this time,

which results in a high-rate absorption of solar radiation by the PV/T collector. However, it should be noted that each element reaches its maximum value depending on its location in the system and its physical characteristics. The temperatures of the front part of the PV/T collector and the upper plate have higher values when compared to the other elements of the PV/T collector. The temperature reaches 60°C at noon of the day for the front of the collector and 50°C for the upper plate, while the temperature of the other components do not exceed 45°C in the same period.

We notice that a temperature difference in the range of about 9°C is achieved between the inlet and outlet during the time interval ranging between 10 and 14 hours. This is explained as follows. Because of the increase in the temperature of the upper plate that is bonded to the PV module,

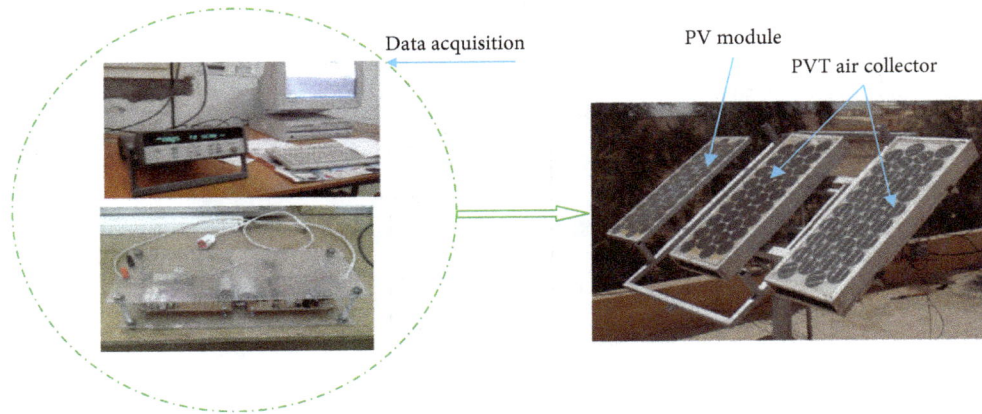

FIGURE 7: Photo of the experimental prototype.

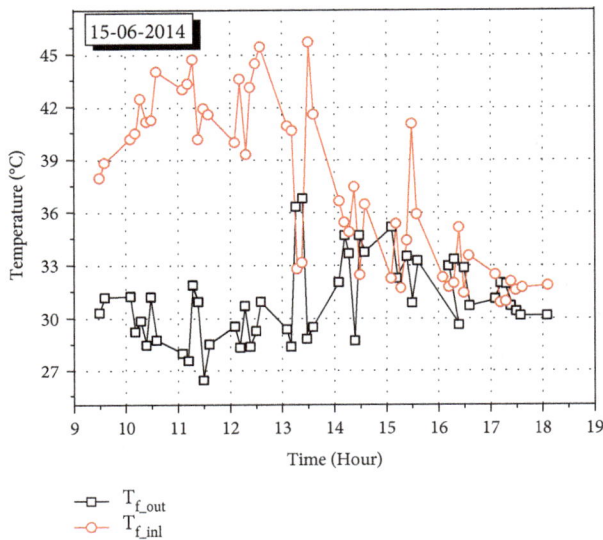

FIGURE 8: Hourly variations of the inlet and outlet air temperatures of the PV/T collector.

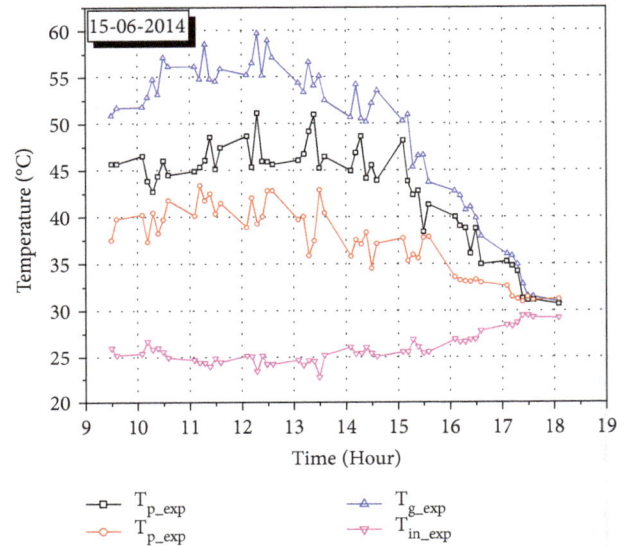

FIGURE 9: Hourly variation of the temperature elements of the PV/T.

the upper plate absorbs the heat generated by the PV module. When the air flows naturally through the channel, a convective exchange occurs between the lower and upper plates with air, and the air takes the heat generated by the two plates which increases their temperature at the outlet of PV/T air collector.

Figure 10 shows the current-voltage characteristic curves obtained by the experimental tests in the morning and evening for the two collectors, that is, the photovoltaic module and the hybrid PV/T air collector. The results showed that for the same surface of the PV and PV/T collector, there is an increase of 7 to 30% of the current intensity of the hybrid PV/T relative to the photovoltaic module.

In Figure 11, the comparison between the results obtained by experimental numerical modeling of the air temperature at the outlet of the PV/T air collector is presented. Polynomial interpolation is implemented to approximate the experimental results. The approximation curve is expressed by (35), in the local time (TL) between 8 hours

and 17 hours and at the temperature between 25°C and 40°C. It is given by

$$T_{fout\_exp} = -659.03457 + 195.48251\text{TL}$$
$$- 20.32914\text{TL}^2 + 0.94632\text{TL}^3 \qquad (35)$$
$$- 0.01688\text{TL}^4.$$

The coefficient of determination (R^2) is the key parameter which serves as the basis of comparison between the simulated and experimental results of the fluid outlet temperatures. The correlation coefficient (R^2) has been evaluated by using the following expression [25]:

$$R^2 = 1 - \frac{\sum_i^k \left(T_{fout-\text{Exp}} - T_{fout-\text{Theo}}\right)^2}{\sum_i^k \left(T_{fout-\text{Exp}} - T_{fout-\text{mav}}\right)^2}. \qquad (36)$$

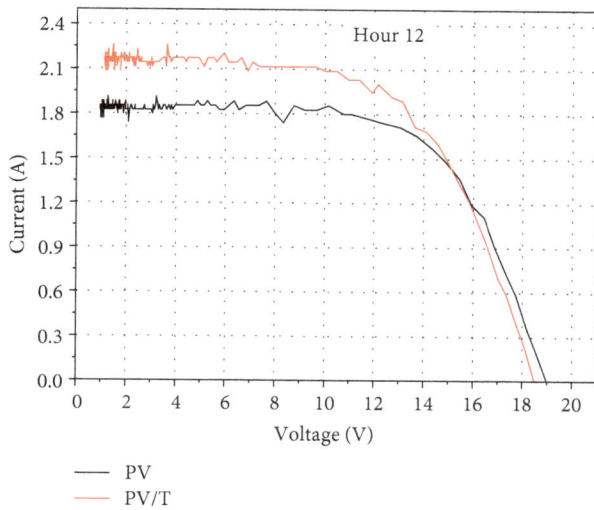

FIGURE 10: Current-voltage characteristic curves of the PV/T air collector and PV module.

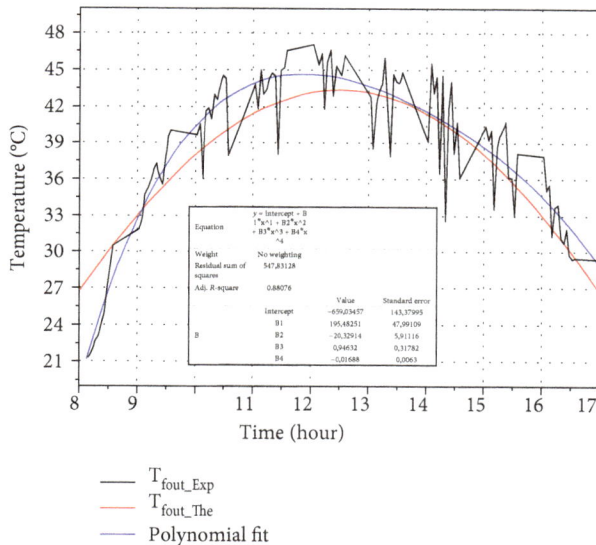

FIGURE 11: Validation of the mathematical model.

The correlation coefficient (R^2) is an indicator for judging the quality of a linear regression, either single or multiple, at a value between 0 and 1. We found that the R^2 value is equal to 0.7, in we find this value acceptable.

We did not study the effect of pressure on electrical and thermal efficiencies. We consider that the temperature is too low for there to be an influence of the pressure on the heat transfer and consequently on the efficiency of the collector.

5. Conclusion

In this work, a hybrid PV/T system was investigated, modeled, constructed, and tested. It was found that by combining a PV module with an air collector, the electrical performances are noticeably enhanced. The resulting thermal energy can be exploited to achieve thermal comfort in a building. A mathematical model based on energy balance was developed. Simulation results of this model obtained by using FORTRAN 90 programming language are plotted. The temperatures in the inlet and outlet are monitored and recorded. It was found that the temperature at outlet is higher than the temperature at the inlet, which suggests that a heat exchange has taken place and this is the reason for the improvement of the PV module performances.

The results found through simulation were compared to those found experimentally. Their comparison does show good agreement. This agreement is more pronounced in the case of the inlet and outlet fluid temperatures.

In conclusion, the combination of the PV module with the thermal collector would enhance the electrical performances of the PV module and the resulting thermal energy can be used for other purposes, for instance, to achieve thermal comfort in a building when this combination is wall integrated. In other applications, when water is used instead of air, the heated water can be used for sanitary purposes.

As a future work to further improve the performances of such systems, the PV module is to be replaced by a high-efficiency PV module and a reflector is used to concentrate solar radiation upon it.

Nomenclature

A: Surface (m^2)
C_p: Specific heat ($J\,kg^{-1}\,k^{-1}$)
D_h: Hydraulic diameter (m)
e: Thickness (m)
h^c: Convective-exchange coefficient ($W\,m^{-2}\,k^{-1}$)
h^r: Radiative-exchange coefficient ($W\,m^{-2}\,k^{-1}$)
h: Channel depth (m)
I: Solar intensity ($W\,m^{-2}$)
l: Length of the collector (m)
$\dot{m}$: Mass flow ($kg\cdot s^{-1}$)
M: Mass (kg)
n: Number of fins
R: Area ratio
Re: Reynolds number
q: Heat flow (W)
P: Power (W)
t: Time (s)
Tl: Local time (hour)
T: Temperature (°C)
v: Speed ($m\,s^{-1}$)
w: Width of the collector (m)
U: Heat transfer coefficient ($W\,m^{-2}\,k^{-1}$)
x: Distance (m).

Greek Symbols

α: Absorbance
δ: Thickness of fin (m)
ε: Emissivity
η: Efficiency
λ: Thermal conductivity ($W\,m^{-1}\,k^{-1}$)
ρ: Density ($kg\,m^{-3}$)

μ: Dynamic viscosity (kg·m^{-1} s^{-1})
σ: Stefane-Boltzmann constant (W m^{-2} K^{-4})
τ: Transmittance
ζ: Convergence criterion.

Subscripts and Abbreviations

a:	Ambient
b:	Lower plate
bp:	Bottom surface of absorber plate
c:	Solar cell
f:	Fluid
inl:	Inlet
g:	Glass cover
in:	Insulator
p:	Upper plate
pv:	Electric
out:	Outlet
s:	Sky
ted:	Tedlar
Exp:	Experimental
t:	Total
th:	Thermal
tp:	Top surface of absorber plate
Theo:	Theory
mav:	Mean value.

Conflicts of Interest

The authors declare that there is no conflict of interest regarding the publication of this paper. This research was performed as part of the requirements during the employment of the authors at the Unit of Applied Research in Renewable Energy (URAER)

References

[1] Z. B. Liu, L. Zhang, Y. Q. Luo, Y. L. Zhang, and Z. H. Wu, "Performance evaluation of a photovoltaic thermal-compound thermoelectric ventilator system," *Energy and Buildings*, vol. 167, pp. 23–29, 2018.

[2] D. Gürlich, A. Dalibard, and U. Eicker, "Photovoltaic-thermal hybrid collector performance for direct trigeneration in a European building retrofit case study," *Energy and Buildings*, vol. 152, pp. 701–717, 2017.

[3] M. Fiorentini, P. Cooper, and Z. Ma, "Development and optimization of an innovative HVAC system with integrated PVT and PCM thermal storage for a net-zero energy retrofitted house," *Energy and Buildings*, vol. 94, pp. 21–32, 2015.

[4] C. Lamnatou and D. Chemisana, "Photovoltaic/thermal (PVT) systems: a review with emphasis on environmental issues," *Renewable Energy*, vol. 105, pp. 270–287, 2017.

[5] M. Piratheepan and T. N. Anderson, "Performance of a building integrated photovoltaic/thermal concentrator, for facade applications," *Solar Energy*, vol. 153, pp. 562–573, 2017.

[6] M. Song, F. Niu, N. Mao, Y. Hu, and S. Deng, "Review on building energy performance improvement using phase change materials," *Energy and Buildings*, vol. 158, pp. 776–793, 2018.

[7] D. Su, Y. Jia, Y. Lin, and G. Fang, "Maximizing the energy output of a photovoltaic–thermal solar collector incorporating phase change materials," *Energy and Buildings*, vol. 153, pp. 382–391, 2017.

[8] W. Lin, Z. Ma, P. Cooper, M. I. Sohel, and L. Yang, "Thermal performance investigation and optimization of buildings with integrated phase change materials and solar photovoltaic thermal collectors," *Energy and Buildings*, vol. 116, pp. 562–573, 2016.

[9] K. Touafek, M. Haddadi, and A. Malek, "Design and modeling of a photovoltaic thermal collector for domestic air heating and electricity production," *Energy and Buildings*, vol. 59, pp. 21–28, 2013.

[10] M. Alobaid, B. Hughes, J. K. Calautit, D. O'Connor, and A. Heyes, "A review of solar driven absorption cooling with photovoltaic thermal systems," *Renewable and Sustainable Energy Reviews*, vol. 76, pp. 728–742, 2017.

[11] M. Hosseinzadeh, A. Salari, M. Sardarabadi, and M. Passandideh-Fard, "Optimization and parametric analysis of a nanofluid based photovoltaic thermal system: 3D numerical model with experimental validation," *Energy Conversion and Management*, vol. 160, no. 15, pp. 93–108, 2018.

[12] J. F. Chen, L. Zhang, and Y. J. Dai, "Performance analysis and multi-objective optimization of a hybrid photovoltaic/thermal collector for domestic hot water application," *Energy*, vol. 143, pp. 500–516, 2018.

[13] L. Sahota and G. N. Tiwari, "Review on series connected photovoltaic thermal (PVT) systems: analytical and experimental studies," *Solar Energy*, vol. 150, pp. 96–127, 2017.

[14] E. Sakellariou and P. Axaopoulos, "Simulation and experimental performance analysis of a modified PV panel to a PVT collector," *Solar Energy*, vol. 155, pp. 715–726, 2017.

[15] D. B. Singh, "Improving the performance of single slope solar still by including N identical PVT collectors," *Applied Thermal Engineering*, vol. 131, pp. 167–179, 2018.

[16] M. Farshchimonfared, J. I. Bilbao, and A. B. Sproul, "Channel depth, air mass flow rate and air distribution duct diameter optimization of photovoltaic thermal (PV/T) air collectors linked to residential buildings," *Renewable Energy*, vol. 76, pp. 27–35, 2015.

[17] V. Delisle and M. Kummert, "A novel approach to compare building-integrated photovoltaics/thermal air collectors to side-by-side PV modules and solar thermal collectors," *Solar Energy*, vol. 100, pp. 50–65, 2014.

[18] J.-H. Kim and J.-T. Kim, "A simulation study of air-type building-integrated photovoltaic-thermal system," *Energy Procedia*, vol. 30, pp. 1016–1024, 2012.

[19] A. K. Athienitis, J. Bambara, B. O'Neill, and J. Faille, "A prototype photovoltaic/thermal system integrated with transpired collector," *Solar Energy*, vol. 85, no. 1, pp. 139–153, 2011.

[20] M. Gholampour and M. Ameri, "Energy and exergy study of effective parameters on performance of photovoltaic/thermal natural air collectors," *Journal of Solar Energy Engineering*, vol. 136, no. 3, article 031001, 2014.

[21] L. M. Candanedo, A. Athienitis, and K.-W. Park, "Convective heat transfer coefficients in a building integrated photovoltaic/thermal system," *Journal of Solar Energy Engineering*, vol. 133, no. 2, article 021002, 2011.

[22] J. A. Duffie and W. A. Beckman, *Solar Engineering of Thermal Processes*, JohnWiley & Sons, New York, NY, USA, 2nd edition, 1991.

[23] E. M. Ali Alfegi, K. Sopian, M. Y. H. Othman, and B. B. Yatim, "Transient mathematical model of both side single pass photovoltaic thermal air collector," *ARPN Journal of Engineering and Applied Sciences*, vol. 2, p. 5, 2007.

[24] K. Touafek, M. Haddadi, and A. Malek, "Experimental study on a new hybrid photovoltaic thermal collector," *Applied Solar Energy*, vol. 45, no. 3, pp. 181–186, 2009.

[25] I. Tabet, K. Touafek, N. Bellel, N. Bouarroudj, A. Khelifa, and M. Adouane, "Optimization of angle of inclination of the hybrid photovoltaic-thermal solar collector using particle swarm optimization algorithm," *Journal of Renewable and Sustainable Energy*, vol. 6, no. 5, article 053116, 2014.

Permissions

All chapters in this book were first published in IJP, by Hindawi Publishing Corporation; hereby published with permission under the Creative Commons Attribution License or equivalent. Every chapter published in this book has been scrutinized by our experts. Their significance has been extensively debated. The topics covered herein carry significant findings which will fuel the growth of the discipline. They may even be implemented as practical applications or may be referred to as a beginning point for another development.

The contributors of this book come from diverse backgrounds, making this book a truly international effort. This book will bring forth new frontiers with its revolutionizing research information and detailed analysis of the nascent developments around the world.

We would like to thank all the contributing authors for lending their expertise to make the book truly unique. They have played a crucial role in the development of this book. Without their invaluable contributions this book wouldn't have been possible. They have made vital efforts to compile up to date information on the varied aspects of this subject to make this book a valuable addition to the collection of many professionals and students.

This book was conceptualized with the vision of imparting up-to-date information and advanced data in this field. To ensure the same, a matchless editorial board was set up. Every individual on the board went through rigorous rounds of assessment to prove their worth. After which they invested a large part of their time researching and compiling the most relevant data for our readers.

The editorial board has been involved in producing this book since its inception. They have spent rigorous hours researching and exploring the diverse topics which have resulted in the successful publishing of this book. They have passed on their knowledge of decades through this book. To expedite this challenging task, the publisher supported the team at every step. A small team of assistant editors was also appointed to further simplify the editing procedure and attain best results for the readers.

Apart from the editorial board, the designing team has also invested a significant amount of their time in understanding the subject and creating the most relevant covers. They scrutinized every image to scout for the most suitable representation of the subject and create an appropriate cover for the book.

The publishing team has been an ardent support to the editorial, designing and production team. Their endless efforts to recruit the best for this project, has resulted in the accomplishment of this book. They are a veteran in the field of academics and their pool of knowledge is as vast as their experience in printing. Their expertise and guidance has proved useful at every step. Their uncompromising quality standards have made this book an exceptional effort. Their encouragement from time to time has been an inspiration for everyone.

The publisher and the editorial board hope that this book will prove to be a valuable piece of knowledge for researchers, students, practitioners and scholars across the globe.

List of Contributors

Simon D. Hodgson and Alice R. Gillett
Department of Mechanical Engineering, University of Chester, Chester, UK

Chong Li
School of Electronic Engineering, Nanjing Xiaozhuang University, Nanjing 211171, China
College of Economics and Management, Nanjing University of Aeronautics and Astronautics, Nanjing 211106, China

Saad Motahhir, Abdelaziz El Ghzizal, Souad Sebti and Aziz Derouich
Laboratory of Production Engineering, Energy and Sustainable Development, Higher School of Technology, SMBA University, Fez, Morocco

Yin Guo, Qibing Liang, Bifen Shu, Jing Wang and Qingchuan Yang
Institute for Solar Energy Systems, Sun Yat-Sen University, Guangzhou, China
Guangdong Provincial Key Laboratory of Photovoltaic Technologies, Guangzhou, China

Shaoliang Wang and Zheng Xu
Institute of Optoelectronics Technology, Beijing Jiaotong University, Beijing 100044, China

Qingsong Huang and Jialiang Qiu
CECEP Solar Energy Technology (Zhenjiang) Co. Ltd., Zhenjiang 212132, China

Xianfang Gou
CECEP Solar Energy Technology (Zhenjiang) Co. Ltd., Zhenjiang 212132, China
Beijing University of Technology, Beijing 100124, China

Su Zhou and Junlin Huang
CECEP Solar Energy Technology (Zhenjiang) Co. Ltd., Zhenjiang 212132, China
Nanjing University of Aeronautics and Astronautics, Nanjing 211106, China

Honglie Shen
Nanjing University of Aeronautics and Astronautics, Nanjing 211106, China

Qi Wang, Bin Lu, Yichao Sun and Jin Liu
School of Electrical & Automation Engineering, Nanjing Normal University, Nanjing 210042, China
Jiangsu Province Gas-electricity Integrated Energy Engineering Laboratory, Nanjing 210046, China

Xiaobo Dou
School of Electrical Engineering, Southeast University, Nanjing 210096, China

Rongrong Zhai, Ying Chen, Hongtao Liu, Hao Wu and Yongping Yang
School of Energy, Power and Mechanical Engineering, North China Electric Power University, Beijing 102206, China

Yu-Pei Huang, Xiang Chen and Cheng-En Ye
Department of Electronic Engineering, National Quemoy University, Kinmen County, Taiwan

Kam Hoe Ong, Ramasamy Agileswari and Chakrabarty Chandan Kumar
Department of Electronics & Communication Engineering, Universiti Tenaga Nasional, Selangor, Malaysia

Biancamaria Maniscalco and Jake W. Bowers
Centre for Renewable Energy Systems Technology (CREST), Wolfson School of Mechanical, Electrical and Manufacturing Engineering, Loughborough University, Loughborough, Leicestershire LE11 3TU, UK

Panagiota Arnou
Laboratory for Energy Materials, University of Luxembourg, L-4422 Belvaux, Luxembourg

Marayati Bte Marsadek
Institute of Power Engineering, Universiti Tenaga Nasional, Selangor, Malaysia

Hongbo Wang, Runzhou Su and Yuanzuo Li
College of Science, Northeast Forestry University, Harbin, Heilongjiang 150040, China

Qian Liu
Department of Applied Physics, Xi'an University of Technology, Xi'an 710054, China

Dejiang Liu
Life Science College, Jiamusi University, Jiamusi, Heilongjiang 154007, China

Jinglin Liu
College of Science, Jiamusi University, Jiamusi, Heilongjiang 154007, China

Mao Mao and Xiao-Lin Zhang
School of Atmospheric Physics, Nanjing University of Information Science & Technology, Nanjing 210044, China

Guo-Hua Wu
College of Science and Technology, Nihon University, 1-18-14 Kanda Surugadai, Chiyoda-ku, Tokyo 101-8308, Japan

Abdullrahman A. Al-Shamma'a and Abdullah M. Noman
Department of Electrical Engineering, College of Engineering, King Saud University, Riyadh 11421, Saudi Arabia
Department of Mechatronics Engineering, College of Engineering, Taiz University, Taiz, Yemen

Khaled E. Addoweesh and A. I. Alolah
Department of Electrical Engineering, College of Engineering, King Saud University, Riyadh 11421, Saudi Arabia

Ayman A. Alabduljabbar
King Abdulaziz City for Science and Technology (KACST), Riyadh 11442, Saudi Arabia

Daniel Tudor Cotfas, Petru Adrian Cotfas and Octavian Mihai Machidon
Electronics and Computers Department, Transilvania University of Brasov, Brasov, Romania

Gloria Zanotti, Giuseppe Mattioli, Anna Maria Paoletti and Giovanna Pennesi
Istituto di Struttura della Materia (ISM), CNR, Via Salaria km 29.300, 00015 Monterotondo, Rm, Italy

Daniela Caschera
Istituto per lo Studio dei Materiali Nanostrutturati (ISMN), CNR, Via Salaria km 29.300, 00015 Monterotondo, Rm, Italy

Nitzan Maman, Iris Visoly-Fisher and Eugene A. Katz
Department of Solar Energy and Environmental Physics, Swiss Institute for Dryland Environmental and Energy Research, Jacob Blaustein Institutes for Desert Research, Ben-Gurion University of the Negev, Midreshet Ben-Gurion 8499000, Israel
Ilse Katz Institute for Nanoscale Science and Technology, Ben-Gurion University of the Negev, 84105 Be'er Sheva, Israel

Ravi K. Misra and Lioz Etgar
Center for Applied Chemistry, the Institute of Chemistry, The Hebrew University of Jerusalem, Jerusalem, Israel

Thelma Beatriz Pavón-Silva
Unidad Académica Profesional-Acolman, Universidad Autónoma del Estado de México, Camino de Caleros No. 11, Ejidos de Santa Catarina, Acolman, MEX, Mexico

Hipólito Romero-Tehuitzil, Gonzálo Munguia del Río and Jorge Huacuz-Villamar
Instituto Nacional de Electricidad y Energías Limpias, Av. Reforma 113, Col. Palmira, 62490 Cuernavaca, MOR, Mexico

Khaled Touafek, Abdelkrim Khelifa and Lyes Boutina
Unité de Recherche Appliquée en Energies Renouvelables (URAER), Centre de Développement des Energies Renouvelables (CDER), 47133 Ghardaïa, Algeria

Ismail Tabet and Salim Haddad
Université de Skikda, Algeria

Index

www.ingramcontent.com/pod-product-compliance
Lightning Source LLC
Chambersburg PA
CBHW050457200326
41458CB00014B/5209

9 781641 163392